Nicotinic Acetylcholine Receptors

Jean-Pierre Changeux and Stuart J. Edelstein

Nicotinic Acetylcholine Receptors

FROM MOLECULAR BIOLOGY TO COGNITION

Copyright © 2005 by Odile Jacob Publishing Corporation, New York.

www.odilejacob.com.

All rights reserved. Printed in the United States of America and distributed by The Johns Hopkins University Press.

No part of this book may be reproduced in any form or by any electronic or mechanical means, including storage and retrieval systems, except in the case of brief quotations embodied in critical articles and reviews.

FIRST EDITION

Printed on acid-free paper.

Library of Congress Control Number: 2005927665

ISBN: 0-9768908-0-1

CONTENTS

PREFACE		vii
1.	Historical Background	1
2.	Purification and Characterization of the Nicotinic Receptor	22
3.	Evolution and Diversification of Pentameric Receptor Channels	36
4.	Chemical Structure of the Agonist-Binding Site	51
5.	Identification and Properties of the Receptor Ion Channel	64
6.	Activation and Desensitization at the Structural Level	83
7.	Three-Dimensional Structure at the Amino Acid Level	108
8.	Inherited Pathologies of the Acetylcholine Receptor	125
9.	Genesis of the Postsynaptic Membrane by Targeted Receptor Gene Transcription at the Motor Endplate	140
10.	Supramolecular Assembly of the Postsynaptic Membrane	164
11.	Molecular Biology of Brain Nicotinic Receptors	177
12.	Nicotinic Receptors and Brain Functions	192
	APPENDIX: Models of Acetylcholine Receptor Dynamics	213
	WORKS CITED	223
	INDEX	277

PREFACE

The aim of this book is to give a clear and straightforward account of the remarkable properties of the nicotinic receptor for acetylcholine, a membrane protein involved in chemical transduction in the nervous system that is also the target of a widely used drug, nicotine. This molecule also happens to be the first pharmacological receptor and ion channel ever to have been identified. The history of its discovery and the advances made over the last forty years in understanding its functional organization constitute a remarkable case study of the emergence of a new field of scientific research. The overwhelming amount of information that must be assimilated by neuroscientists today places a premium on timely and succinct summaries of the current state of research, but such progress reports do not critically examine the process by which the results they describe were acquired. Our purpose, by contrast, is not only to review the most recent experimental and theoretical breakthroughs in the study of the nicotinic receptor, but also to give the reader a sense of the intellectual excitement and adventure that accompanied the various stages of discovery, each one having its own evolutionary dynamic, its own challenges and trials, and its own successes and failures. We have therefore adopted a historical and multidisciplinary approach in order to trace the consequences of "variation and selection" in a key field of neurobiology.

Communication between neurons takes place primarily at the level of specialized contacts in the nervous system called synapses. Signal transmission at the synapse is mainly mediated by a chemical substance, the neurotransmitter, that is stored in the nerve ending and released upon arrival of a nerve impulse. After diffusion through the space that separates the cells, the neurotransmitter is recognized by a specific receptor molecule, and the resulting binding event is transduced into an electrical signal. Such receptors are now known to be the target of many drugs active on the nervous system, including nicotine, curare, neuroleptics, and tranquilizers. The story of the receptor for acetylcholine, the first membrane receptor for a neurotransmitter to be isolated and characterized at the molecular level, provides a unique perspective on the development of a branch of science at the crossroads of four major domains: biochemistry, neurophysiology, pharmacology, and the behavioral sciences. It is rare for the full trajectory of a scientific subject to be presented in a single book. This volume furnishes an exceptional opportunity for scientists and students to follow the course of a major advance in our understanding of the molecular basis of brain functions.

In the first chapter we examine the series of events by which research in several distinct fields in the first half of the twentieth century came to converge upon the study of the receptor. The influence of these approaches, which we regard as authentic scientific "cultures," can still be discerned in contemporary research programs concerning four broad topics: receptor pharmacology and chemical transmission at synapses; enzyme biochemistry and stereochemical specificity of binding sites; electric potentials and ion channels; and allosteric transitions and elementary mechanisms of signal transduction. In the second chapter we consider the early history of receptor purification and characterization, and go on in the three following chapters to describe the course of research into the various aspects of the receptor structure and function: the chemical structure of the binding site; the identity and properties of the ion channel; and the mechanism of signal transmission, including activation of the ion channel and its regulation.

In the second half of the book, starting with Chapter 6, we survey the results of recent studies on the three-dimensional structure of the receptor molecule, particularly with regard to the valuable insights they offer into inherited pathologies such as congenital myasthenia and epilepsies. After considering the integration of the receptor into its synaptic membrane environment at the junction between motor nerve and skeletal muscle and at the surface of neurons in the brain, we conclude with a look at still ongoing work on the distribution, physiology, and regulation of the nicotinic receptors in brain functions and cognition.

Much of the material contained in this book has been the subject of prior reviews and articles. A number of previously published accounts deal with aspects of the analysis presented here, notably Changeux (1981, 1990), Cartaud and Changeux (1993), Duclert and Changeux (1995), Changeux et al. (1998), Changeux and Edelstein (1998), Schaeffer et al. (2001), and Champtiaux and Changeux (2003).

We would like to express our thanks to colleagues who read portions of the draft manuscript, especially Marc Ballivet, and to Pierre-Jean Corringer, Thomas Grutter, and Antoine Taly for generously supplying figures. We owe thanks as well to Malcolm DeBevoise for his skillful assistance in putting the text into finished form; to Alice Calaprice for her expert copyediting; to Lynn Edelstein for her attentive proofreading; and to Mike Burton for his elegant page design. Most of all we are grateful to Odile Jacob and Bernard Gotlieb for their role in producing this book and for their continued support and understanding.

Paris
February 2005

Nicotinic Acetylcholine Receptors

CHAPTER 1

Historical Background

The word "neuroscience" is relatively new, referring as it does to a field of scientific inquiry that developed only recently. Until the late 1960s, investigation of the nervous system, and the brain in particular, dealt mainly with its anatomy (observed by means of light or electron microscopy), its physiology (using electrical recordings), and its chemistry and pharmacology. Psychology and animal behavior were still seen as belonging to the humanities rather than the physical sciences, despite the attempts of physiologists to establish causal relationships between brain anatomy and behavior. In 1971, the first annual meeting of the newly created Society for Neuroscience attracted 1,100 scientists. Was this event evidence of a major paradigm shift resulting from the formulation of a new body of explanatory hypotheses? Or was it simply the consequence of a development and strengthening of local modes of interaction between disciplines, with novel techniques being introduced from different domains of research, extending and combining concepts from distinct, apparently unrelated schools of thought? It is still too early to decide which of these interpretations is correct.

At nearly the same moment, convergence among a number of different scientific disciplines had led to the identification of the first neurotransmitter receptor. Four distinct research traditions—or what we may refer to as "cultures"—contributed to this achievement, and to modern work on receptors generally, focusing respectively on receptor pharmacology and chemical transmission at synapses; enzyme activity and stereochemical specificity of binding sites; electric potentials and ion channels; and allosteric transitions and elementary mechanisms of signal transduction. We begin by examining these four scientific cultures in historical perspective. In subsequent chapters we will consider the current understanding of acetylcholine receptors with respect to genes and their regulation, quaternary structures, agonist-binding sites, ion channels, allosteric regulation, genetic diseases, synaptic membrane assembly, and higher cognitive functions.

Receptor Pharmacology and Chemical Transmission

Early Investigations of Drug Action

The action of drugs on the human organism was first observed in prehistoric times, in connection with shamanic practices based on primitive experiments involving natural substances derived mainly from plants. The word "pharmacology" itself comes from the Greek *pharmakon,* signifying the magical action of a substance either in curing disease or causing death. The emergence of rational medicine with the Hippocratic school soon eliminated the term's magical content, and it gradually came to designate a substance that modifies the condition of the organism, whether healthy or sick, in a definite manner. With the introduction of the notions of "active principle" and "active dose," due to Paracelsus in the sixteenth century, and the insight that these substances were chemical entities having distinct structures, due to Lavoisier in the late eighteenth century, the modern idea of *pharmacological agents* was established. Our concern here is with the specific effects and mode of action of such agents, particularly certain compounds extracted from plants such as nicotine and curare.

The systematic experimental investigation of the effects of medicines and poisons made considerable progress in the mid-nineteenth century with the pioneering work of the French physiologist Claude Bernard. Several of his 1857 lectures at the Collège de France on toxic and medicinal substances were devoted to the physiological effects of curare (a substance used by native South Americans to poison the tips of arrows) and the target of its paralytic action. Bernard sought to understand its physiological effect by "localizing" the site of this action. Curare, he said, acts as a "chemical lancet" that "dissects the motor system," by which he meant that curare blocks motor (but not sensory) activity. In particular, he demonstrated that curare does not alter muscle contraction, but instead affects the action of the motor nerve on the muscle. Surprisingly, however, he concluded that curare acted on the motor nerve ending and from there traveled through the central nervous system. His disciple Alfred Vulpian correctly reported in 1886 that curare does not act on the motor nerves themselves—more precisely, the ventral roots of the spinal cord—but causes "communication between nerve fibers and muscle fibers to be interrupted" (see discussion in Changeux, 1979).

Early Conceptions of "Receptors"

The theoretical notion of a *pharmacological receptor* as it is used today in the neurosciences can be traced to the work of the English physiologist John Newport Langley during the period 1905–1908, though a closely related concept had already emerged independently from studies in immunology carried out by the German bacteriologist Paul Ehrlich (1897). Inspired by the work of Claude Bernard (to whom he gave ample credit), Langley sought to localize and further specify the actions of curare and nicotine on neuromuscular preparations from fowl. Like curare, nicotine is native to the New World. It was first isolated by the French chemist Nicholas-Louis Vauquelin in 1809 from tobacco plants that had been introduced in Europe in the sixteenth century, in part owing to the interest of the French ambassador to Lisbon, Jean Nicot de Villemain. Langley showed that nicotine blocks transmission in a manner different from that of curare. Applied to neuromuscular preparations, it first causes a contraction and then produces a blocking effect. In modern terms, nicotine (like acetylcholine) behaves as an *agonist,* and the response *desensitizes.* Curare, on the other hand, does not cause contraction, but instead blocks the effect of nicotine; that is, it behaves as an *antagonist* of nicotine action.

It remained unclear from the work of Bernard and Vulpian, reviewed by Langley (1906) in his Croonian Lecture of that year, as well as from studies by Wilhelm Kühne in Germany, whether curare (and nicotine) acted on the neural or muscular side of the motor endplate. To resolve this issue, Langley (1905, 1906) denervated the frog gastrocnemius muscle. Several weeks later, despite degeneration of the nerve terminal, the muscle tissue still responded to nicotine and curare, "the only difference apparently observed being an increased response to small doses of the poison." Since neither of these compounds prevented the contraction of the muscle, he concluded that "the muscle substance which combines with nicotine and curare is not identical with the substance which contracts. It is convenient to have a term for the especially excitable constituent, and I have called it the *receptive substance.* It receives the stimulus and by transmitting it causes contraction" (1906).

In an important further observation, Langley (1907) noted that "a small drop of nicotine placed near, but not touching a nerve ending is without effect, whilst placed on the nerve ending produces local contraction." The receptive substance therefore appeared to be concentrated under the nerve terminal. Comparing the effect of nicotine on striated muscle with that of adrenaline on "unstriated muscle and glands" (Elliott, 1904, 1905), Langley concluded that "the different systems of efferent nerves would chiefly, at any rate, owe their difference to the different characters of the receptive substances of the cells with which they have become connected." This physiological concept of a *receptor*, whose normal function is to receive "special substances secreted by the end of the nerve"—what we now call *neurotransmitters*—complemented the structural views of stereochemical specificity developed by Ehrlich and Fischer (see below). Even though Langley's "receptor theory" was dismissed as overly speculative, even "unnecessary," by some of the most distinguished pharmacologists of the period (Dale, 1943), it nonetheless provided a useful and productive interpretation of experimental data on drug action in both peripheral and central synapses. The distinction by Dale (1914) between the "muscarine action" of "certain esters and ethers of choline" upon smooth muscles (blocked by atropine) and the "nicotine action" of these esters and ethers upon striated voluntary muscles (blocked by curare) appears in retrospect to have been in agreement with Langley's receptor theory, and indeed may have been influenced by it.

Discovery of the Synapse

Anatomical studies of brain tissue in the nineteenth century wrestled with the question whether its component cells form a reticular continuum or a discontinuous network. A turning point was reached in 1889 with the publication by the Spanish anatomist Santiago Ramón y Cajal of a paper entitled "Conexión general des los elementos nerviosos," in which he defended the idea—developed and extended in his authoritative *Textura del sistema nervioso del hombre y de los vertebrados* (1899)—that nerve cells form independent units, or "neurons" (the term introduced by Heinrich Waldeyer in 1891), which are juxtaposed in "contiguity" rather than in "continuity" with one another. The neuron was supposed to mediate the propagation of electric signals—for which Cajal used characteristic arrows to specify the "marche des courants," or movement of currents—throughout the neural network. According to his concept of dynamic polarization, the nerve impulses collected by the dendrite and the cell body propagate in a polarized manner toward the axon. However, Cajal's neuron doctrine did not clearly specify the function, if any, of the "articulation" between nerve cells (today called the *synapse*) in the polarity of neuronal transmission; indeed, Cajal favored assigning responsibility for this process to the neurofibrillary network within the nerve cell. It was the English physiologist Sir Charles Sherrington who explicitly argued on behalf of the synapse's role in this regard in 1897. Trained by Langley, with whom he had published an article in 1884 on the consequences of excising portions of the brain in dogs and monkeys, Sherrington met Cajal on a visit to Spain in 1885 and his subsequent experimental work on spinal reflexes owed much to Cajal's influence. On the basis of *in vivo* experiments, Sherrington conceived of the synapse as a kind of intracellular "valve" that creates the polarity and threshold for transmission underlying what he called the "discontinuous chain of neurons." More than two centuries earlier, in his *Traité de l'homme* (1648), Descartes had proposed nearly the same word—"valvule"—for structures that orient the unidirectional flow of "animal spirits" from the nerve to the muscle.

Confirmation of the status of the synapse as a morphological entity had to await the development of the electron microscope (Palade and Palay, 1954; Robertson, 1956). The images demonstrate that the juxtaposed cell membranes do not fuse, but are separated by a gap about 50–100 nm wide, with vesicles 30–60 nm in diameter visible on the side of the nerve ending and characteristic densities and folds (Couteaux, 1958) on the postsynaptic side (see Figure 1.1). Although the function of the synapse as a valve in the propagation of nerve signals had already largely been established by physiological data at the beginning of the century, controversy persisted until the 1950s with regard to the actual mechanism of synaptic transmission. Was it electrical or chemical?

A chemical basis for the propagation of signals in the nervous system had initially been suggested by the German physiologist Emil Du Bois-Reymond, and later was explicitly formulated by one of Langley's students, Thomas R. Elliott (1904, 1905). The neurotransmitter synthesized and stored in the nerve ending is released upon the arrival of an electrical impulse; after diffusion in the synaptic cleft,

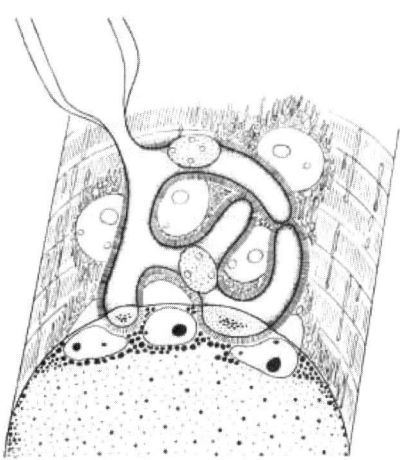

FIGURE 1.1 Diagram of the motor endplate showing the junction between a motor nerve and a striated skeletal muscle fiber. The nerve ending is in light gray and the fringe under the nerve represents the folds of the muscle postsynaptic membrane. The gray nuclei with small nucleoli belong to the Schwann cells of the nerve; the larger paler ones with large nucleoli are specialized muscle nuclei known as "fundamental nuclei," which play a critical role in the biosynthesis of the postsynaptic membrane. Mitochondria (black dots) also accumulate in the vicinity of these nuclei in the postsynaptic domain. (From Couteaux, 1978.)

it serves as a chemical relay that creates a converse chemoelectrical transduction at the level of the postsynaptic membrane. As a result of subsequent studies on the chemistry and pharmacology of neurotransmitters by Otto Loewi, Henry Dale, Ulf von Euler, and their students (see review in von Euler, 1981), the role of chemical transmission in the peripheral nervous system came to be accepted. With regard to activity in the brain itself, however, the electrical hypothesis continued to receive support from a number of distinguished scientists (among them Erlanger and Gasser, Lorente de Nó, and Eccles), until finally it was recognized that chemical and electrical synapses coexist in the nervous system (see review in Eccles, 1964).

Enzyme Activity and Stereochemical Specificity

The Forerunners of Modern Enzymology

A second research tradition, focusing on enzymes and related processes, played a crucial role in the history of receptors. Beginning in the nineteenth century, it developed alongside, and at times together with, the pharmacological tradition. In this context it should be recalled that Louis Pasteur's early work as a chemist, establishing the relationship between the crystalline form of tartrate isomers and their action on polarized light, led him to the seminal observation that living organisms such as molds and bacteria selectively utilize—and therefore recognize—a particular isomer in an asymmetrical (or "stereospecific") manner. The extension of this concept to the fermentation of organic substances subsequently gave rise to the concept of enzyme stereospecificity. Inquiring into the difference in taste between isomeric forms of asparagine, Pasteur had already postulated in 1886 that the "asymmetric active body which would play a role in the nervous impression, translated by a sugar taste in the one case, and a nearly insipid one in the other, would be nothing other, in my view, than the nervous matter itself, an *asymmetric matter*, as for all primordial substances of life" (see review by Debru, 1983).

The German chemist Emil Fischer (1894), in his extensive studies of the chemical synthesis of sugars, their isomeric forms and stereochemistry, and their selective transformations by yeast through respiration and fermentation, went further than this. He proposed that cells contain "chemically active agents" displaying a configuration complementary to that of the sugars on which they act, and concluded: "It is known that invertin and emulsin show some similarity to proteic substances and undoubtedly, like them, possess a molecule that is constructed in an asymmetric manner. Their limited action on glucoside may then be explained by the hypothesis that an association of molecules necessary to trigger the chemical process cannot take place without an analogous geometrical construction. To use a picture, I would like to say that enzyme and glucoside have to fit each other like a *lock and key* in order to exert a chemical effect on each other" (Fischer, 1894).

Several years later Ehrlich (1897) extended the *stereochemical concepts* applied by Fischer to fermentation to the immune reaction. Referring to the interaction of toxins with "antitoxic antibodies," Ehrlich stated that the "capacity to bind the antibodies must be related to the existence of specific atomic groupings that belong to the toxic complex, display a maximal specific affinity for a given atomic grouping of the antitoxic complex, and easily insert themselves in it, like a key and a lock, to use Emil Fischer's well-known analogy" (reprinted in Ehrlich, 1957).

In order to go beyond the qualitative notions of Fischer and Langley, quantitative analysis of experimental data was needed. Progress in this area could not be achieved without the development of a mathematical formalism. Interestingly, almost identical models were independently proposed for enzymes (Henri, 1903) and for drug action (Hill, 1910); later models utilized in pharmacology (Clark, 1926; Ariëns, 1954; Gaddum, 1957) were inspired by work on enzyme kinetics (Michaelis and Menten, 1913; Haldane, 1930). The equations devised to account for the observed dependence of enzyme rates (or pharmacological effect) on substrate (or agonist) concentration were based on the hypothesis, directly influenced by Fischer's model, of a reversible, *stereospecific complex* between the substrate (S) and a limited population of sites on the enzyme (E):

$$E + S \rightleftharpoons ES \rightleftharpoons E + P.$$

The basic equations, adapted from the mathematical description of the adsorption isotherm (Langmuir, 1918), yielded results consistent with the experimental data, and the formalism was readily extended to treat compounds that act analogously as *competitive* inhibitors (I) of enzyme activity:

$$E + I \rightleftharpoons EI.$$

A similar formulation was applied to *antagonists* of receptor activity. These structural analogues of the substrate, or agonist, are assumed to bind to the same site but without being transformed or causing a response (Michaelis and Menten, 1913; Gaddum, 1957; Jenkinson, 1960).

The enzyme model for receptors in an electrophysiological context (see following section) was most clearly enunciated by Del Castillo and Katz (1957a). They explicitly based their analysis on similarities observed between the acetylcholine receptor and acetylcholinesterase, citing publications by Augustinsson (1948, 1950), and assumed that a complex forms between receptor and agonist (A), as in the case of the ES complex, with channel opening occurring only for the bound state (R*), analogously with product formation in an enzyme reaction:

$$R + A \rightleftharpoons RA \rightleftharpoons R^*A \text{ (open)}.$$

This formalism is still widely used in the electrophysiological tradition (Colquhoun and Sakmann, 1998).

Enzymes subsequently were isolated as pure proteins and then crystallized, leading to the first description of an enzyme's atomic structure by David Phillips and his colleagues (Blake et al., 1965). Early on, the substrate-binding sites of enzymes naturally provided models for the still hypothetical receptor sites of pharmacological agents. David Nachmansohn (1959) had already suggested that the macromolecule bearing the receptor site for the acetylcholine neurotransmitter was a protein. The catalytic site of acetylcholinesterase, an enzyme present in abundant quantities at the motor endplate, was thought to be identical with the site of the physiological action of acetylcholine (Roepke, 1937; Zupancic, 1952; Belleau, 1964), but pharmacological evidence (Nachmansohn, 1959; Podleski and Nachmansohn, 1966; Karlin, 1967b; Podleski, 1967) demonstrated that the two sites were distinct.

Nonetheless, the enzyme model appeared to be adequate to support an empirical analysis of the effect of the large number of acetylcholine analogues and curare derivatives that had so far been synthesized—in particular the widely used flaxedil (Bovet et al., 1949) and methonium compounds (Barlow and Ing, 1948; Paton and Zaimis, 1949; see also Khromov-Borisov and Michelson, 1966)—as well as a precise definition of the structural features of molecules that acted either as "depolarizing" nicotinic agonists or as "stabilizing" competitive antagonists.

Electric Potentials and Ion Channels

How the brain propagates and processes the signals it receives from the outside world remained an enigma for centuries. In the seventeenth century, "animal spirits" were thought to mobilize "pneumatic" forces; magnetic, or electrical, forces became popular in the eighteenth century. The investigations of Luigi Galvani and his wife Lucia at the turn of the nineteenth century led to general acceptance of the view that "nervous fluid" was identical to electricity and that there existed an "electrical principle in the contracting muscle" and in the nerve. This idea furnished the basis for a third research tradition that relies almost exclusively on electrical measurements. A crucial step forward in the development of this culture occurred with the demonstration that electric signals propagate at a measurable speed (Helmholtz, 1850) that is smaller than the speed of sound (see review by Corsi, 1986). The nature of the electromotive force nonetheless remained mysterious. But the German physiologist Julius Bernstein (1902; see also Bernstein and Tschermak, 1904) proposed that, in the resting cell, the membrane potential originates from a difference in ion concentrations (potassium) across the cell membrane, which was subsequently found to result from the active transport of Na^+ and K^+ ions catalyzed by a membrane-bound enzyme known as the Na^+, K^+-ATPase (Skou, 1965). In this view, the nerve signal would result from a transient change of passive ionic permeability across the cell membrane (Lillie, 1909), with selective ion transport leading to electrical phenomena that were subsequently incorporated into a mathematical description in the model of Hodgkin and Huxley (1952).

In the case of neuromuscular transmission, the initial electrical response of skeletal muscle to stimulation of the motor nerve (later referred to as the *postsynaptic potential*) was first recorded with *extracellular* electrodes by Göpfert and Schaefer in 1938. The following year, Feldberg, Fessard, and Nachmansohn carried out an experiment at the Arcachon marine biological station, in France, that was to be one of the first syntheses of the electrophysiological, pharmacological, and enzymological traditions (Feldberg et al., 1940). Using the electric organ of *Torpedo marmorata* (Figure 1.2), which David Nachmansohn earlier had shown (see review in Nachmansohn, 1959) to be an extremely rich source of acetylcholine and its degradative enzyme acetylcholinesterase, and hence of cholinergic synapses, they demonstrated that neural stimulation of the electric organ elicits the release of acetylcholine and that "close range arterial injections of acetylcholine in the perfused organ connected with an amplifier caused potential changes in the same directions as those of the discharge." In other words, the electric organ mediates chemoelectrical transduction (Feldberg et al. 1940).

Valuable information was also obtained from recordings of the postsynaptic potential at the neuromuscular junction with intracellular electrodes (Eccles et al., 1941; Fatt and Katz, 1950, 1951, 1952) and from stimulation of the neuromuscular transmission junction by the iontophoretic application of acetylcholine to the subsynaptic membrane (Nastuk, 1953; Del Castillo and Katz, 1955a, 1955b, 1957a, 1957b; Kuffler and Yoshikami, 1953). Elicited by electrical stimulation of the nerve ending and recorded intracellularly, the postsynaptic potential (or current) has a simple wave shape, with a rapid rise time of less than 0.2 msec and a half-decay time of a few milliseconds. The delay

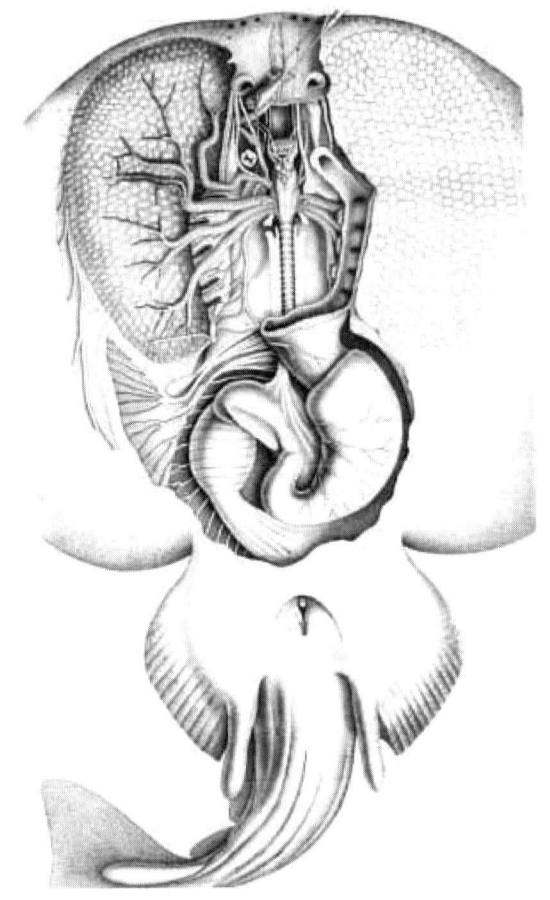

FIGURE 1.2 This romantic lithograph shows a dissection of the ventral face of *Torpedo narce, Risso* from Mediterranean waters. It depicts the electric organs as paired lobes, with a hexagonal pattern of elementary cells (or electroplaques) on both sides of the fish; on the left the skin has been removed to reveal the innervation of the organ by the electric nerves. The electric organs from fish such as *Electrophorus* and *Torpedo* contain large numbers (millions, even billions) of cholinergic synapses and thus constitute exceptionally rich sources of cholinergic receptors, which played a crucial role in the identification of the receptor protein. (From Savi, 1844.)

between stimulus and response of about 0.75 msec (Katz and Miledi, 1965) is explained by the time required for the release of acetylcholine packets or quanta (about 300 in frog endplates) and for their diffusion across the cleft. The synaptic cleft is so narrow that, in spite of the relatively small number of acetylcholine molecules released by a single vesicle (~5000), the local concentration of acetylcholine in the cleft rises transiently (in <1 msec) to approximately 3×10^{-4} M (Katz and Miledi, 1973b; Kuffler and Yoshikami, 1975; 1977) compared to a background release (when acetylcholinesterase is blocked) of approximately 10^{-8} M (Katz and Miledi, 1977). In other words, at the motor endplate, the intercellular communication signal is a transient *chemical pulse* involving a very high local concentration of neurotransmitter. A similar mode of chemical communication may take place at central synapses, for example in transmissions involving glutamate at large numbers of excitatory synapses (Trussell and Fischbach, 1989). Additionally, a mode of chemical communication characterized by more diffuse paracrine release—known as "volume transmission" (Fuxe and Agnati, 1991)—causing a spillover between synapses may be an important mechanism of neuromodulation in the brain, particularly in the case of acetylcholine (Descarries and Mechawar, 2000). In this respect, rapid electrochemical communication ("wiring transmission") would differ from endocrine or paracrine communication.

As we noted in the previous section, the early quantitative description of receptor action was based on an analogy with enzymes. The extension of the enzyme model to the quantification of concentration-effect curves, however, is not as unproblematic as it may appear at first sight. In the case of enzymes, the formalism directly applies to a molecular phenomenon: under steady-state conditions, a linear relation typically exists between the rate of product formation and the concentration of intermediate ES complexes. This is not, of course, what one finds in the majority of the standard pharmacological experiments conducted with isolated organs (Hill, 1910; Gaddum, 1957; Furchgott, 1966; Paton, 1970; see Neubig and Cohen, 1980), for at least two reasons: first, a complex sequence of reactions usually links the postulated drug-receptor (RA) complex to the measured effect; second, the effect of a neurotransmitter at the synapse under physiological conditions transiently occurs within milliseconds. To obtain a standard concentration-effect curve, a pharmacological agent is added to the bath. Generally speaking (see the "rate theory" described by Paton, 1961), a steady state is reached within seconds or minutes, and it is this state that is taken as a measure of the response, even when desensitization occurs. All the constants determined under these conditions are "phenomenological," which is to say they cannot be identified with intrinsic molecular parameters of the receptor.

An important step forward in the biophysical study of the acetylcholine receptor occurred with the observation, first made by Langley (1905), of a gradual decrease in the permeability response to acetylcholine when contact between the neurotransmitter and the membrane is maintained for several minutes. To account for this experimental evidence of "desensitization," Katz and Thesleff (1957) proposed a kinetic schema in which the receptor was postulated to exist in two forms, "effective" (A) and "refractory" (B):

$$\begin{array}{ccc} S + A & \xrightleftharpoons[\text{(fast)}]{a} & SA \\ \text{(slow)}\ k_2 \updownarrow k_4 & & k_3 \updownarrow k_1\ \text{(slow)} \\ S + B & \xrightleftharpoons[\text{(fast)}]{b} & SB. \end{array}$$

The experimental data made it necessary to assume, the authors noted, that the "affinity of the drug to receptor B [is] much higher than to receptor A." Moreover, their model was designed in such a way that "the free receptors are distributed, even in the absence of a drug, between states A and B; that is, a proportion of receptors is present in a refractory form, and on account of its very high affinity will preferentially absorb small quantities of applied acetylcholine." In other words, the preferentially populated form of the receptor would be "selected" by acetylcholine.

Empirical determination of the intrinsic values of the various parameters governing activation and desensitization was refined by reducing the complexity of the system and the time scale of the measurements. Improved electrophysiological methods made it possible to record elementary voltage levels at the neuromuscular junction (Katz and Miledi, 1970, 1972, 1973a) and fluctuations in current (Anderson and Stevens, 1973) observed in the presence of acetylcholine. The discovery of individual, discrete, all-or-none "molecular events" occurring under patch-clamp conditions was interpreted as evidence of molecular transitions representing the opening and closing of single-ion channels associated with the acetylcholine receptor (Neher and Sakmann, 1976). Desensitization was supposed to consist in periods of inactivity between active phases of rapid, discrete channel opening and closing (Sakmann et al., 1980).

With the recording of electric phenomena at the molecular level, the electrophysiological tradition had finally reached the limits of what it could achieve on its own, unaided by other disciplines. In the meantime, biochemists and molecular biologists had succeeded in obtaining unexpected results that were to dramatically change the course of research.

Allosteric Interactions

Hemoglobin and Regulatory Enzymes

A basic limitation of electrophysiological experimentation, and of the models to which it gave rise, is that the interaction of acetylcholine with its site can only be inferred indirectly from electrical recordings of the open state of the ion channel. Nor have attempts to infer molecular structure from function been successful. To do this, entirely different types of data (from biochemistry or X-ray crystallography, for example) are required (for discussion of this point, see Changeux and Connes, 1995; Edelstein et al., 1997b). From the start, the fundamental challenge was to explain how acetylcholine creates a transient "short circuit," or "puncture," allowing ions to pass through the membrane (Katz, 1966). In 1953, Nachmansohn (then working on the transformation of the enzyme-substrate complex of acetylcholinesterase) wrote: "The free ester (acetylcholine) acts upon a receptor, presumably a protein, and this action upon the receptor is responsible for the change of ionic permeability and thus the generation of the bioelectric potential. Some facts suggest the possibility that the effect of acetylcholine is to produce a change in configuration (conformation) of the protein. Rearrangements of acidic and basic groups by folding or unfolding of the protein chains would be one possibility to account for the increased sodium permeability affected by the system."

The nature of the structural link between the acetylcholine-binding site and the ion channel was difficult to conceptualize, however. Early studies of hemoglobin and, later, of bacterial regulatory enzymes suggested one possibility, namely that the activity of enzymes such as threonine deaminase (Umbarger, 1956; Changeux, 1961) or aspartate transcarbamylase (Yates and Pardee, 1956; Gerhart and Pardee, 1962) is regulated in a reversible manner by specific metabolites: the end product of the biosynthetic pathway initiated by a particular enzyme would inhibit that enzyme ("feedback inhibition"). Moreover, the fact that this regulatory property is conserved under enzymatic action indicates that it is an intrinsic feature of the protein molecule. The analysis of the kinetic behavior of these enzymes shows, in particular, that certain physical transformations (such as heating) and chemical transformations (such as a modification of sulfhydryl groups) may uncouple regulatory interactions while leaving enzymatic activity unaffected (Changeux, 1961; Gerhart and Pardee, 1962). This phenomenon was referred to by Monod et al. (1963) as "densensitization" (the authors were unaware of the prior use of the word in pharmacology). As Changeux (1961) put it, the "classical model of competitive inhibition involving binding of substrate and (regulatory) inhibitor at the same site and the same groups could not possibly account for these results. It seemed inevitable to assume the existence of two distinct sites" (see Figure 1.3). The term "allosteric" was introduced to describe this state of affairs by Monod and Jacob during the general discussion that took place at the Cold Spring Harbor Symposium at which these data were presented (Monod and Jacob, 1961).

The two sites were subsequently described by Monod et al. (1963) as being the "active site . . . responsible for the biological activity of the protein [and the] *allosteric* site complementary to the structure of . . . the allosteric effector, which it binds specifically and reversibly. The formation of the enzyme-effector complex does not activate a reaction involving the effector itself: it is assumed only to bring about a discrete reversible alteration of the molecular structure of the protein, or *allosteric transition,* which modifies the properties of the active site, changing one or several of the kinetic parameters that characterize the biological activity of the protein." Further biochemical and X-ray

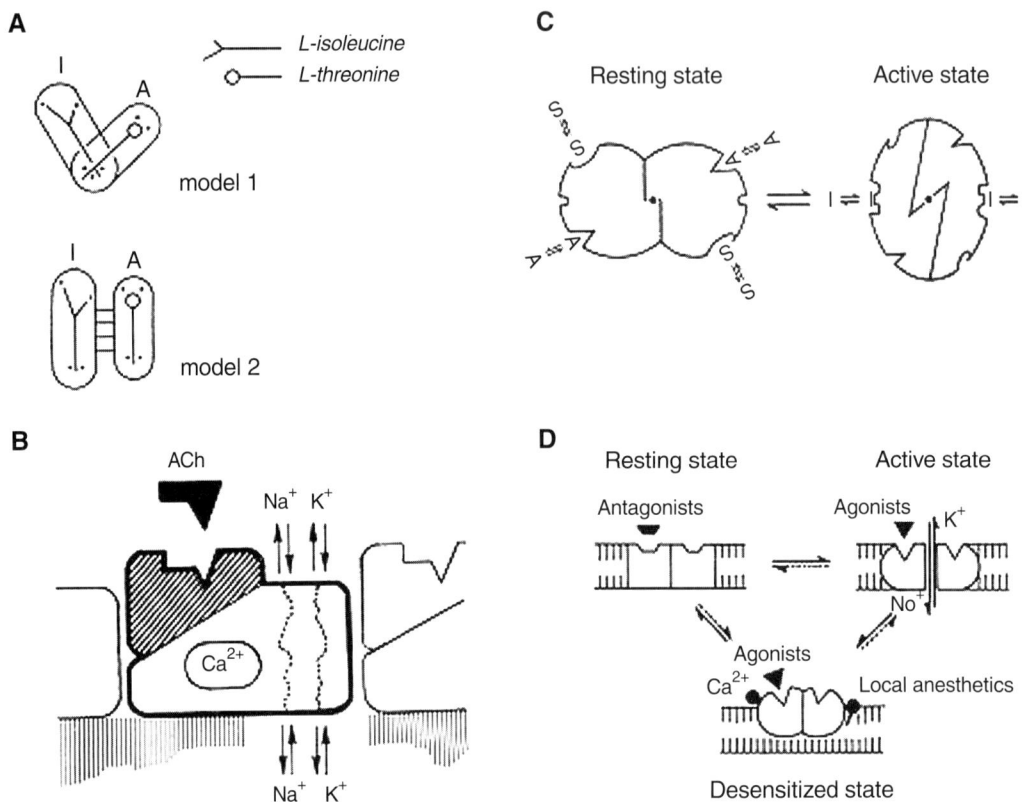

FIGURE 1.3 Schematic representation of models successively proposed to account for the allosteric properties of bacterial regulatory enzymes and of the acetylcholine receptor. (A) To explain the feedback inhibition of L-threonine deaminase (the first enzyme of isoleucine biosynthesis [*Escherichia coli*]) by L-isoleucine, the end product of the pathway, two models were considered: *direct* interaction between feedback inhibitor and substrate at the level of *overlapping sites;* and *indirect* interaction between topographically distinct, *nonoverlapping* sites. Experimental evidence supports the second "allosteric" model positing interaction between regulatory and catalytic sites. (From Changeux, 1961, courtesy of Cold Spring Harbor Laboratory Press.) (B) The model for regulatory enzymes was extended to the regulation of ion permeability by acetylcholine: the acetylcholine-binding site is the "regulatory" site; the site of ion translocation—the ion channel, or "ionophore"—is the "biological active" site. (From Changeux, 1969.) (C) A conformational transition between a small number of discrete states (the two-state Monod-Wyman-Changeux model) that preserves the symmetry of the protein molecule was postulated to mediate both heterotropic interactions, between regulatory effector and substrate, and homotropic interactions, between substrate (or effector) molecules. (From Changeux, 1965.) (D) The same idea was extended to the activation of the subsynaptic membrane receptor by acetylcholine. (From Changeux et al., 1976, courtesy of Cold Spring Harbor Laboratory Press.)

diffraction studies of aspartate transcarbamylase (Gerhart and Schachman, 1965; Lipscomb, 1994), glycogen phosphorylase (Weber et al., 1978; Fletterick and Madsen, 1980; Johnson and Barford, 1990), phosphofructokinase (Schirmer and Evans, 1990), bacterial L-lactate dehydrogenase (Iwata et al., 1994), and threonine deaminase (Gallagher et al., 1998) supported the theory's basic claim that the allosteric and catalytic sites are topographically far apart (up to 30 Å in the case of phosphorylase) on the surface of the molecule. In the case of hemoglobin, a site that binds the regulatory ligand 2,3-diphosphoglycerate has also been discovered in the central cavity of the molecule between the beta chains, at a significant distance ($\sim$20 Å) from the oxygen-binding sites (Arnone, 1972). It is interesting to note, however, that in most cases, including hemoglobin, the allosteric and substrate-binding sites are carried by the *same* polypeptide chain, although in the case of aspartate transcarbamylase they are carried by *different* chains.

By analogy with these enzymes, it was proposed with respect to the acetylcholine receptor and other ligand-gated ion channels that the ion channel corresponds to the "active" site, with the agonist-binding site being the homologue of a regulatory "allosteric" site (see Changeux, 1965; Changeux et al., 1967b; Changeux, 1969).

Another important feature of allosteric proteins is the manner in which they cooperate to bind ligands. Studies of hemoglobin provided a unique opportunity for understanding this phenomenon. Like many regulatory proteins, including enzymes and gene repressors, hemoglobin exhibits ligand-binding curves that significantly deviate from the classical Langmuir isotherm. The sigmoidal curve for the binding of oxygen by hemoglobin had intrigued biochemists early in the twentieth century and led to models (for example, by Haldane, Hill, Adair, Pauling) based on interactions between individual oxygen-binding sites (see reviews by Edelstein, 1975; Edsall, 1980). It came as a surprise, however, to discover that once a crystal structure had been achieved (for a review of early structural studies, see Perutz, 1969) the sites are not in direct contact, but more than 20 Å apart from one another. The indirect nature of the cooperative interaction between O_2 molecules prompted theorists to look in new directions. Work being done at about this time on regulatory enzymes revealed that they resembled hemoglobin in also displaying a sigmoidal curve for catalytic activity (and binding) as a function both of substrate and of regulatory ligand concentration (Umbarger, 1956; Changeux, 1961; Gerhart and Pardee, 1962).

Attempts to interpret this striking property in terms of protein structure had a critical impact on the understanding of the conformational changes that account for allosteric coupling (Monod et al., 1965). These *homotropic* interactions were attributed to a particular aspect of the protein molecule's design, namely the cooperative and symmetrical assembly of several identical subunits into a finite "microcrystal" or oligomer (an idea already suggested for hemoglobin by Wyman in 1948). As a result, the protein molecule was assumed to possess two or more sets of identical binding sites, and it was proposed, as the simplest possible mechanism, that these sites interact via a *concerted all-or-none transition* of the subunits between discrete symmetrical states (R and T) with different ligand-binding properties (Monod et al., 1965; Edelstein, 1971):

$$R \rightleftharpoons T.$$

The two states were postulated to exist prior to ligand binding, and a novel intrinsic parameter for the protein, $L = T/R$, was introduced to describe their equilibrium constant for isomerization.

The homotropic interactions between identical sites and the "heterotropic" interaction of an allosteric effector with substrate binding, for instance, thus result from the stabilization by the ligand in question of the state for which it displays the highest binding affinity. As a consequence the $T \rightleftharpoons R$ equilibrium shifts in favor of this state. The conservation of molecular symmetry is thought to be a condition for cooperative transition. Because of the stringent limitations imposed by symmetry requirements, only a few different conformations are available in principle for purposes of ligand binding.

In a more general formulation of the theory (Rubin and Changeux, 1966), the claim that the ligands bind to only one of the two R and T states was abandoned and the hypothesis proposed instead that any given ligand can bind to the same site on the R *and* T conformations, only with different affinities. One of the consequences of this *nonexclusive binding* is that a saturating level of effector may produce an incomplete shift of the $T \rightleftharpoons R$ equilibrium in favor of one of the two states. The *partial* activation and inhibition observed in several experimental systems could be accounted for in these terms (Rang and Dale, 1987). Furthermore, cooperativity is seriously diminished when either state is excessively favored, as has been observed in the case of several mutant hemoglobins (Edelstein, 1971).

A critical feature of the mathematical formalism developed in this context is the distinction between two functions, the *binding* function ($\bar{Y}$) and the *state* function ($\bar{R}$), which express the proportion of sites occupied by the ligand and the proportion of molecules in the R conformation, respectively. This departure from the classical enzyme model could therefore be used to test the theory (see Figure 1.4A and Chapter 6).

An alternative schema, initially suggested in the case of hemoglobin (Adair, 1925; Pauling, 1935) and subsequently formalized by Koshland et al. (1966), is based on the view that proteins in general, and regulatory proteins in particular, are flexible, plastic objects that can exist in multiple conformations "induced" by the ligand (Koshland, 1963) and therefore are directly shaped to its structure. The relevant formalism unavoidably includes a significant number of parameters, some of which have no evident physical significance. General models

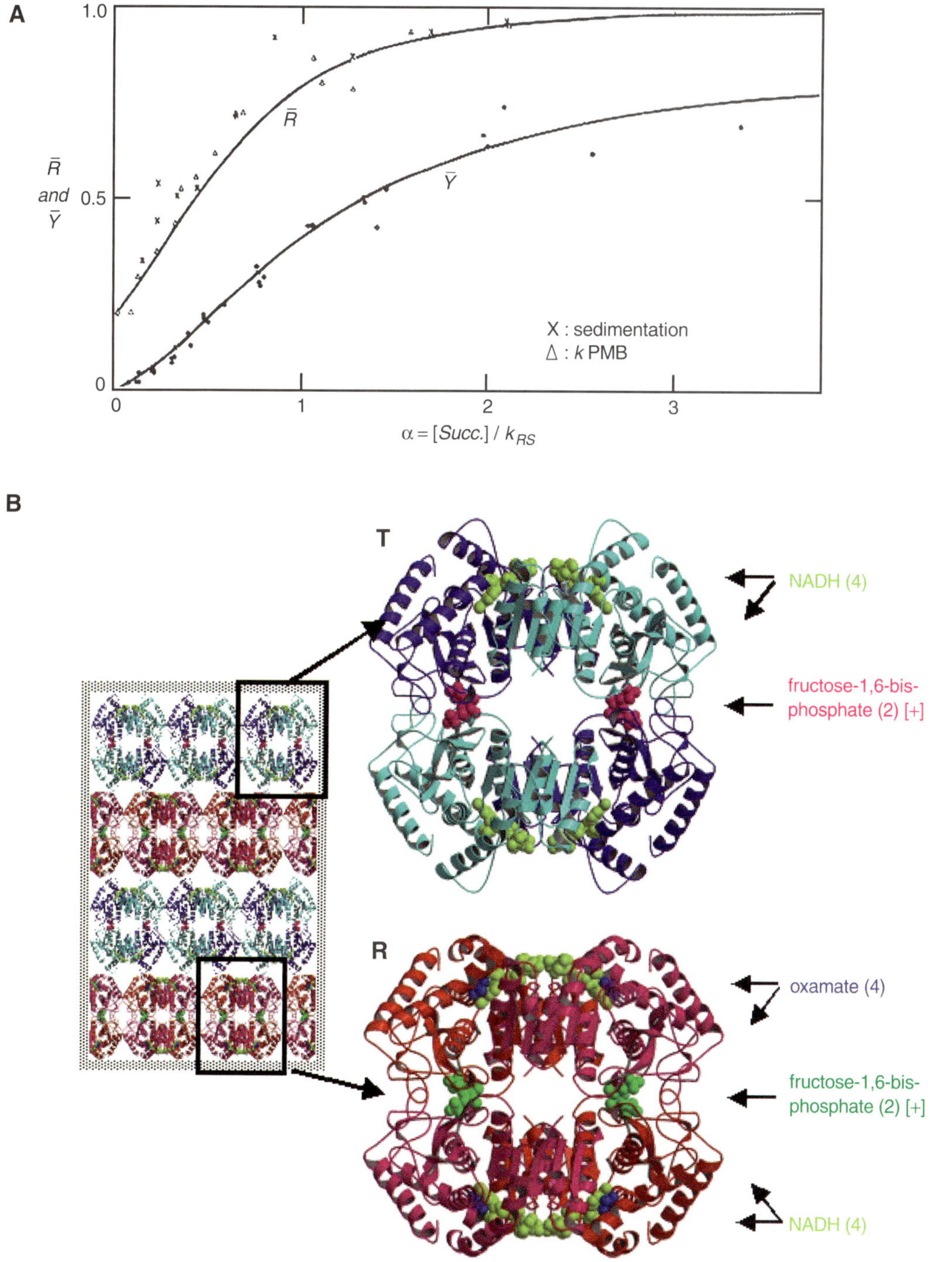

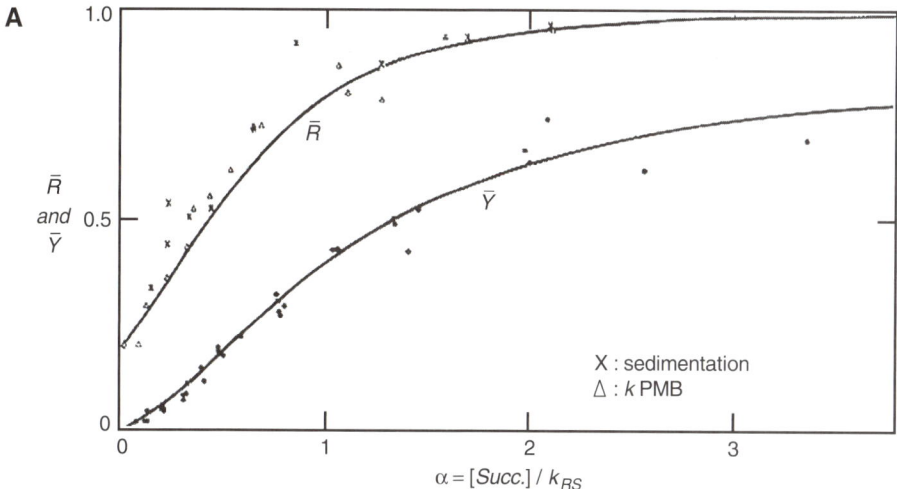

FIGURE 1.4 Functional and structural evidence supporting the Monod-Wyman-Changeux model. (A) Functional evidence: distinct progressions of the state ($\bar{R}$) and binding ($\bar{Y}$) functions measured with aspartate transcarbamylase as a function of succinate—a substrate analogue—concentration; in these experiments succinate binding was compared with the change in conformation as monitored by the sedimentation coefficient (X) and by the rate of reaction of sulfhydryl groups with p-mercuribenzoate (PMB) (Δ). (From Changeux and Rubin, 1968.) (B) Structural evidence: X-ray structure of bacterial L-lactate dehydrogenase from *Bifidobacterium longum* showing the conservation of the molecular symmetry during the course of the R-T transition, together with the associated ligand-binding sites located at subunit interfaces. (From Iwata et al., 1994.)

have subsequently been proposed that incorporate features of both the two-state and induced-fit approaches (Eigen, 1967; Hammes and Wu, 1974; Jardetzky, 1996).

Applied to real systems, the induced conformational change model is sometimes difficult to distinguish from elaborate versions of the two-state model. But one of its critical and experimentally testable predictions—a prediction not made by the two-state model—is that state and binding functions should coincide.

Experiments carried out under equilibrium conditions with aspartate transcarbamylase (Changeux and Rubin, 1968; Gerhart and Schachman, 1968; Hensley and Schachman, 1979) and phosphorylase b (Buc and Buc, 1968), and by rapid kinetic methods with glyceraldehyde-3-phosphate dehydrogenase (Kirschner et al., 1966) or hemoglobin (Henry et al., 1997), unambiguously showed that in these systems the binding function differs from the state function (see Figure 1.4A). In the case of hemoglobin, certain aspects of these conclusions have been challenged (Ackers et al., 1997), but the objections may be based on questionable data (Marden et al., 1998), at least in part (Ackers et al., 2000). The bulk of the evidence indicates, however, that the major conformational transitions of the protein molecule are neither "induced" by the ligand nor "tertiary-linked"; instead they are "quaternary-linked," exist prior to ligand binding, and undergo modification as a function of the energy of ligand binding (Shulman, 2001). Results recently obtained concerning aspartate transcarbamylase provide new evidence in support of this conclusion (Macol et al., 2001). In a remarkable investigation by Iwata et al. (1994) involving bacterial L-lactate dehydrogenase, both R and T conformations were observed in the same crystal (see Figure 1.4B).

The two-state model surely does not account for all modes of interaction between sites in regulatory proteins. It was not designed to do so. It does not, for instance, provide a simple explanation for the so-called negative cooperativity encountered with rabbit muscle glyceraldehyde-3-phosphate dehydrogenase (see Henis et al., 1979). In a typical regulatory enzyme like aspartate transcarbamylase, interactions between heterotropic ligands have also been reported to exist without significantly affecting their homotropic (positive) cooperative effects (Kerbiriou and Hervé, 1977). Even the X-ray diffraction studies of hemoglobin by Perutz reveal that, in addition to major transitions affecting primarily the quaternary structure of the molecule, local conformational changes may also occur as a consequence of the actual bonding of the ligand (oxygen) to its site (the heme). Baldwin and Chothia (1979) nonetheless concluded in a general review of work on hemoglobin that "the change in geometry at one site is not produced by *direct* propagation of structural changes induced by the binding of ligand at another site. Only very small structural changes occur on ligand binding to a subunit unless the quaternary structure changes." This is essentially the same conclusion recently reached by Henry et al. (2002) with a generalized two-state model.

The controversies surrounding this issue led to the unambiguous distinction, now validated by a great many experimental observations (see Jardetzky, 1996), between the main regulatory interactions of physiological relevance, which are primarily mediated by intrinsic, major ligand-independent transitions of the quaternary structure, and the actual binding of the ligand, which may cause local rearrangements specific to its particular structure. In other words, standard allosteric molecules may undergo conformational transitions on two distinct levels, one *global* and the other *local*. In the first case, the transitions involve the quaternary structure of the allosteric molecule and are associated with a regulatory function; in the second case, they involve the tertiary structure of individual subunits, a circumstance that is found in the classical monomeric enzymes, but that has also been confirmed by precise functional measurements of certain voltage-gated channels (Changeux and Edelstein, 1998), for example.

The first explicit applications of an early version of the two-state model to receptors (Changeux et al., 1967b; Karlin, 1967a; Changeux, 1969; Edelstein, 1972) opened the way to testing of more robust versions that successfully accounted for the full range of properties of the receptor, including single channels, desensitized states, and complex mutant phenotypes (see review by Changeux and Edelstein, 1998). The analysis of quaternary structures in soluble enzymes with mutually perpendicular axes of symmetry was able to be generalized to membrane receptors with only a single rotational axis of symmetry (or pseudosymmetry in the case of the muscle nicotinic receptor, for example) perpendicular to the plane of the membrane. The basic concept of a T $\rightleftharpoons$ R concerted transition was shown to be sufficient to account for both homotropic and heterotropic interactions, and was readily extended to desensitized forms of the receptor by incorporating high-affinity states with a closed ion channel. Certain additional modifications are nevertheless required to account for the loss of symmetry accompanying the specialization of subunits in the heteropentameric forms of the receptor. Moreover, the evolutionary development of desensitized states requires that multiple transitions be incorporated into complete allosteric models for receptors, giving rise to a "network of transitions" (Galzi et al., 1996a). The various features of the model, and their experimental verification, are described in detail in Chapter 6 and in the Appendix.

Some Epistemological Remarks

With regard to models in general, the opinion is commonly found among experimental scientists, physiologists, and biochemists that preliminary theorizing is not necessary for experiments to be carried out. In the view famously associated with Francis Bacon, theory should follow, not precede, the collection of data. From its inception, however, receptor research has followed Descartes in constructing *a priori* theoretical hypotheses and formal models. Langley, as we noted earlier, was often criticized for his "unnecessary" proposal of a "receptive substance" (see Dale, 1943). Yet very often nature answers only in the language one uses to ask questions of it. In this respect the receptor concept turned out to be a fruitful one.

In the early stages of research on a new problem, a model furnishes, at the very least, the basic language and the nomenclature required to raise and give precise expression to critical issues. A model is an attempt to conceive and formulate a "representation," in simplified form, of a real object or phenomenon (see Thom, 1979; Suppes, 2002; Changeux, 2004). It is not expected to exhaust all the properties or features of the object or phenomenon being investigated: reality will be always be different and, in any case, more complex. What is sought are "mathematical objects that are compatible with natural objects" (Changeux and Connes, 1995). Models, in other words, are mental representations produced in the brains of scientists that, in the best case, closely describe relevant features of real objects or phenomena. Furthermore, because models have been designed, through extensive "tinkering" and selection in the conscious workspace of scientists' brains, to be parsimonious, noncontradictory, and internally consistent (see Changeux, 2004), they are frequently expected to reveal certain features of reality that had not been anticipated on the basis of straightforward empirical inspection alone. In any case, they yield predictions that may be compared quantitatively to experiments. This "interpretation" of the theory, and conversely of the data by the theory, requires that the theoretical variables of the model be placed in precise correspondence with the experimental observations (Changeux et al., 1973).

Sometimes this correspondence cannot be established in any simple way—for instance, when the model posits the existence of a mechanism that does not occupy the same *level* of description as the observations. Experimentalists then are apt to feel, and sometimes correctly, that what they judge to be an unanticipated but essential complexity has been missed by the oversimplified formalism of a model. Occasionally they claim that experimentation is more fruitful when it is conducted without reference to explicit models, unaware that they are nonetheless implicitly drawing upon previously established models, often very crude and rudimentary ones. What matters in the case of a given model at a particular moment, of course, is that it adequately fit the experimental data. Whether it is right or wrong, it may also serve a *heuristic* function by opening up new avenues of theoretical and experimental inquiry. In this respect, the foregoing discussion of receptor models illustrates what might be called the epistemic evolution of models throughout the past history of the research. One notes, for example, that the mechanisms underlying the biological evolution of antibody diversity suggested by early models are more closely related to immediate "empirical" perception by the senses than those suggested by later versions. Here, as in other chapters in the history of biology, instructive (Lamarckian) models have been abandoned in favor of selective (Darwinian) models (see Changeux, 2004).

This brief survey of the progress made in understanding neurotransmitter receptors also points to another aspect of the dynamics of scientific research, namely the presence of separate, often competing traditions or cultures, even in fields as specialized as the neurosciences. These cultures typically possess a distinctive set of founding doctrines, exploit highly specialized technologies, employ their own vocabularies and concepts, publish results in their own journals, and embrace their own value systems, which, though they are seldom acknowledged publicly, make themselves felt in scientific politics at the national as well as the international level. The historical confrontation of the four scientific cultures that have contributed to the development of the receptor field has been rich in consequences. Multicultural contacts (or "coalitions," to use Claude Lévi-Strauss's term) have yielded fresh experimental approaches, which in turn have led to the formation of evolving cognitive networks, with the result that various lines of research have converged to produce new "bridging concepts," or "cognitive nodes," linking work in adjacent disciplines (see Changeux, 2004).

An important cognitive node emerged in the 1970s with the introduction of a mode of thinking and a set of techniques peculiar to molecular biology in a field that until then had been dominated by electrophysiology, anatomy, and pharmacology. The harnessing of these disciplines enabled researchers to share a coherent perspective as they moved from the physiological level (characterization of the synapse) to the structural level (three-dimensional organization of receptors and channels and the regulation of their gene expression). Thanks to advances in the biochemistry and molecular biology of receptors, previously inaccessible questions of functional organization could now be posed in increasingly reductionist terms, making it possible to analyze events over shorter and shorter times on smaller and smaller scales—ultimately, on the scale of individual amino acid residues of single molecules.

CHAPTER 2

Purification and Characterization of the Nicotinic Receptor

In the 1970s, the convergence of the various cultures we have just discussed within the newborn field of neuroscience resulted in the identification of a few discrete molecular objects as determinants of critical functions in the nervous system. The isolation of the nicotinic receptor marked one turning point, since prior to that time the chemical nature of the receptor was a mystery—so much so that no less distinguished a scientist than Sir Henry Dale (1943) expressed doubts that it would ever be identified, and considered efforts to do so a vain enterprise. Nevertheless, the description of a neurotransmitter receptor and its associated ion channel as well-defined biochemical entities provided a first concrete link between physiology and biochemistry. It opened up the way to detailed structure-function studies of novel molecules specialized in interneuronal communication and signal transduction, as well as of new aspects of gene expression in neurogenesis and synapse development. The story of how receptors came to be isolated is enlivened by the fact that research focused on what we regard as two of the greatest wonders of the animal world: the fish electric organ, an extremely rich source of receptors (Nachmansohn, 1959); and the venom of snakes of the genus *Bungarus*, which contains an α-toxin (Lee and Chang, 1966) having an exquisite sensitivity and a high affinity for the receptor (see Changeux, 1981; Grutter and Changeux, 2001).

Electric Organs

Electric organs in fish (see reviews in Nachmansohn, 1959; Changeux, 1981) played a critical role in the identification and purification of the acetylcholine receptor in a functional state for several reasons. First, the electric organs from *Electrophorus* and *Torpedo* (Figures 1.2 and 2.1A) are exceptionally rich in a single class of cholinergic (nicotinic) synapses (Marnay, 1937, cited in Nachmansohn, 1959). The total number of synapses in an electric organ of *Electrophorus electricus,* for example, is 10^{10} or more. The component elements of the cholinergic synapse and, in particular, the nicotinic receptor therefore make up a large share of the content of such organs (as we shall see, up to several hundred milligrams per kilogram of fresh tissue in *Torpedo marmorata*).

Second, the elementary unit of the electric organ, or electroplaque, is a highly polarized and giant multinucleated structure, one face of which is innervated and excitable, the other specialized in active transport and rich in Na^+, K^+-ATPase (see reviews in Fessard, 1958; Changeux, 1981). The dissection of a single electroplaque from the tissue of *E. electricus* permitted electrophysiological recordings and quantitative *in vivo* pharmacological assays (Schoffeniels and Nachmansohn, 1957; Higman et al., 1963; see Figure 2.1B). Isolation of the electroplaque heralded the development of ligands that were subsequently used to characterize the acetylcholine receptor *in vitro* (Figure 2.1C, D), and made it possible to distinguish agonists, competitive antagonists, and noncompetitive blockers, establish their relative potencies, and test for the reversibility of their effects (Higman et al., 1963; Podleski and Bartels, 1963; Karlin and Bartels, 1966; Changeux and Podleski, 1968; Lester et al., 1975; Langenbuch-Cachat et al., 1988). The effects of SH reagents on the *in vivo* response to cholinergic ligands was noted as well, and used as an additional argument for distinguishing between receptor and catalytic sites of acetylcholinesterase (Karlin and Bartels, 1966). In short, the pharmacology of the electroplaque response was found to be largely similar, though not identical, to that of the postsynaptic membrane of the vertebrate neuromuscular junction—which is to say, to the response typical of a *nicotinic* receptor (Dale, 1914). Desensitization was also shown to occur upon application of nicotinic agonists *in vivo* to the electroplaque of *E. electricus* (Larmie and Webb, 1973; Lester et al., 1975) and of *Torpedo marmorata* (Bennett et al., 1961; Moreau and Changeux, 1976).

Third and finally, fragmentation of the electroplaque into subcellular membrane preparations that could be purified (Changeux et al., 1969; Cartaud et al., 1971) brought physiology into fruitful contact with biochemistry.

Response of Microsacs to Acetylcholine

A key step toward characterizing the receptor at the molecular level was the isolation and development of functionally competent mem-

CHAPTER 2

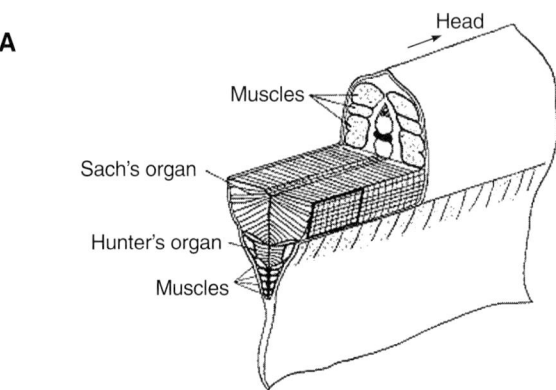

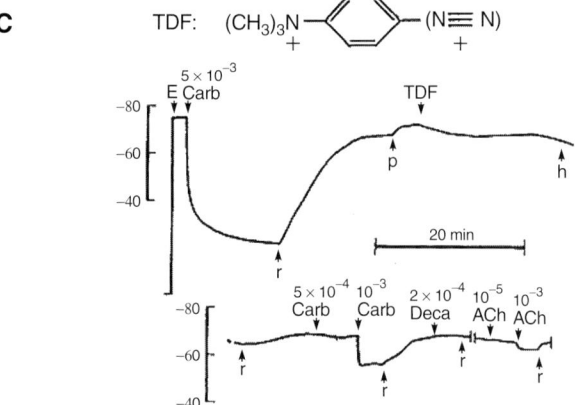

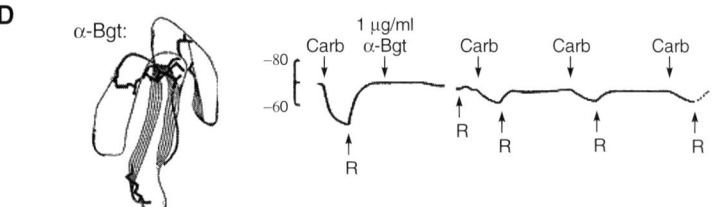

FIGURE 2.1 Electrical responses of a single electroplaque from *Electrophorus electricus*. (A) Anatomy of the electric organ displaying the rows of electroplaques from which individual electroplaques are selected for dissection. (From Massoulié et al., 1970.) (B) Electrical activity recording of a single electroplaque by means of an intracellular electrode. (Modified from Nachmansohn, 1959.) (C) Irreversible blocking of the electrical response of an individual electroplaque to carbamylcholine (Carb) following exposure to the affinity-labeling reagent p-(trimethylammonium) benzene diazonium fluoroborate (TDF). ACh: acetylcholine; Deca: decamethonium. (From Changeux et al., 1967a.) (D) At low concentrations the snake venom toxin α-bungarotoxin (α-Bgt) almost irreversibly blocks the electrical response of the isolated electroplaque *in vivo*. R: Ringer's solution. (From Changeux et al., 1970b.)

brane fragments *in vitro* from an electric organ. Suspensions of membrane fragments from the electric organ of *E. electricus* were selected for purification on the basis of their high acetylcholinesterase content. Observations made possible by electron microscopy subsequently revealed that a majority of the fragments reseal into closed vesicles, or "microsacs" (Changeux et al., 1969; Cartaud et al., 1971). Such microsacs may then retain permeant molecules. The permeability of these microsacs to radioactive cations such as ^{22}Na$^+$, ^{42}K$^+$ was able to be determined (Kasai and Changeux, 1970, 1971) using a simple filtration technique (Figure 2.2A, B) The suspension was first equilibrated overnight in a medium containing the radioactive ion, and the *efflux* was monitored as a function of time. In the presence of a nicotinic agonist, the "apparent volume" of the microsac suspension (0.7–2.0 µl/mg membrane protein) did not change, but the rate of Na$^+$ efflux increased three- to fourfold by comparison with a significant "leak flux." This increase was reversible, and susceptible to blocking by *d*-tubocurarine and other competitive antagonists, including snake venom α-toxins. Furthermore, channel blockers such as tetracaine inhibited the permeability response in a noncompetitive manner. The response of the microsacs to nicotinic agonists thus closely reproduced the *in vivo* process of "activation" by the neurotransmitter, as recorded by electrophysiological methods, yet with the significant advantages associated with a completely cell-free and chemically controlled situation marked by a defined environment on both sides of the membrane (Kasai and Changeux, 1970, 1971). The same tissue could thus be used for convergent physiological, pharmacological, and biochemical studies (Changeux et al., 1970a, 1970b).

The finding that microsacs retained virtually all of the properties of intact receptors was confirmed by several groups working on *E. electricus* microsacs (Hess et al., 1979, 1982), and repeated with receptor-rich membrane preparations initially purified from the *Torpedo marmorata* electric organ (Cohen et al., 1972; Hazelbauer and Changeux, 1974; Popot et al., 1974; Miller et al., 1978; Deleglane and McNamee, 1980; Moore and Raftery, 1980; Neubig and Cohen, 1980). This novel method for measuring the activity of ligand-gated ion channels was subsequently improved from both the theoretical and practical point of view (Bernhardt and Neumann, 1978, 1981) to allow rapid flux measurements in the millisecond-to-minute time range (Hess et al., 1979; Aoshima et al., 1980; Moore and Raftery, 1980; Neubig and Cohen, 1980, 1982; Heidmann et al., 1983; see also the review in Coombs and Hess, 1984). Its utilization played a decisive role in the functional reconstitution of acetylcholine-regulated ion fluxes with the purified receptor protein (as we will see in the final section of this chapter).

CHAPTER 2

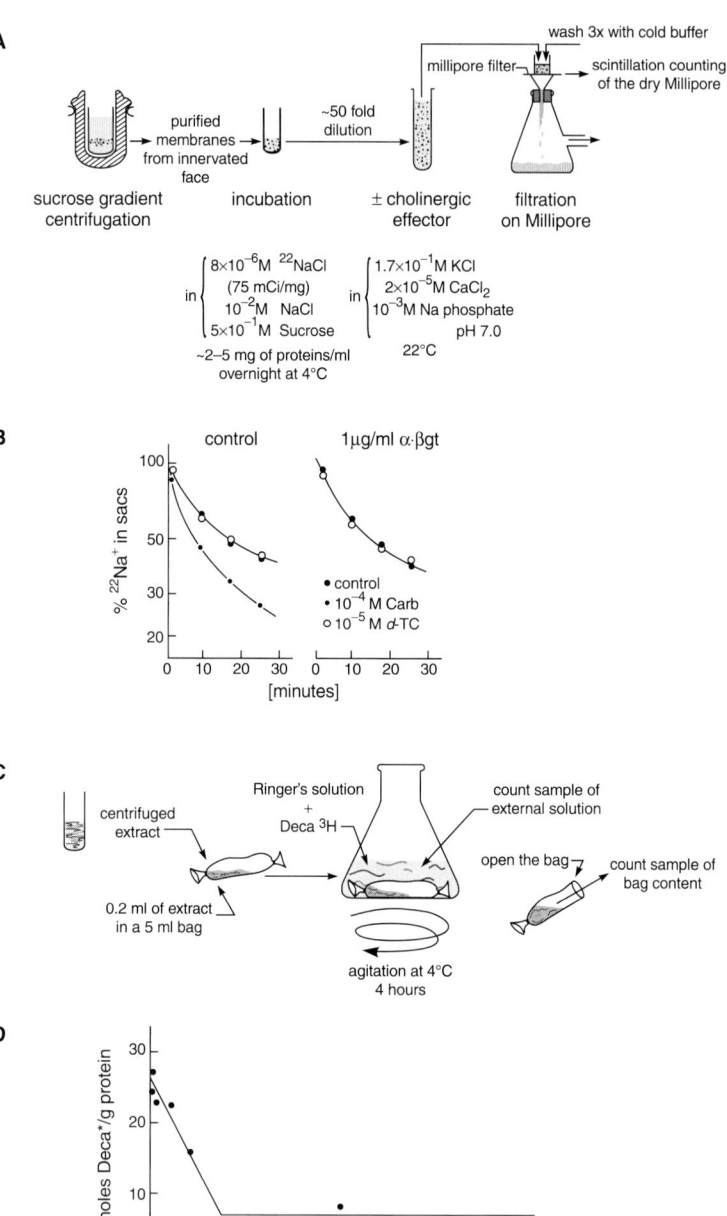

FIGURE 2.2 The use of α-bungarotoxin to identify the acetylcholine receptor. The blocking effect of α-bungarotoxin (α-Bgt) is preserved when the ion permeability response is measured *in vitro* with membrane fragments purified from *E. electricus* electric organ. (A) This method is based on the observation that the membrane fragments reseal into closed vesicles, or "microsacs," which retain permeant molecules and make it possible to measure ion fluxes using a simple filtration technique. (B) Nonreversible blocking by α-bungarotoxin of the micro-sacs' response to carbamylcholine (Carb), which is also reversibly inhibited by *d*-tubocurarine (*d*-TC). (C) Equilibrium dialysis method used to measure the binding of a radioactive ligand (Deca ^{3}H) to sodium deoxycholate extracts of microsacs prepared from the *E. electricus* electric organ. (D) α-bungarotoxin displaces the radioactive agonist decamethonium (Deca) bound to proteins in a sodium deoxycholate extract. (All panels from Changeux et al., 1970b.)

Receptor-Specific Ligands

Early attempts (Chagas et al., 1957; Ehrenpreis, 1960) to characterize the nicotinic receptor by radioactive cholinergic ligand-binding assays, for instance as a proteolipid (De Robertis, 1971), did not appear to be reliable. An attractive method of covalent labeling by a high-affinity ligand with a reactive group, already used with success to label the immunoglobulin-binding site (Singer and Doolittle, 1966), was developed and applied initially to the pharmacology of the muscarinic acetylcholine receptor (Gill and Rang, 1966). The compound first used with the nicotinic receptor was p-(trimethylammonium) benzene diazonium fluoroborate (TDF) (Fenton and Singer, 1965), which was selected (Changeux et al., 1967) for its homology with the nicotinic agonist phenyltrimethylammonium (Changeux and Podleski, 1968) and for its highly reactive diazonium group (Figure 2.1C). Indeed, TDF irreversibly blocks the response to carbamylcholine, and this effect is abolished in the presence of the reversible competitive antagonist *d*-tubocurarine (Changeux et al., 1967). TDF was then used to radiolabel *in vitro* the acetylcholine-binding site (Weiland et al., 1979). Subsequently, its dimethyl photoactivatable homologue p-N, N-dimethylamino benzene diazonium fluoroborate (DDF) (Langenbuch-Cachat et al., 1988) led to the successful identification of several sets of labeled amino acids that participate in the acetylcholine-binding site (Dennis et al., 1988; see Chapter 4).

An improvement in affinity labeling of the acetylcholine receptor site occurred with the development of a compound directly inspired by TDF, but with a maleimide group instead of a diazonium group as the reactive group (Karlin and Winnik, 1968). This compound, 4-(N-maleimido)-phenyl-trimethylammonium iodide (MPTA), does not react covalently with the acetylcholine site unless the electroplaque has been exposed to the disulfide bond-breaking agent dithiothreitol. This treatment, which was previously shown to alter the response of the electroplaque to agonists (Karlin and Bartels, 1966), enhanced MPTA labeling. Ultimately, it led to the identification of cysteinyl residues belonging to the acetylcholine-binding area (Kao et al., 1984). In the late 1960s, however, when these affinity-labeling methods were first used with cellular or subcellular preparations from the electric organ, the specificity of the labeling data was insufficient to allow efficient characterization of the receptor protein from crude detergent extracts.

Small toxic polypeptides purified from the venom of poisonous snakes such as *Bungarus multicinctus* (Lee and Chang, 1966) made it possible to characterize the acetylcholine-binding site, and so to give a chemical description of the macromolecule carrying this site, thereby dramatically shifting the course of research on neurotransmitter receptors. These toxins—compact polypeptides of 61 to 74 amino acids having four to five disulfide bonds and a net positive charge—block neuromuscular transmission in vertebrate tetrapods at the postsynaptic level; *d*-tubocurarine was found to provide effective protection against this block (Chang and Lee, 1963; Lee and Chang, 1966; Lester, 1970). It was not clear, however, whether such molecules would affect the response of electroplaque from organisms as primitive as *Electrophorus* or *Torpedo*. A critical moment in the history of acetylcholine receptor research was therefore the demonstration that α-bungarotoxin blocks the response of *E. electricus* electroplaque to nicotinic agonists, as well as the ion-flux response of the excitable microsacs (Changeux et al., 1970b; Kasai and Changeux, 1970, 1971) to such agonists, at concentrations of less than 1 µg/ml and in an almost irreversible manner (Figure 2.1D). Moreover, the same result was rapidly extended to *Torpedo marmorata* (Miledi et al., 1971; Moreau and Changeux, 1976), thus making possible a reduction from *in vivo* physiology and pharmacology to *in vitro* biochemistry. In combination with the electric organ, these toxins became essential and widely used tools throughout the world for identifying, isolating, and purifying the acetylcholine receptor protein from this tissue *in a physiologically relevant conformation, competent for agonist binding* (Changeux et al., 1970a, 1970b).

Receptor Extraction without Denaturation

Initial attempts to identify the acetylcholine-binding site were hindered by the fact that several of the effectors that are active on electroplaque also bind to several classes of saturable sites on molecules distinct from the receptor and/or have high partition coefficients in lipidic compartments. In order to discriminate between the receptor and other classes of molecules, and to facilitate its purification, it was essential to be able to obtain samples of the receptor in solution without loss of its binding capacity. How to go about doing this was not obvious. Many of the detergents used in protein solubilization were known to cause denaturation of such integral membrane proteins (Changeux, 1966). A few nondenaturing detergents, however, had been found both to disperse the membrane *and* to preserve the α-toxin binding capacity. They included sodium deoxycholate, in association with α-toxin-sensitive radioactive decamethonium binding (Changeux et al., 1970a, 1970b; see Figure 2.2C, D), and Triton X-100, in association with direct binding of the iodinated radioactive α-toxin (Miledi et al., 1971). The solubilized molecule was retained by dialysis membranes with pores selective for molecules smaller than 50,000 daltons; it was sensitive to heating at 100°C for two minutes, to incubation at pH 4.75 for five minutes, and to digestion by pronase (Changeux et al., 1970a, 1970b). As anticipated by Nachmansohn (1959), the physiological receptor site for acetylcholine is carried by a protein.

For detergent-solubilized receptors in solution, it was then possible to show that snake venom α-toxins, which do not have any significant effect on the catalytic activity of acetylcholinesterase (Lee and Chang, 1966), displace in nearly stoichiometric amounts the radioactive agonist *decamethonium* (or acetylcholine), bound to one major class of membrane sites but not to others (Changeux et al., 1970b). On the other hand, the decamethonium not displaced by α-toxins could be largely displaced by compounds known to be powerful inhibitors of acetylcholinesterase (Changeux et al., 1970a, 1971; Kasai and Changeux, 1971). The possibility had been considered that the catalytic site of acetylcholinesterase might also serve as the pharmacological receptor site (Zupancic, 1952; Belleau, 1964; see, however, Podleski and Nachmansohn, 1966; Karlin, 1967). It was therefore possible to unambiguously distinguish between binding to the catalytic site of acetylcholinesterase and binding to another major class of sites that interact with both cholinergic ligands *and* snake α-toxins.

Several of the snake α-toxins were then labeled with radioisotopes (Lee and Tseng, 1966; Barnard et al., 1971; Miledi et al., 1971; Cooper and Reich, 1972), yielding chemically different but still pharmacologically active forms. One particularly elegant procedure consists of iodination followed by dehalogenation in the presence of tritium gas (Menez et al., 1971). This produces a molecule almost identical to the native α-toxin, except that on one histidine residue a hydrogen atom is replaced by its radioactive isotope, tritium. It is interesting to note that cholinergic effectors such as carbamylcholine and *d*-tubocurarine prevented binding of the radioactive α-toxin (Miledi et al., 1971; Meunier et al., 1972a). Moreover, the same rank order was found for the ability of a compound to protect against α-toxin binding and for its *apparent* affinity with the isolated electroplaque (Meunier et al., 1972a; Weber et al., 1972, 1974a, 1974b, 1974c). On the basis of the known antagonistic action of the α-toxins, it was proposed (Changeux et al., 1970b, 1971) that the site at which cholinergic ligands *and* α-toxins bind in a mutually exclusive manner *is* the physiological target site for the electrogenic action of acetylcholine. This suggestion has been confirmed by all subsequent studies of electric organ and skeletal muscle cells (see reviews in Karlin, 1980; Changeux, 1981; Conti-Tronconi and Raftery, 1982).

First Steps toward Receptor Purification

Direct and reliable assays of the physiological receptor site for acetylcholine, based on the use of these toxins, were developed and used to monitor the extraction of the receptor protein from membranes prepared from electric tissue. The solubilized protein was then separated physically from the enzyme acetylcholinesterase by selective adsorption to snake venom α-toxins covalently bound to sepharose (Changeux et al., 1971; Meunier et al., 1971a; see also Changeux et al., 1970b). This experimental observation eliminated the possibility—suggested by Changeux (1966) on the basis of the experimental demonstration of "peripheral sites" for curare-like agents on acetylcholinesterase—that the physiological receptor site was an allosteric site present on a macromolecular complex along with acetylcholinesterase.

Standard sedimentation measurements for the deoxycholate-solubilized protein from *E. electricus* revealed a sedimentation coefficient slightly smaller than that of catalase on sucrose gradients (about 9S) and a Stokes radius close to that of β-galactosidase on Sepharose columns (about 7 nm) (Meunier et al., 1971b; see Figure 2.3A). These values were confirmed by other groups working with *Electrophorus* or *Torpedo californica*, with the same (or different) detergent(s) (Raftery et al., 1971; Berg et al., 1972; Klett et al., 1973; review by Changeux, 1981). Moreover, apparent inconsistencies among the values observed (if one assumes that the receptor is a globular water-soluble protein) were assigned to the association of significant amounts of detergent to the receptor protein in solution (Meunier et al., 1972a, 1972b), for example, up to 0.35–0.49 g Triton X-100 per g protein (Wise et al., 1979).

The receptor protein was thus sufficiently well characterized in crude extracts to contemplate the further steps of fractionation and purification. In the period 1972–1973, several groups working on the acetylcholine receptor achieved a significant degree of purification of the α-toxin binding material using crude detergent extracts from *Electrophorus* electric organs (Olsen et al., 1972; Biesecker, 1973; Karlin and Cowburn, 1973; Klett et al., 1973; Chang, 1974; Lindstrom and Patrick, 1974; Meunier et al., 1974) or *Torpedo* electric organs (Karlsson et al., 1972; Schmidt and Raftery, 1972; Eldefrawi and Eldefrawi, 1973; Potter, 1973). The most common method was affinity chromatography, which previously had been employed for a number of different proteins, including insulin receptors (Cuatrecasas, 1971, 1972). Application to the acetylcholine receptor relied on columns with immobilized quaternary ammonium agonists or antagonists (Figure 2.3B) (Olsen et al., 1972; Schmidt and Raftery, 1972) and snake α-toxins (Karlsson et al., 1972; Klett et al., 1973; Lindstrom and Patrick, 1974).

The most highly purified preparations of receptor protein had specific activities of 100,000–125,000 daltons per α-toxin site, and gave a symmetrical protein peak on sucrose gradients that coincided with α-toxin binding as well as a single band after cross-linking with bifunctional reagents. These preparations, when examined by electron microscopy, revealed for the first time 90 Å ringlike particles of identical size and shape (Cartaud et al., 1973; Meunier et al., 1974; Cartaud et al., 1978; see Figure 2.4A)—observations that established the

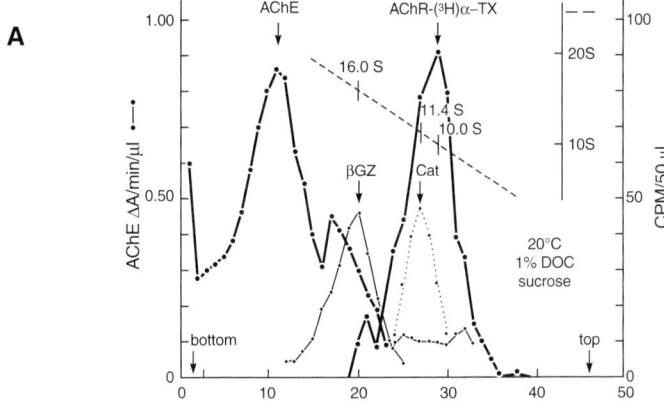

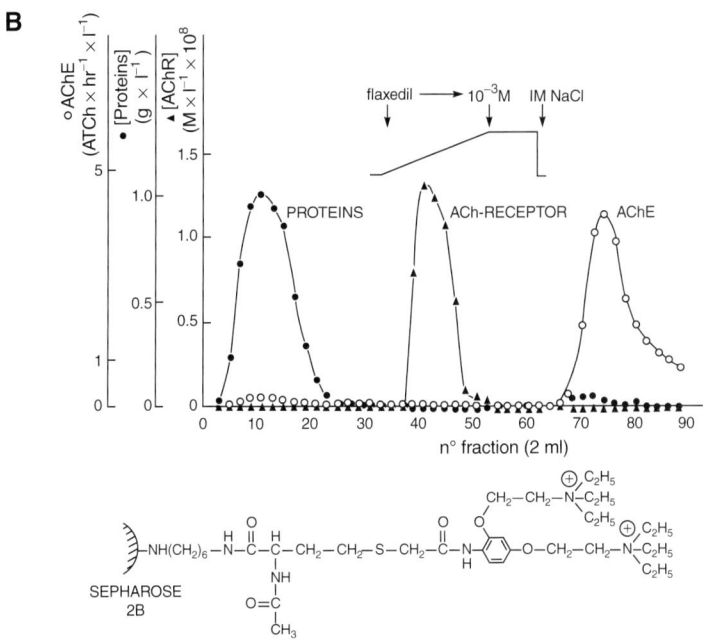

FIGURE 2.3 Physical properties of detergent-extracted acetylcholine receptor proteins from *E. electricus*, assayed with tritiated *Naja nigricollis* α-toxin (α-TX). (A) Sedimentation profile in a sucrose gradient established in the presence of 1% sodium deoxycholate (DOC). Cat: catalase. (From Meunier et al., 1971b). (B) Affinity chromatography on a column made from an immobilized derivative of the cholinergic competitive antagonist flaxedil (formula indicated). The receptor protein was assayed by following radioactive α-toxin binding after selective elution by flaxedil. It is clearly separated from acetylcholinesterase (AChE), which is released in the presence of IM NaCl. (From Olsen et al., 1972.)

homogeneity of the purified preparations, obtained in milligram amounts (Changeux, 1975; Heidmann and Changeux, 1978; Fambrough, 1979; Karlin, 1980; Lindstrom and Dau, 1980; Changeux, 1981).

Identification of the Purified Protein with the Physiological Receptor

The characterization and assay of the receptor protein, before and during purification, were based on the ability of the protein to bind snake venom α-toxins. As expected, the purified protein also bound cholinergic ligands. All the tested agonists and antagonists that were pharmacologically active on the isolated electroplaque interacted with the purified receptor with the same rank order observed in tests on live cells within a given category (see Eldefrawi and Eldefrawi, 1973; Klett et al., 1973; Meunier and Changeux, 1973; Moody et al., 1973; Meunier et al., 1974). These binding measurements provided further proof that the purified protein was indeed the pharmacological receptor of acetylcholine.

Decisive evidence for the physiological role of the purified protein was supplied by immunological experiments (Heilbronn et al., 1973; Patrick and Lindstrom, 1973; Sugiyama et al., 1973, 1974;). When injected into various animals (rabbits, rats, mice, monkeys, etc.), the receptor

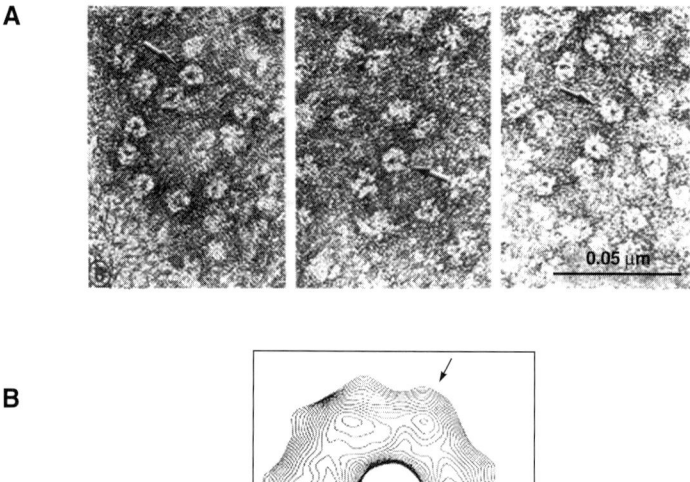

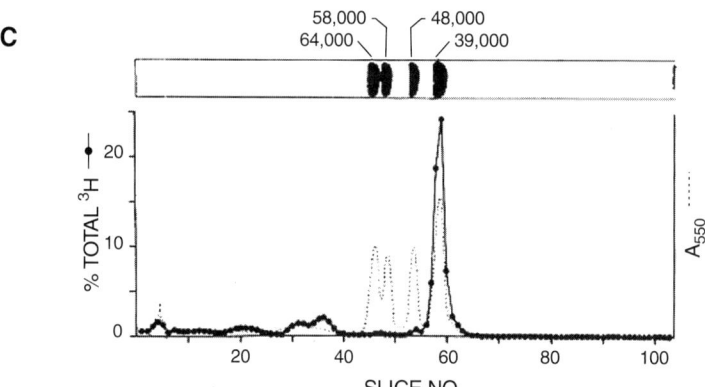

FIGURE 2.4 Structural observations of early purified preparations of acetylcholine receptor protein. (A) Electron micrographs of one of the first preparations from *E. electricus*. The rosettes have a diameter of about 9 mm, with a central depression about 1.5 nm wide, and display a structure composed of five to six subunits. (From Cartaud et al., 1973.) (B) Image obtained by computerized analysis after reorientation and summation of negatively stained individual rosettes. The image displays five unequal peaks of electron density distributed around the central pit. The two peaks marked by an arrow are reinforced by a snake venom α-toxin and therefore assigned to the α-subunits, which are nonadjacent. (From Bon et al., 1982, 1984.) (C) Denaturing polyacrylamide gel electrophoresis of a receptor molecule from *T. californica* that yields four polypeptide chains: α (39,000), β (48,000), γ (58,000) and δ (64,000); only the α-chain is labeled by [^{3}H]-MPTA, an affinity-labeling reagent of the binding site. (From Weill et al., 1974.)

protein from fish causes an immune reaction associated with a flaccid paralysis, which resembles the human disease myasthenia gravis (see reviews by Fuchs, 1979; Lindstrom and Dau, 1980; Lindstrom, 1986). Serum from immunized animals precipitates the purified fish receptor protein as well as the receptor protein from the animal recipient (rat, rabbit)—proof of the receptor's remarkable evolutionary stability (Sugiyama et al., 1973; see review by Lindstrom and Dau, 1980). Such antisera did not react against the cholinergic proteolipid (De Robertis, 1971) transferred to detergent solution, nor did antisera raised against the proteolipid react against the detergent-purified receptor protein (Barrantes et al., 1975; Heilbronn, 1975). Moreover, the serum of immunized animals blocks the response of the isolated electroplaque to bath-applied cholinergic agonists (Patrick et al., 1973; Sugiyama et al., 1973). This blocking effect demonstrates that the protein used in immunization plays a critical role in the electrogenic action of acetylcholine—which had been expected in the case of the actual acetylcholine receptor.

The most compelling evidence that the purified receptor encompasses all functional features of the native receptor was obtained by two sets of reconstitution experiments that achieved a functional acetylcholine-gated ion channel in an artificial lipid environment from chemically defined components.

Receptor Reconstitution

The first method to give successful results (Kagawa and Racker, 1971) took as its point of departure the physical reintegration of the cholate solubilized receptor from *E. electricus* into a membrane structure by extensive dialysis without loss of α-toxin binding activity (Changeux et al., 1972). It is interesting to note that the particulate material thus obtained appeared under the electron microscope as a set of closed vesicles resembling the native microsacs (Figure 2.5A). In a second stage, such microsacs reformed from *T. marmorata* receptor-rich membranes (after cholate solubilization and supplementation with *Torpedo* lipids and divalent cations) were shown to have recovered their initial function. They were sufficiently sealed to retain Na$^+$, and the rate of ^{22}Na$^+$ efflux was accelerated by carbamylcholine (producing a value of $K_{app} = 5 \times 10^{-5}$ M, close to that obtained with native microsacs). This increase was blocked by α-toxin. It had therefore been shown possible to reconstitute an acetylcholine-sensitive membrane from a detergent extract of high-specific-activity membranes containing the acetylcholine receptor as the dominant protein (Hazelbauer and Changeux, 1974). This method was significantly improved by the addition of a complex lipid mixture from soybeans, known as asolectin, at all stages of the reconstitution process (Epstein and Racker, 1978; Changeux et al., 1979; Huganir et al., 1979; Wu and Raftery, 1979; Anholt et al., 1980; Popot et al., 1981).

The reproducibility obtained with highly purified preparations of receptor protein permitted a detailed analysis of the pharmacology (Changeux et al., 1979), ionic selectivity (Popot et al., 1981), and desensitization of the response (Figure 2.5B). In short, the functional properties of the receptor oligomer appeared to be essentially the same as those of the original membranes (Heidmann et al., 1980; Sobel et al., 1980; Popot et al., 1981). Fast kinetic experiments were subsequently performed by following Tl^+ (Wu and Raftery, 1981a, 1981b) or Rb^+ (Walker et al., 1982; Kaldany and Karlin, 1983, 1984) influx into vesicles in a rapid-mixing apparatus. Once again, the results were similar to those obtained with the native microsacs (see review in Montal, 1986).

The success of these experiments did not seem to be due to any *absolute* requirement involving a particular class of lipid, but both cholesterol and negatively charged lipids were important in reconstituting ion-gating activity (Kilian et al., 1980; Fong and McNamee, 1986, 1987; McNamee et al., 1986; Jones and McNamee, 1988). Analysis of the ion-influx response to carbamylcholine and ligand binding under conditions of controlled progressive delipidation revealed a minimum requirement of 45 lipid molecules per receptor for stabilization in a fully functional state. Progressive irreversible inactivation occurs as the lipid-to-protein ratio decreases below 45. Complete inactiva-

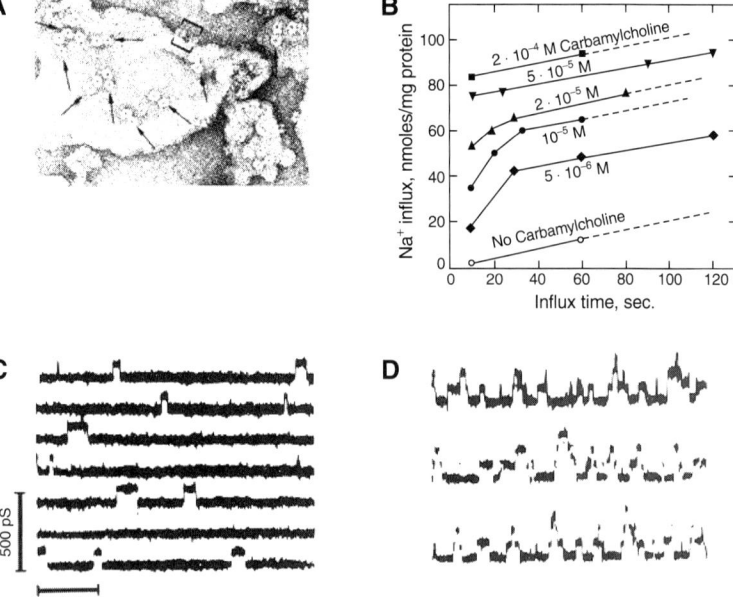

FIGURE 2.5 Reconstitution of the nicotinic acetylcholine receptor into lipid vesicles or bilayers. (A) Electron micrograph of the preparation analyzed in B. The arrows point to doubles of acetylcholine receptor rosettes. (From Popot et al., 1981). (B) Reconstitution into vesicles composed of soybean phospholipids. The time course of $^{22}Na^+$ influx is measured as a function of carbamylcholine concentration. (From Epstein and Racker, 1978.) (C, D) Reconstitution into lipid bilayers. Single-channel current fluctuations recorded (C) 1 minute and (D) 10 minutes after addition of 10 μM carbamylcholine. (From Schindler and Quast, 1980.)

tion takes place below a ratio of 20—a result consistent with a functional requirement for a single shell of lipids around the perimeter of the receptor molecule (Jones et al., 1988; see reviews in Baenziger and Chew, 1997; Barrantes, 1997; Corbin et al., 1998; Burger et al., 2000).

A second method for reconstituting the acetylcholine-gated channel consisted in incorporating the receptor protein into planar black lipid films. This method was based on the observation that material from isolated membranes (or liposomes) assembles spontaneously into lipid-protein monolayers at the solution surface, which may be combined into planar bilayers (Schindler and Quast, 1980). Openings and closings of single acetylcholine-gated channels, which closely resemble those observed *in vivo* (Neher and Sakmann, 1976), were recorded in such artificial bilayers after incorporation of receptors from native receptor-rich membranes (Schindler and Quast, 1980) or from purified preparations (Suarez-Isla and Hucho, 1977; Nelson et al., 1980; Boheim et al., 1981; Montal, 1986; Montal et al., 1986) (Figure 2.5C, D), as well as transmembrane peptides (Montal et al., 1993; Montal and Opella, 2002).

Conclusion

Purification of the receptor was a critical step in the experimental investigation of nicotinic receptors. It definitively resolved the uncertainty surrounding the chemical nature of the receptor, demonstrating that it is neither an enzyme nor a soluble binding protein, but rather an integral membrane protein. The fact that it could be solubilized by detergents without becoming denatured made it possible to specify detailed structure-function relations. Purification of the 9S receptor protein demonstrated the feasibility of reconstituting functional receptors, and established that the acetylcholine-binding site and the ion channel belong to the same biochemical species. Indeed, the oligomer not only carries acetylcholine-binding sites and contains an ionic channel; it also possesses all the structural elements required for the activation (and desensitization) of the ionic response to nicotinic agonists on physiological timescales and over acetylcholine-concentration ranges. Further evidence for the detailed roles of various portions of the receptor in each of these functional aspects was obtained by subsequent studies involving chemical modification and recombinant DNA-gene expression techniques. We will examine these in the chapters that follow.

CHAPTER 3

Evolution and Diversification of Pentameric Receptor Channels

The isolation of the nicotinic acetylcholine receptor opened a new chapter of research in the life sciences. It provided the means for a mechanism that had previously been studied by physiologists as a transducing "black box" (see Eccles, 1964; Katz, 1966; Hille, 1992) to be investigated as a biochemical entity, using the methods of protein chemistry, recombinant DNA technologies, and molecular biophysics. In the wake of the pioneering achievement we have just described in isolating the nicotinic receptor, thousands of other receptors for neurotransmitters, hormones, second messengers, and related internal signals were identified, as well as receptors for odorant and taste substances. They were found to compose several protein families, with genes that account for about 5% of the human genome (~1500–2000 genes) (Lander et al., 2001; Venter et al., 2001). Owing to its availability in large quantities from fish electric organs, however, the nicotinic receptor remained a favored candidate for detailed analyses of protein architecture. As a membrane-bound regulatory protein, the nicotinic receptor molecule displays a number of specific features that were progressively elucidated: its subunit composition; the sequences of amino acid residues of each subunit; the organization of the subunits into an oligomeric structure; and, finally, the three-dimensional structure of the folded polypeptide chains (now largely worked out—see Chapter 7) and of their molecular assembly.

Shortly after the first sequencing results were obtained by chemical methods, the advent of cloning technologies permitted identification of a set of additional nicotinic receptor genes and their corresponding proteins, together with certain features of gene organization (particularly introns and promoters), thereby casting light upon the evolution of nicotinic receptor genes and the manner of their regulation.

The Determination of Subunit Composition

At the beginning of a new millennium dominated by the sequencing of the human genome and related developments in biotechnology, one is apt to forget how difficult it was to grasp the structure of what was then an entirely new type of membrane protein. Indeed, it took more than a decade. First of all, it was unclear which form represented the functional receptor. The early preparations of detergent-solubilized receptor from *Torpedo* displayed, in addition to the predominant 9S form (Meunier et al., 1971a), an α-toxin binding component with a 13S sedimentation coefficient and an 85 Å Stokes radius (Raftery et al., 1972). Reducing agents such as 2-mercaptoethanol or dithiothreitol converted this "heavy" form into a state indistinguishable from the "light" form (Chang and Bock, 1977; Hamilton et al., 1977; Sobel et al., 1977a). The molecular weight of the heavy form was found to be exactly twice that of the light form (Reynolds and Karlin, 1978), establishing that the heavy form represents a dimer of two light forms linked covalently by an intermolecular disulfide bridge. The functional properties of the *Torpedo* heavy form have generally been found to be virtually identical to those of the light form (Anholt et al., 1980; Montal, 1986), but some minor specific differences have been reported (Schindler et al., 1984; Yeramian et al., 1986; Schmidt et al., 1998). Certain authors (Schindler et al., 1984) regard the dimer as the predominant active form. The heavy form has not been found in extracts of muscle from higher vertebrates, however, nor in the brain. In what follows, when reference is made to the nicotinic receptor, only the "light" form is implied.

Early studies of the purified protein attempted to determine the molecular weight of the 9S receptor, but the initial hydrodynamic measurements were subject to uncertainty because of ill-defined amounts of bound detergent (Meunier et al., 1972a). Various other analytical methods were applied (Biesecker, 1973; Hucho and Changeux, 1973; Reynolds and Karlin, 1978; Wise et al., 1979), and ultimately light scattering and hydrodynamic measurements (Doster et al., 1980; Lo et al., 1982) produced estimates close to the value of ~290 kDa computed from the amino acid sequences of the subunits, plus the contribution (~20,000) of covalently linked sugars (Vandlen et al., 1979). The shape of the molecule is characterized by a radius of gyration of 46 ± 10 Å observed with low-angle neutron scattering (Wise et al., 1979). It carries two sites for α-bungarotoxin and nicotinic ligands (Reynolds and Karlin, 1978; Weiland et al., 1979).

Soon after purification of the receptor had been achieved, the idea that it is composed of subunits emerged from observation of an apparent molecular weight in the presence of sodium dodecylsulfate (SDS) several times smaller than that of the native molecule (Meunier et al., 1972b; Reiter et al., 1972). Elucidating the correct quaternary structure required several years of effort on the part of a number of different laboratories, principally because proteolytic cleavage had distorted the subunit pattern. Once the proteases were inactivated, consensus was finally reached with regard to *four subunits* (indicated by Greek letters), each with the following *apparent* molecular weight: α, 40,000; β, 50,000; γ, 60,000; δ, 66,000 (Raftery et al., 1974; Weill et al., 1974; Raftery et al., 1975; Lindstrom et al., 1979; Karlin, 1980; Raftery et al., 1980; Saitoh et al., 1980) for both electric organ and muscle acetylcholine receptor (see Figure 2.4C).

For many years, only one chain (α, the smallest) had a known functional significance. The α-subunit remains associated with the radioactive α-toxin after partial dissociation of the receptor protein in SDS (Meunier et al., 1972b). As we noted in Chapter 2, the α-subunit is labeled by affinity-labeling reagents for the acetylcholine-binding site (Weill et al., 1974; Kao et al., 1984). Moreover, α-toxin binding can be recovered after partial renaturation of α-subunits purified in the presence of SDS (Haggerty and Froehner, 1981; Gershoni et al., 1983; Tzartos and Changeux, 1983b; Tzartos and Changeux, 1984; Wilson et al., 1984). In the early stages of research, however, the evidence that the β-, γ-, and δ-subunits were intrinsic components of the receptor protein was structural only (Witzemann and Raftery, 1978; Karlin, 1980). These polypeptides might well have represented distinct membrane components that become, or remain associated with, the α-subunit in the course of membrane solubilization, as in the case of certain membrane-bound molecules from the cytoskeleton that interact with the glycine receptor (Betz, 1987) or the voltage-sensitive ion channel (Goldin et al., 1986). Primary structural homologies between the four subunits (Raftery et al., 1980; Tzartos and Lindstrom, 1980) nonetheless strongly suggested an intrinsic functional role. Moreover, one of the chains, the δ-subunit, was covalently labeled by a photoaffinity derivative of the noncompetitive blocker trimethisoquin, and this labeling was *positively regulated* by the binding of agonist to the acetylcholine receptor site (Oswald et al., 1980; Saitoh et al., 1980), thus supporting the view of a subunit contribution to receptor function.

The exact stoichiometry of the four chains that comprise the receptor was also debated for years. In an initial report (Hucho and Changeux, 1973), which appeared at a moment when the identity of the subunits had not yet been established with certainty, partial cross-linking experiments with suberimidate followed by SDS gel electrophoresis yielded a five-band pattern (with some of the bands as doublets). These data were correctly interpreted in terms of a *pentameric* molecular structure (Hucho and Changeux, 1973). Various stoichiometries were subsequently suggested (Raftery et al., 1975; Karlin et al., 1976; Sobel et al., 1979). A 2:1:1:1 subunit stoichiometry ($2\alpha:\beta:\gamma:\delta$) was finally sustained on the basis of molecular weight measurements (Reynolds and Karlin, 1978) and demonstrated by direct quantification (Lindstrom et al., 1979) and sequence analysis (Raftery et al., 1980). This stoichiometry also accounts for the observed two acetylcholine-binding sites per oligomer (Reynolds and Karlin, 1978), and remains universally accepted today.

Subunit Organization within Receptor Oligomers

One of the most convincing lines of evidence for the identification of a novel molecule continues to be its visualization by electron microscopy. In the 1970s, observations made using this technique on negatively stained preparations of purified or membrane-bound receptor from *Electrophorus electricus* (Cartaud et al., 1973) and *Torpedo marmorata* receptor-rich membranes (Cartaud et al., 1973; Nickel and Potter, 1973) revealed ringlike particles 80–90 Å in diameter, with a prominent phosphotungstate-accumulating axial pit (Cartaud et al., 1973; Nickel and Potter, 1973). These were the first images ever obtained of a neurotransmitter receptor. Numerous investigations subsequently confirmed and extended these initial observations, achieving progressively higher resolution of the subunit organization and three-dimensional structure of the individual subunits (Reed et al., 1975; Ross et al., 1977; Schiebler, 1977; Brisson, 1978, 1980b; Cartaud et al., 1978, 1981; Klymkowsky and Stroud, 1979; Zingsheim et al., 1980, 1982; Kistler and Stroud, 1981b; Bon et al., 1982, 1984; Brisson and Unwin, 1984, 1985; Kubalek et al., 1987; Toyoshima and Unwin, 1988; Unwin et al., 1988; Miyazawa et al., 1999, 2003).

These early structural studies brought direct confirmation of the subunit organization of the electric organ and muscle-type nicotinic receptors in *Torpedo*. Computerized-image analysis at ~18 Å resolution after reorientation and summation of up to 132 individual rosettes confirmed the established stoichiometry in disclosing the expected *five* unequal peaks of electron density distributed around a central pit (Bon et al., 1982, 1984), as presented in Figure 2.4B. Two of the peaks, located at 160° from each other, were reinforced by α-toxin binding (Bon et al., 1984) and thus were assigned to the α-subunits. These images of nonadjacent α-subunits were consistent with earlier biochemical labeling experiments using biotinylated α-toxin (Holtzman et al., 1982; Karlin et al., 1983) and monoclonal antibody fragments (Fairclough et al., 1983).

The conclusion that the two α-subunits are not adjacent within the oligomer was confirmed by cryoelectron microscopy on flattened tubular vesicles crystallized in two dimensions (Brisson, 1978, 1980; Brisson and Unwin, 1984, 1985). Studies of these ordered arrays yielded a pseudopentagonal structure of the receptor molecule in three dimensions with an axis of quasisymmetry perpendicular to the plane of the membrane. The location of each individual polypeptide chain was also determined by means of monoclonal antibodies specific for each of the individual subunits (Kubalek et al., 1987). An αβαγδ arrangement of the subunits was inferred from these observations (see also Kistler and Stroud, 1981; Chatrenet et al., 1990), and until recently this ordering of the subunits was the interpretation favored by Unwin and his colleagues (Miyazawa et al., 1999). Cross-linking experiments with labeled α-bungarotoxin (Holtzman et al., 1982; Oswald and Changeux, 1982; Karlin et al., 1983) and *in vivo* expression experiments of different pairs of α- and non-α-subunits (Blount and Merlie, 1989) favor the order αγαδβ, however (see discussion in Popot and Changeux, 1984; Blount and Merlie, 1989; Hucho et al., 1996). This interpretation, now accepted by Unwin et al. (2002), is also consistent with observations indicating that the β- and δ-subunits in a myasthenic mutant are adjacent (Quiram et al., 1999). Moreover, the handedness of the circular subunit arrangement in the order αγαδβ was initially reported as clockwise, when viewed from the synaptic cleft (Machold et al., 1995), but the X-ray structure of a related molecule, the acetylcholine-binding protein from *Lymnaea stagnalis* (see Chapter 7), indicates a counterclockwise arrangement (Brejc et al., 2001). Numerous observations concerning the contributions of the γ- and δ-subunits to the two ligand-binding sites at their respective interface with the α-subunits (see Chapter 4) lend further support to the αγαδβ arrangement. On balance, the evidence clearly suggests that the γ-chain lies between the two α-chains. The receptor molecule thus displays a rather unconventional kind of hidden pseudosymmetry.

Primary Structure and Transmembrane Organization

The main chemical properties of the purified receptor, such as amino acid composition, state of glycosylation, isoelectric point, and so on, did not display any notable features, except perhaps a hydrophobic character slightly greater than that of acetylcholinesterase but less than that of Na^+, K^+ ATPase (Vandlen et al., 1979). Analysis of the primary structure of the subunits was required for a detailed understanding of structure-function relationships, and the historical record shows that relatively rapid progress was made in this area. In the late 1970s, the availability of large amounts of protein and improvements in microsequencing techniques made it possible to take the first step. Thus the NH_2-terminal sequence of the MPTA-labeled α-subunit (Devillers-Thiéry et al., 1979) from *T. marmorata* was identified as follows:

$$NH_2\text{-Ser-Glu-His-Glu-Thr-Arg-Leu-Val-Ala-Asn-Leu-Leu-Glu-Asn-Tyr-Asn-Lys-Val-Ile-Arg-} \ldots$$

The same partial sequence was subsequently reported for the α-chain purified from *T. californica* (Hunkapiller et al., 1979) and was confirmed by all cloning and sequencing data for the α-subunit (Noda et al., 1982; Sumikawa et al., 1982; Devillers-Thiéry et al., 1983). These were, in fact, the first reports concerning the primary structure of a receptor for neurotransmitter since the proposal of the receptor theory by Langley in 1905. The "substance" of the receptor was now unambiguously established as a protein.

The search for structural analogies between the subunits was then carried on with the development of monoclonal antibodies against each individual subunit. It is interesting that extensive immunological cross-reactivity was initially discovered for the chains (Tzartos and Lindstrom, 1980). Microsequencing of the first 54 amino acids of the four purified subunits by Raftery et al. (1980) revealed striking sequence identities (ranging between 35% and 50%) at the amino termini of the four subunits, interpreted as a divergence from a common ancestral gene by fourfold duplication (Raftery et al., 1980). These data suggested that the receptor oligomer possesses authentic (albeit imperfect) structural symmetry, thus reconciling the rather baroque quaternary structure of the protein (four subunits having different molecular weights) with the regular pentameric structure that normally characterizes a symmetrical allosteric oligomer (Changeux, 1981).

Additional insight came from the complete deciphering of the entire primary structure of all four receptor subunits by molecular cloning. About 2.4% of the total messenger RNA (mRNA) molecules in the electric organ of *Torpedo californica* were found to encode for the receptor polypeptides (Mendez et al., 1980), a discovery that facilitated the constitution of cDNA libraries. Clones coding for the receptor subunits were successfully selected from these libraries by two methods: screening for electric organ specificity, selection of mRNA from those clones using the hybridization-selection technique, and identification of the products obtained *in vitro* by immunoprecipitation (Ballivet et al., 1982; Giraudat et al., 1982); and direct hybridization with two sets of oligodeoxyribonucleotides corresponding to known fragments of subunit sequences (Noda et al., 1982, 1983a, 1983b; Sumikawa et al., 1982).

Complete cDNA coding sequences of *T. californica* were established for the precursors of the subunits α (Noda et al., 1982), β (Noda et al., 1983b), γ (Claudio et al., 1983; Noda et al., 1983a), and δ (Noda et al., 1983b), as well as partial (Sumikawa et al., 1982) and complete nucleotide sequences (Devillers-Thiéry et al., 1983) for the α-subunit from *T. marmorata* (Figure 3.1). The corresponding amino acid sequences include the partial amino-terminal peptide sequences previously identified from purified α-subunit of *T. marmorata* (Devillers-Thiéry et al., 1979) and α-, β-, γ-, and δ-subunits from *T. californica* (Raftery et al., 1980). They are preceded by a signal peptide containing 20 (α and β), 17 (γ), or 21 (δ) amino acid residues, all of them largely hydrophobic, that is cleaved off in the mature subunits.

The coding sequences of the mature subunits in *T. californica* are 437 (α), 469 (β), 489 (γ), and 501 (δ) amino acids long, with exact molecular weights of 50,116 (α), 53,681 (β), 56,279 (γ), and 57,565 (δ), corresponding to a total for $\alpha_2\beta\gamma\delta$ of 267,757. As with the partial amino acid sequence data (Raftery et al., 1980), homology among chains was observed to range from 10% to 60% amino acid sequence identity, depending on the region, with an average of 40%. The fact that closer homologies were found between subunits α and β, on the one hand, and between γ and δ on the other (Noda et al., 1983b), suggests that a common ancestral gene for the α- and β-subunits diverged from that of the γ- and δ-subunits in the acetylcholine receptor phylogenetic tree (see below). In the receptor from the muscle of higher vertebrates, the γ-subunit present in the embryo is replaced by an ε-subunit in the adult (Takai et al., 1985; Witzemann et al., 1989).

Comparison of the aligned sequences of the four receptor subunits by different laboratories (Claudio et al., 1983; Devillers-Thiéry et al., 1983; Noda et al., 1983a) disclosed similar hydrophobicity profiles. A subdivision of the homologous chains into four domains was proposed by these investigators: a *large hydrophilic* amino-terminal domain of 210 to 224 amino acids; a *compact hydrophobic* region of 68 residues subdivided into three segments of 19 to 27 uncharged amino acids (numbered M1, M2, and M3); a *small hydrophilic* domain of 109 to 146 amino acids; and a carboxyl-terminal segment of 20 *hydrophobic* residues (numbered M4) (Figure 3.1). The various models of transmembrane organization inferred from this assignment of domains will be discussed in Chapter 7.

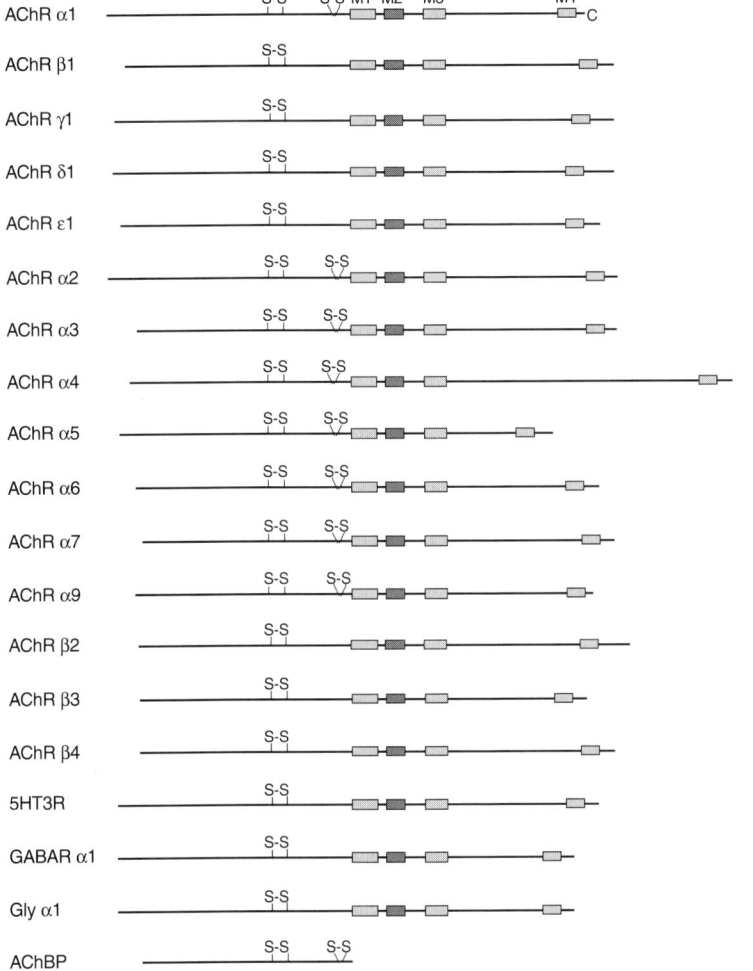

FIGURE 3.1 Schematic representation of the primary structure of the subunits of the nicotinic acetylcholine receptor and of several other members of the acetylcholine receptor family. The four transmembrane regions are shown, as well as disulfide bonds. The linear representations, proportional to the lengths of their respective sequences (obtained from the SwissProt database), are aligned at the N-terminal extremity of M1. The subunits for the human forms of muscle and neuronal acetylcholine receptors are shown, as well as subunits from several additional human channel receptors for other neurotransmitters (serotonin [5HT3R], GABA, and glycine) from the same "nicotinoid" superfamily. *Bottom:* the molluscan acetylcholine-binding protein (AChBP).

Molecular Phylogeny

A fundamental result of applying recombinant DNA techniques to the nicotinic receptor in the early 1980s was the demonstration of close homologies between nicotinic receptor subunit sequences, first in the fish electric organ and skeletal muscle from higher vertebrates, including humans (Numa et al., 1983), and subsequently between muscle and brain nicotinic receptors (Boulter et al., 1986a; Couturier et al., 1990a; review in Role and Berg, 1996). Together these subunits constitute a multigene family (Le Novère and Changeux, 1995; Ortells and Lunt, 1995). The sequences of 12 types of neuronal (and epithelial) nicotinic receptor subunits have so far been determined in higher vertebrates (Le Novère and Changeux, 1999). Nine are designated as α-subunits (α2–α10) and share with the electric organ α1-subunit a pair of adjacent cysteines at positions 192 and 193 in *Torpedo* (Karlin, 1993), while the others are referred to as non-α- or β-subunits (β2–β4). Moreover, systematic and comparative sequence analyses demonstrate that the acetylcholine receptor genes constitute a superfamily of ligand-gated channels that includes the 5-HT$_3$, GABA$_A$, GABA$_C$, glycine, and some invertebrate anionic glutamate receptors (Ortells and Lunt, 1995) (Figure 3.1). No apparent structural relation exists with ligand-gated ion channels belonging to the family of cationic glutamate receptors, nor with ATP, inositol phosphate, or cyclic nucleotide-gated channels.

Within the acetylcholine family, the various vertebrate subunits can be related by time of divergence from a common ancestral gene, as well as by the time of divergence from the most closely related member outside the acetylcholine receptor group, the 5-HT3 receptor (see Figure 3.2). A close relationship for acetylcholine- and serotonin-gated channels is further supported by the production of a functional chimeric protein with the N-terminal domain from α7 that renders the 5-HT$_3$ channel responsive to acetylcholine (Eiselé et al., 1993). Coassembly of 5-HT$_3$ and α4-subunits into functional receptors has also been reported (van Hooft et al., 1998). Interspecies combination of subunits from the acetylcholine family has been observed to produce functional receptors, including complexes involving subunits of invertebrate and vertebrate origin (Bertrand et al., 1994; Schulz et al., 1998).

The complete annotation of the human genome sequence brought with it no reports of additional genes encoding nicotinic receptor subunits. For invertebrates, the complete DNA sequence of the nematode *C. elegans* contains more than 40 genes tentatively assigned to nicotinic receptor subunits (Bargmann, 1998), several of which assemble into functional nicotinic receptors upon expression in *Xenopus* oocytes (Ballivet et al., 1996; Treinin et al., 1998). These include unc-29 and unc-38, whose elimination leads to loss of the levamisole-sensitive activation of an acetylcholine receptor at the neuromuscular junction (Richmond and Jorgensen, 1999). The unc-38 gene codes for an α-subunit and the unc-29 gene codes for a non-α-subunit. The surprisingly large number of nicotinic receptor genes may correlate with defined distributions and specialized roles at particular classes of neurons and synapses (see below). The sequencing of the fruit fly *Drosophila* led to the preliminary identification of only about 10 nicotinic receptor subunit genes (Littleton and Ganetzky, 2000; Rubin et al., 2000), significantly fewer than in *C. elegans*. Three of them most closely resemble α7 (Littleton and Ganetzky, 2000), whereas the others are more closely related to each other than to known mammalian α- or β-subunits. The lack of α1 and other muscle-subunit homologues in *Drosophila* may be attributed to the fact that the neurotransmitter for the neuromuscular junction in fruit flies is glutamate rather than acetylcholine.

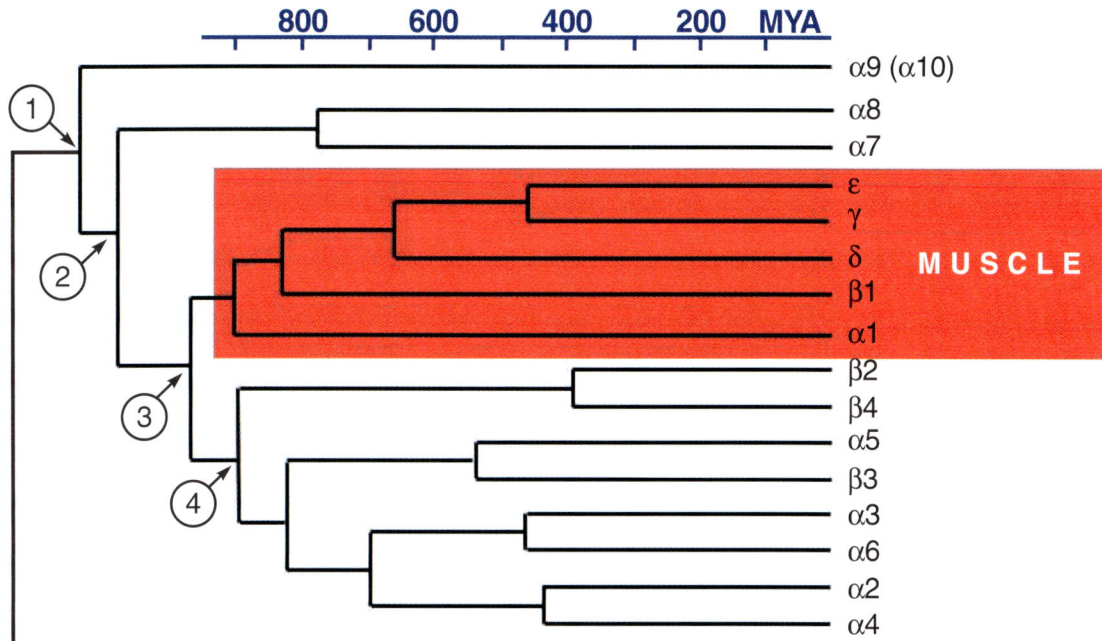

FIGURE 3.2 Evolutionary relationship of amniote acetylcholine receptor subunits. The subunits carry the principal component of an acetylcholine-binding site. The circled numbers refer to groups defined by (1) activation by acetylcholine; (2) activation by nicotine; (3) participation in a heteropentamer; (4) insensitivity to α-bungarotoxin. (Modified from Le Novère, 1998; see also the Ligand-Gated Ion Channels [LGIC] Web site: http://www.ebi.ac.uk/compneur-srv/LGICdb/.)

Analysis of the sequence of known acetylcholine receptor subunits indicates that these subunit genes share a common origin and have a long phylogenetic history. The reconstituted phylogenetic tree (Le Novère and Changeux, 1995; Ortells and Lunt, 1995; Changeux et al., 1998) posits an early divergence from the α9-subunit (see Figure 3.2), which in *Xenopus* oocyte reconstitution experiments forms homo-oligomeric receptors that are able to respond to acetylcholine. Yet receptors composed of the α9-subunit are inhibited by α-bungarotoxin, as well as by both nicotine and muscarine (Elgoyhen et al., 1994; Rothlin et al., 1999), and thus exhibit atypical nicotinic/muscarinic pharmacology. As we noted in Chapter 2, the major muscarinic class of acetylcholine receptors was also affinity-labeled at an early stage of receptor research (Gill and Rang, 1966) and belongs to the G-protein-coupled class of receptors exhibiting seven transmembrane helices. As for the homopentameric nicotinic receptors, divergences subsequently took place involving the vertebrate neuronal α7- and α8-subunits, which also form "low-affinity" homopentamers inhibited by α-bungarotoxin.

The vertebrate muscle and heterooligomeric α-bungarotoxin-insensitive neuronal subunits comprise a large group that, around 1000 million years ago (MYA), may have diverged from their invertebrate ancestors. Structural gene duplications and tissue-specific promoter switches then led to muscle α1-, β1-, and δ/γ/ε-subunit clades, which form $(\alpha)_2\gamma\delta\beta$ complexes, and to the α-bungarotoxin-insensitive neuronal subunit subfamily, which emerged at the beginning of the chordate phylum. Segregation of the ancestral gene for the β2/β4-subunits probably occurred more than 800 MYA, and that for α5/β3 and the rest of the subunit genes slightly later. Finally, the duplications yielding α2 and α4, in one branch, and α3 and α6, in another branch, may have taken place more than 620 MYA and 500 MYA, respectively, followed by the most recent β2/β4 bifurcation a little less than 420 MYA. This last duplication coincides with the split between the teleost and the tetrapod lineages, and thus occurred during the same period that saw an increase in the complexity of the brain associated with the advent of terrestrial life, together with a correlative diversification of the cholinergic pathways. It is quite remarkable that no known duplication of acetylcholine receptor genes accompanied the dramatic increase of brain complexity that characterizes the recent evolution of mammals, particularly primates and the human lineage—with the possible exception of human α7 (Gault et al., 1998).

Sequence analysis at the gene level has also revealed the existence of various introns, providing further insight into the evolution of the genes (Jonas et al., 1990; Le Novère and Changeux, 1995; see Figure 3.3). The genes coding for subunits found in the muscle form of the receptor contain 11 or 12 exons (family IV), whereas the genes for neuronal subunits that participate in heteropentamer formation (family III) retain the first four and the last exon of family IV, while possessing one large central exon that codes for a major portion of the pro-

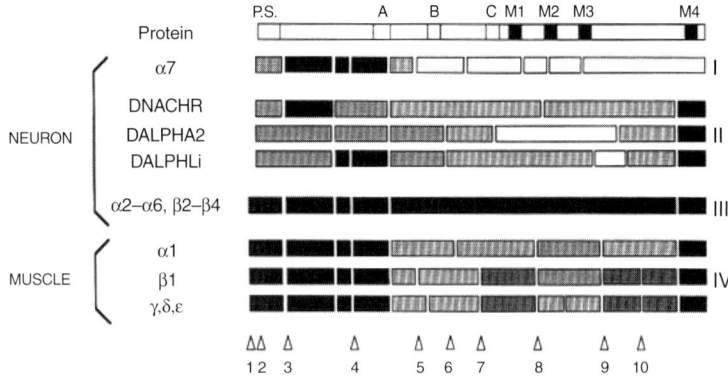

FIGURE 3.3 Structure of nicotinic receptor subunit genes. Only exons are represented: the gray level reflects the exon's overall conservation throughout the family (i.e., an exonic frontier is present at the same place in different subunits independent of the degree of sequence similarity among the exons themselves). A, B, C: binding site loops. M1, M2, M3, M4: transmembrane segments. The arrowheads mark the informative limits. (Adapted from Jonas et al., 1990, with the help of Alain Bessis.)

tein sequence that encompasses about half of the N-terminal domain, transmembrane regions M1, M2, and M3, and most of the cytoplasmic domain. Distinct exon distributions were observed for the α7 gene and for *Drosophila* genes. As in the case of most other proteins (Stoltzfus et al., 1994), there is no clear-cut systematic relationship between exons and functional domains. In addition, alternatively spliced gene products, which have been reported for subunits α1 (Talib et al., 1993) and γ (Mileo et al., 1995), may be particularly prevalent in the case of subunits from *Drosophila* (see Rubin et al., 2000).

The diversity of the nicotinic receptor subunit genes was therefore established nearly 400 MYA and has not changed significantly since then. The conservation of nicotinic receptor subunit gene structure stands in sharp contrast to the explosive development of brain organization and complexity in the course of mammalian evolution.

Chromosome Localization and Patterns of Expression

The distribution of neuronal acetylcholine receptor subunits in distinct portions of the brain varies strikingly from one subunit to another and from one species to another. Early studies of motor endplate development revealed an *embryonic* extrajunctional form and an *adult* subjunctional form, which in some species differ with respect to γ- versus ε-subunit contribution to the muscle heteropentamer (Mishina et al., 1986). The situation is more complicated for nicotinic receptors in different regions of the brain, which vary considerably in architecture and connectivity, but also differ strikingly in the pattern of genes they express (see Chapter 10). In general, each subunit displays a distinct pattern of expression: some are highly restricted, as in the case of α2 in the chick brain (Daubas et al., 1990); others are widely distributed, as in the case of β2 in the rat brain (Hill et al., 1993); many others exhibit intermediate distributions. In the rhesus monkey, α2 is widely expressed to the extent that it may replace α4 as the ubiquitous partner of β2 (Han et al., 2000). In effect, each gene encoding a given nicotinic receptor subunit displays an almost unique pattern of expression throughout the central and peripheral nervous system.

In a first attempt to relate the patterns of neuronally expressed nicotinic receptor subunit genes to the organization of the genome, the chromosomal location of these genes was investigated in several animals. In the mouse, the genes coding for the four subunits of the nicotinic receptor were located on three chromosomes by analyzing restriction-fragment-length polymorphisms between the domestic mouse and *Mus spretus* (Robert et al., 1985). The γ- and δ-subunit genes were found to cosegregate in this species, as in the chick (Nef et al., 1984), and are allocated to chromosome 1, whereas the β-subunit gene is located on chromosome 11 (Heidmann et al., 1986). The α-subunit gene was found to be closely linked to the Actc-1 gene located on chromosome 2 (Heidmann et al., 1986). The genes coding for the neuronal nicotinic receptor subunits were also localized on mouse chromosomes (Bessis et al., 1990). The α2-, α3-, and β2-subunits (for references see Heinemann et al., 1989) were found on chromosomes 14, 9, and 3, respectively. The α4-subunit gene is located on chromosome 2, but it is not genetically linked to the α1-subunit gene (Bessis et al., 1990). However, a conserved cluster encompasses the α3-, α5-, and β4-genes in the rat (Boulter et al., 1990), the chick (Couturier et al., 1990b), and all other vertebrate species that have been examined, including humans (Raimondi et al., 1992). The high proportion of linkages conserved over the course of evolution suggests that strong selective pressures must have operated in their favor. The simplest explanation is that the linkages gave rise to coordinated transcriptional control, but there is no direct evidence for such control. Additional analysis of the promoter and regulatory sequences for each individual subunit gene is required (see Chapter 11). The corresponding positions on human chromosomes are chromosome 1: β2; chromosome 2: α1, γ, δ; chromosome 4: α9; chromosome 8: α2, α6, β3; chromosome 11: α10; chromosome 15: α3, α5, α7, β4; chromosome 17: β1, ε; chromosome 20: α4 (see the NCBI Genome data base).

The diversity of the patterns of expression of nicotinic receptor oligomers in central neurons derives from the variety of combinatorial processes operating not only at this level, but also at the level of posttranslational targeting to the somatodendritic, preterminal, and terminal axonal domains of the neuronal membrane (see Chapter 11 and Hill et al., 1993; Sargent, 1993). It is also important to keep in mind that a quantitative variation of pharmacological properties could have been tolerated within a physiological envelope as a consequence of near-neutral genetic drift (Le Novère et al., 2002a). Under these circumstances, the fact that *Caenorhabditis* has a larger number of nicotinic receptor genes (~40)—compared to estimates of 10 in *Drosophila* or 17 in *Homo*—could be related to a large array of different neuron-specific promoters (in addition to fewer contributions from splice variants, as already noted). Also, *Caenorhabditis* receptors are principally homooligomeric, reflecting a predominance of subunits closely related to vertebrate α7. In this case the absence of functional diversity arising from varied subunit combinations could be compensated by a greater number of distinct homopentameric subunits. This possibility is supported by an observed predominance of α- over non-α-subunit genes (Mongan et al., 1998), although at least two *Caenorhabditis* nicotinic receptor genes are linked in an operon, and both gene products are required for formation of functional (heterooligomeric) receptors *in vitro* (Treinin et al., 1998). Investigation of the genetic basis of human evolution by means of sequencing and bioinformatic analysis has revealed a pattern of accelerated evolution among nervous system genes—among them, interestingly, the α2- and α5-subunit genes of the nicotinic receptor (Dorus et al., 2004).

A great deal more information is needed in order to understand the origin and functional consequences of nicotinic receptor diversity in different locations within an organism and its distinct forms among different species. Nevertheless, from a functional point of view, knowledge of an essential aspect of the contribution of nicotinic receptors to brain physiology may be expected to come from their precise distribution rather than from their particular physiological properties (see Le Novère et al., 2002a).

Conclusion

The evolution of the multigene family that includes nicotinic receptors reveals several trends that hold important implications for a deeper understanding of their structure-function relations. In general, the question of molecular symmetry has been a much-debated issue in the case of allosteric proteins (Monod et al., 1965) and, in particular, of allosteric membrane proteins (Changeux, 1969). The presence of nonidentical subunits in nicotinic receptor oligomers results in a pseudosymmetrical structure having a rotational axis coincident with the central axis of the molecule and perpendicular to the membrane. This arrangement represents a novel departure from the perfect symmetry encountered with soluble allosteric enzymes (Changeux and Edelstein, 1998). Yet it presumably does not represent a primitive state, but rather a highly specialized form. Indeed, the most archaic neuronal subunits form symmetric homooligomers, apparently with five equivalent sites, as demonstrated for α7 (Palma et al., 1996). For a homopentameric protein exhibiting strong interactions between its binding sites, a highly cooperative dose-response curve might be expected, with values of the Hill coefficient (n_H) approaching the theoretical upper limit of $n_H = 5$. Although the wild-type chick α7-receptor expressed in *Xenopus* oocytes exhibits low cooperativity ($n_H \sim 1.3$), certain mutated forms of α7 do display very high cooperativity (Bertrand et al., 1993), close to the theoretical limit. Specialization of the subunits has been observed to cause a loss of symmetry, as in the case of muscle and heterooligomeric neuronal nicotinic receptors, and to reduce the number of binding sites to two, with the consequence that cooperativity must also be reduced (to $n_H < 2$).

For heteropentameric neuronal nicotinic receptors with complex subunit compositions, the exact number of each type of subunit per oligomer, despite its important physiological and pharmacological consequences, remains to be established in many cases (see Champtiaux et al., 2002, and Chapter 11).

The possibilities resulting from the combination of subunits with different properties in the same receptor molecules lead to a marked diversification of functional properties and support a capacity to form diverse patterns of receptors in the brain. The evolutionary advantages of the extensive array of properties and distributions produced in this way have clearly outweighed the loss of cooperativity for the heterooligomeric forms that have diverged from the homooligomeric ancestral genes. In subsequent chapters we will examine the contribution of varied subunit combinations to functional diversity with respect to four topics: pharmacological specificity at the agonist binding site; ion selectivity in the channel; modes of activation, desensitization, and up-regulation; and sensitivity to allosteric effectors. These combinations also contribute to the differential distribution and targeting of nicotinic receptor subunit mRNAs and protein oligomers throughout the brain. The overall diversity of these subunit genes generates a broad range of distinct molecular species, marked by a multiplicity of spatial distributions and differential expression patterns during development that enrich the functional complexity of brain networks. It also poses a considerable challenge for molecular pharmacology.

CHAPTER 4

Chemical Structure of the Agonist-Binding Site

In the early phases of research on nicotinic receptors beginning with Langley, Dale, and others, pharmacological analysis of various chemical compounds implied specific structure-function relationships in terms of the chemical form of the agonist (or antagonist) and the nature of the response it elicits. But the exact nature of the relationships was ill-defined. Chemical studies of pharmacologically active substances, including nicotinic and muscarinic ligands (Dale, 1914), synthetic forms of curare, such as flaxedil (Bovet et al., 1949), and other ligands (Khromov-Borisov and Michelson, 1966) suggested possible models for the three-dimensional structure of the ligand (Rang et al., 2003). However, this work in what was called "therapeutic chemistry" did not yield direct information about the complementary receptor site.

The situation changed dramatically once the receptor became available in pure form, permitting the structure of the binding site to be examined directly and the actual binding to be compared with *in vivo* responses to the same ligand. These results inspired a great many studies of the chemistry of the binding site and produced surprisingly accurate deductions of both the structural elements that make up the site and their spatial arrangement (confirmed by structural studies described in Chapter 7). Historically, the investigations that led to the characterization of the ligand-binding site proceeded in discrete steps, from the initial labeling of the binding site and its location at the boundary between subunits to the identification of the chemical interactions responsible for ligand-binding diversity.

CHAPTER 4

Early Studies on the Chemical Identification of Amino Acids Belonging to the Acetylcholine-Binding Site

Because of their highly selective binding and very slow reversibility, α-toxins from snake venoms (Lee and Chang, 1966) became powerful tools that helped to generate many of the initial insights into the nature of the acetylcholine-binding site (see Chapter 1). The binding of the α-toxins was initially studied directly or indirectly by measuring the protection against reversible binding that occurs in the presence of varying concentrations of cholinergic ligands (Weber and Changeux, 1974a, 1974b, 1974c). These ligands decreased the initial rate of radioactive α-toxin binding to a defined population of sites present in receptor-rich membranes from *Torpedo marmorata,* and apparent dissociation constants and rank order were closely related, if not identical, to the α-toxin-sensitive binding of radioactive nicotinic agonists such as acetylcholine and decamethonium (Cohen et al., 1974; Weber and Changeux, 1974a, 1974b, 1974c) and antagonists such as *d*-tubocurarine (Neubig and Cohen, 1979, 1980). The reversible interaction of cholinergic ligands with the receptor site was also monitored by various methods including fluorescence spectroscopy (Cohen and Changeux, 1973a; Martinez-Carrion and Raftery, 1973; Heidmann and Changeux, 1979) and proton magnetic resonance (Miller et al., 1979). The relative stoichiometries of cholinergic ligand and α-toxin were found to be 1.0 ± 0.1 (Weber and Changeux, 1974a, 1974b, 1974c; Damle et al., 1976; Weiland et al., 1976; Heidmann and Changeux, 1979; Neubig and Cohen, 1979, 1980), thus supporting the hypothesis (Changeux et al., 1970a, 1970b) that, despite their many differences in structure, nicotinic agonists and antagonists (including snake venom α-toxins) interact within a common (or overlapping) binding area(s) on the receptor surface.

In order to clarify the mechanism underlying structural recognition of nicotinic ligand by the receptor protein, a variety of experimental approaches have been employed to identify the individual amino acids that form the acetylcholine-binding area. As we noted earlier in Chapter 2, disulfide-bond reduction disturbs the pharmacological response to agonists (Karlin and Bartels, 1966b) and permits the labeling of the α-subunit by cholinergic-affinity ligands specific for sulfhydryl groups (such as the maleimide reagents MBTA and MPTA) (see Karlin, 1969, 1983). Kao et al. (1984) demonstrated by means of peptide mapping and sequencing of *Torpedo* receptors that the residues cysteine 192 and possibly cysteine 193, a tandem unique to the α-subunit, represent the incorporation sites of MBTA. The incorporation of MBTA occurs exclusively on the *reduced* receptor, however. In the native receptor, this pair of cysteines forms a disulfide bridge (Kao and Karlin, 1986; Mosckovitz and Gershoni, 1988), thus implying the presence of a rather rare *cis* peptide bond (see Mosckovitz and Gershoni, 1988).

Subsequent studies based on the binding of snake α-toxins to fragments of the α-subunit (Wilson et al., 1984, 1985; Neumann et al., 1985, 1986; Oblas et al., 1986), synthetic peptides (Wilson et al., 1985; Mulac-Jericevic and Atassi, 1986; Ralston et al., 1987; Radding et al., 1988), deletion mutants (Barkas et al., 1987), or α-subunit fragments expressed in *E. coli* (Barkas et al., 1987), and the consequences for receptor channel response and α-toxin binding of cysteine to serine mutations (Mishina et al., 1985), confirm that the region containing cysteine 192 and cysteine 193 is involved in the interaction of cholinergic ligands. Yet evidence of a high affinity for toxin binding—and, in most experiments, of competition by nicotinic ligands—was never found in isolated α-subunits or their fragments, suggesting that other domains of receptor primary structure participate in the functional organization of the acetylcholine-binding area. Accordingly, the adjacent cys-cys residues of the α-subunit are not by themselves sufficient to achieve binding.

The Binding Site at the Boundary between Subunits

The first indications that the binding site may not be contained exclusively within the α-subunit were obtained after photolabeling of the [^{3}H]α-toxin-receptor complex by ultraviolet irradiation. This technique unambiguously revealed an incorporation of radioactivity into the γ- and δ-subunits in addition to the α-subunits (Oswald and Changeux, 1982; Figure 4.1). Subsequently, the antagonist [^{3}H]*d*-tubocurarine was found to label the γ- and δ-subunits, together with the α-subunits (Pedersen et al., 1986; Pedersen and Cohen, 1988). Significant carbamylcholine-sensitive incorporation of an affinity-labeling reagent, DDF (see below) also takes place in the γ-subunit (Langenbuch-Cachat et al., 1988). Blount and Merlie (1989) then succeeded in expressing different pairs of mouse muscle α- and non-α-subunits in fibroblasts and showed that the γ- and the δ- (but not the β-) subunits readily combined with an α-subunit to form complexes with different high-affinity binding sites for the competitive antagonist *d*-tubocurarine. These data are consistent with the notion, initially suggested by Oswald and Changeux (1982), that the ligand-binding sites overlap at least two subunits and may plausibly be located at the interface between distinct subunits. This situation is not unique. Indeed, generally speaking, ligand-binding sites are also found at subunit interfaces

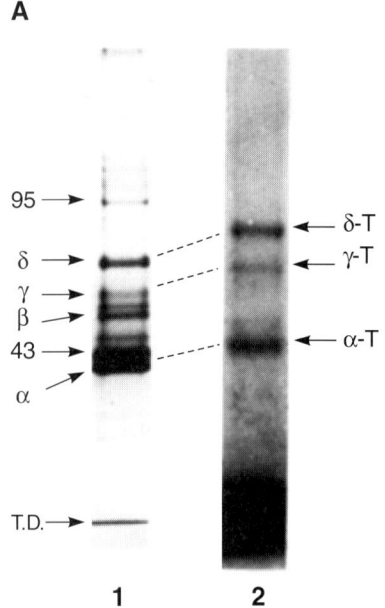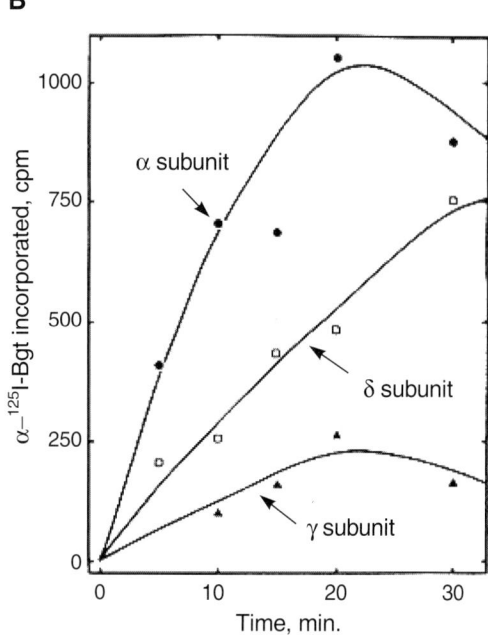

FIGURE 4.1 Ultraviolet light-induced cross-linking of α-^{125}I-bungarotoxin to several subunits of the membrane-bound nicotinic receptor from *Torpedo marmorata* indicating that the toxin binds to interfaces involving different subunits. (A) Denaturing SDS gel of the membrane preparation cross-linked with α-^{125}I-bungarotoxin and stained for protein by Coomassie blue (*panel 1*) with an autoradiogram of the dried gel showing the positions of the labeled cross-linked subunits (*panel 2*): α-^{125}I-bungarotoxin covalently attaches to the α-, γ- and δ-subunits, but in different amounts. (B) Time course of the incorporation of α-^{125}I-bungarotoxin into the α-, γ-, and δ-subunits. (From Oswald and Changeux, 1982.)

in the case of allosteric enzymes such as aspartate transcarbamylase (Lipscomb, 1994), L-lactate dehydrogenase (Iwata et al., 1994), and phosphofructokinase (Schirmer and Evans, 1990); and of hemoglobin (Arnone, 1972) with respect to binding of 2,3-diphosphoglycerate (but not of oxygen) (Perutz, 1989). Unambiguous confirmation of this state of affairs in the case of the nicotinic receptor was obtained with the determination of the crystal structure of an acetylcholine-binding protein (AChBP) (Brejc et al., 2001) found in the snail that is related to the acetylcholine receptor (see Chapter 7).

Multiple Loops of the Acetylcholine-Binding Pocket

A more specific inquiry into the organization of the acetylcholine-binding pocket at the amino acid level required the identification of the individual side chains that form the complementary binding area. This was achieved by systematic photolabeling mapping. The photolabeling reagent DDF (p-N,N-dimethylamino benzene diazonium fluoroborate) (Goldner et al., 1997) was used to probe the topology of the acetylcholine-binding sites on the *native* receptor in detail. This compound is derived from TDF (p-[trimethylammonium] benzene diazonium fluoroborate) (see Figure 4.2), the first affinity-labeling reagent described for the *Electrophorus* electric organ (Changeux et al., 1967a; Weiland et al., 1979), as we noted in Chapter 2. DDF appeared to be a more suitable choice (Langenbuch-Cachat et al., 1988) than sulfhydryl-directed affinity ligands (Karlin, 1983) for two reasons: the photogenerated species possesses a high degree of reactivity with a wide variety of amino acid side chains; and it reacts with the receptor protein without prior reduction, a condition that is known drastically to alter functional properties of the receptor (Karlin, 1969; Chang and Bock, 1977; Barrantes, 1980; Walker et al., 1984). Moreover, because the diazonium ion of DDF is fairly stable in the dark, it was possible to demonstrate that DDF behaves as a reversible competitive antagonist of the *E. electricus* electroplaque response and of the acetylcholine-gated single channel currents recorded in the C_2 mouse cell line (Langenbuch-Cachat et al., 1988). Direct illumination (λ = 320 nm) was found to convert the diazonium of DDF into a highly reactive aryl cation (Bergstrom et al., 1976; Angelini et al., 1984), which irreversibly blocked the agonist response both in electroplaque and in microsacs. The efficiency and selectivity of labeling were significantly improved by energy transfer between a tryptophan of the receptor protein and the photosensitive ligand (Goeldner and Hirth, 1980; Goeldner et al., 1982). Appropriate experimental conditions were then defined whereby DDF labels the agonist/competitive antagonist-binding site with a stoichiometry of one DDF incorporated per α-bungarotoxin-binding site (which is fully prevented by prior incubation with unlabeled α-bungarotoxin). DDF therefore behaves as an efficient photoaffinity label of the native acetylcholine-binding site (Langenbuch-Cachat et al., 1988).

The acetylcholine-binding sites on native membrane-bound acetylcholine receptors from *T. marmorata* were covalently labeled with [^{3}H]DDF and the α-subunits were isolated and cleaved with cyanogen bromide. Three labeled peptides were produced. The amino acids labeled by [^{3}H]DDF were identified as tyrosine 93, tryptophan 149, tyrosine 190, and cysteines 192 and 193 (in addition, the α-subunit residues tryptophan 86, tyrosine 151, and tyrosine 198 were weakly labeled) (Figure 4.2). Both α-toxin and carbamylcholine decreased DDF-labeling of the three peptides in a parallel manner, thus supporting the conclusion that *several loops* (at least three) of the NH_2-terminal large hydrophilic domain contribute to the acetylcholine-binding site (Dennis et al., 1986; Dennis et al., 1988; Galzi et al., 1990). The loops—referred to as A, B, and C—are associated with the portions encompassing residues 86–94, 149–151, and 190–198, respectively (see Figure 4.3). These amino acids were also found to be labeled by the other probes that were subsequently used: tyrosine 93 by acetylcholine mustard (Cohen et al., 1991); tyrosine 190, cysteine 192, and tyrosine 198 by nicotine (Middleton and Cohen, 1991) and *d*-tubocurarine (Pedersen and Cohen, 1990); and tyrosine 190 by lophotoxin (Abramson et al., 1989).

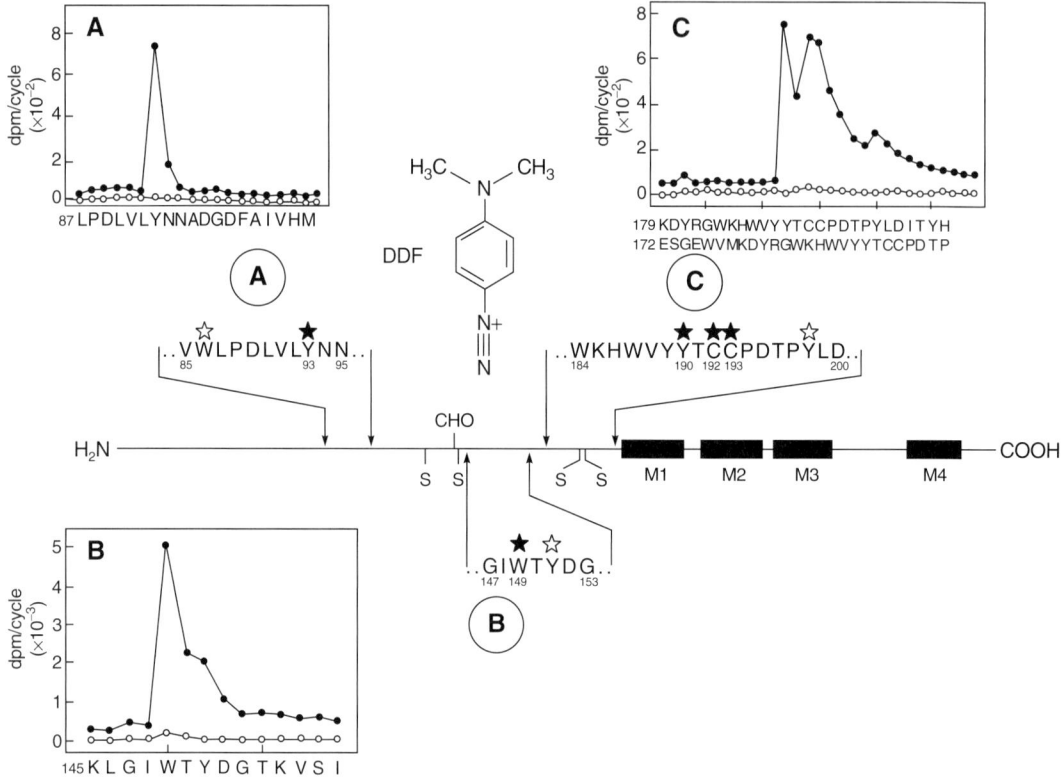

FIGURE 4.2 Photoaffinity labeling of the acetylcholine-binding site by DDF (formula indicated) for the principal components of the *T. marmorata* α-subunit. *Insets:* sequence analysis using high-performance liquid chromatography of three fragments (A, B, and C) of the amino-terminal domain after labeling in the absence (closed circles) or presence (open circles) of carbamylcholine or α-toxin. Positions of photolabeled amino acids on loops A, B, and C are indicated by stars. (From Dennis et al., 1988; and Galzi et al., 1990.)

All probes labeled primarily the α1-subunits, and to a lesser extent the γ- and δ-subunits (see Figure 4.1 for α-bungarotoxin labeling). The much stronger labeling of α1 compared with that of the γ- and δ-subunits argues in favor of an asymmetric location of the binding site with respect to the interface. It was thus proposed that the α-subunits be referred to as carrying a "principal component" of the nicotinic binding site, and the δ- or γ-subunits as contributing to a "complementary component" (Corringer et al., 1995).

Three additional loops (D, E, and F) have been identified in the complementary component. Homologous positions on the γ- and δ-subunits (of loop D), tryptophan 55 and tryptophan 57, respectively, were found to be labeled by nicotine and *d*-tubocurarine (Chiara and Cohen, 1997; Corringer et al., 1999), while the homologous positions tyrosine 111 and arginine 113 (loop E) were weakly but specifically labeled by *d*-tubocurarine (Chiara and Cohen, 1997; Chiara et al., 1998, 1999; Wang et al., 2000). In order to identify negatively charged residues contributing to the stabilization of the cationic ligands, a probe 0.9 nm long, grafted onto the reduced 192–193 disulfide bridge and reacting with aspartates and glutamates, was found to label the δ-subunit aspartic acid residues 165 and 180, and glutamate 182 (Czajkowski and Karlin, 1995). Mutations of aspartic acid → asparagine at position

180 of the δ-subunit (loop F) and of the homologous D174 position of the γ-subunit were found to decrease the affinity for acetylcholine (Martin et al., 1996).

Sequence comparison indicates a high degree of conservation of the loop A, B, C, and D motifs in the binding site of neuronal nicotinic receptors. The labeled residues from loops A, B, and C are indeed present in all α-subunits with the exception of α5, and the labeled residue from loop D in the β2-, β4-, α7-, and α8-subunits. In the homooligomeric α7-receptor, as well as in the α7-V201-5HT3 chimera, which carries the α7-binding site (Eiselé et al., 1993), mutations of the corresponding residues (tryptophan 54, tyrosine 92, tryptophan 149, and tyrosine 188) alter the apparent affinities of binding and activation by nicotinic agonists, establishing their contribution to the acetylcholine-binding site (Galzi et al., 1991b; Tomaselli et al., 1991; Corringer et al., 1995). In this case, the same subunit carries both the principal and the complementary binding components, yet they are located on different domains of the subunit (Figure 4.3). In contrast to this conserved core of amino acids, the labeled residue from loop E is highly variable, whereas the aspartate of loop F is conserved in all γ-, δ-, ε-, and α7-subunits—where it also contributes to acetylcholine binding (Galzi et al., 1996b)—but not in β2 or β4.

The acetylcholine-binding domain thus comprises both highly conserved "canonical" amino acids and variable ones. We now go on to examine their respective functions at the atomic level in the course of ligand binding.

Chemical Interactions Responsible for Binding

It has been widely held that carboxylate *anions* (side chains of aspartic and glutamic acids) in the acetylcholine-binding site form a negative subsite responsible for interaction with the cationic head group of acetylcholine (reviewed in Luyten, 1986). This view was indirectly supported by crystallographic analysis of phosphorylcholine-specific antibody M603 (Padlan et al., 1976), and by molecular recognition studies of unsubstituted ammonium ions using macrocyclic compounds substituted with carboxylate groups (Behr et al., 1976, 1982). In fact, neither glutamyl nor aspartyl residues were among the major [^{3}H]DDF-labeled residues identified (Dennis et al., 1988; Galzi et al., 1990). One cannot rule out the possibility that orientation and/or reactivity restrictions upon DDF within its binding site may prevent labeling to a detectable extent. On the other hand, it is rather interesting to note that [^{3}H]DDF becomes predominantly incorporated into *aromatic* amino acids, particularly tyrosine residues α93 and α190 (and possibly α151 and α198).

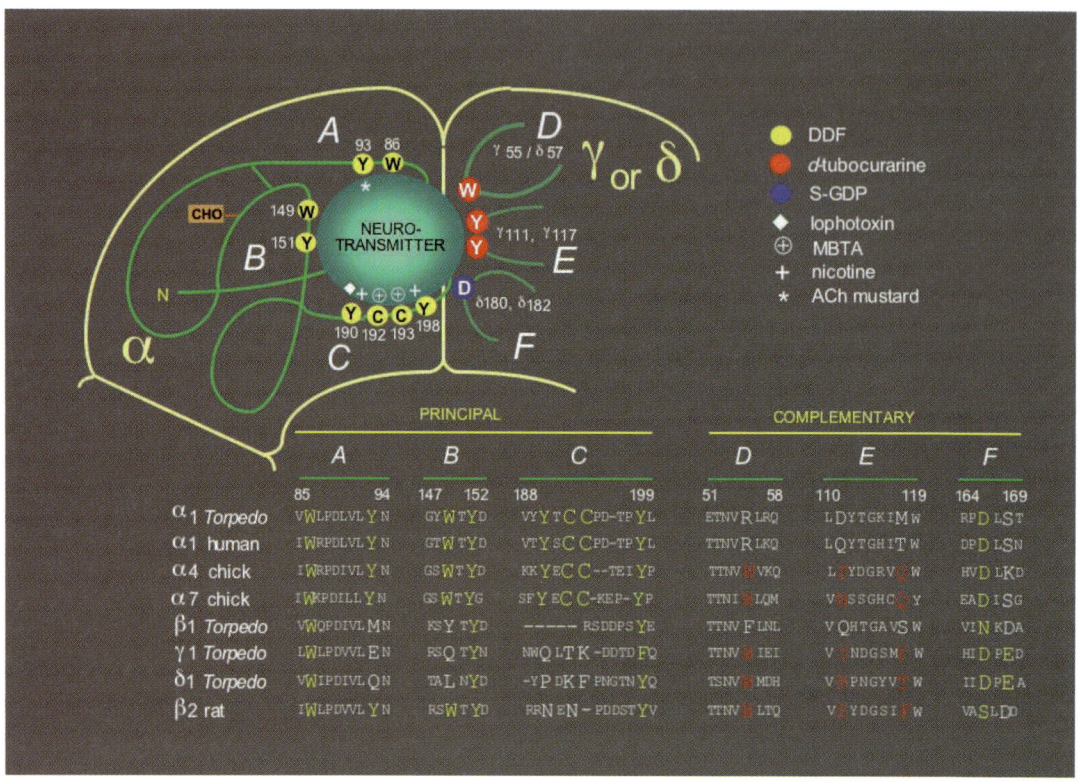

FIGURE 4.3 Principal and complementary portions of the acetylcholine-binding site overlapping the boundary between α- and non-α-subunits, as identified by photoaffinity labeling with a variety of nicotinic ligands. Circles depict residues forming the principal component of the interface (loops A, B, and C); other shapes depict residues belonging to the complementary component of the interface (loops D, E, and F). Aligned sequences of identified loops in the acetylcholine-binding sites of receptors from various species are shown. (From Corringer et al., 2000, courtesy of Annual Reviews.)

This finding is consistent with a model in which the electronegative character of the quaternary ammonium-binding domain consists, at least in part, of a lone pair of electrons associated with the phenolic oxygen of these tyrosines, the sulfur atoms forming the cysteine α192–α193 disulfide, and the nitrogen atom of tryptophan 149 (Dennis et al., 1988; Galzi et al., 1990). Such a view is also consistent with the observation that binding of methyl-substituted ammonium groups by macrocyclic compounds requires not only electrostatic but also hydrophobic interactions (see references in Galzi et al., 1990). These could be provided by noncarboxylate-containing entities such as macrobicyclic polyphenoxides (Schneider et al., 1986; Schneider et al., 1988).

Evidence that the binding site of acetylcholine and nicotinic ligands includes a conserved core of aromatic residues, whose electron-rich side chains may provide stabilizing interactions with the cationic ligands, is further strengthened by mutations of α-subunit tyrosines 93, 190, and 198. All of them affect the apparent affinities of acetylcholine and tetramethylammonium in the same way, suggesting that these tyrosine residues help to stabilize the quaternary ammonium portion of acetylcholine (Sine et al., 1994). Further analysis of the function of these amino acids relying on the incorporation of unnatural amino acids assigned a prominent role to both the hydroxyl group and the aromatic ring of α-subunit tyrosines 93 and 198, whereas tyrosine 190 could not be satisfactorily analyzed (Nowak et al., 1995). With respect to the tryptophan at position 149, the 50% effective concentration (EC_{50}) for acetylcholine correlates with the cation-π-binding capability of a series of fluorinated tryptophan derivatives (Zhong et al., 1998), which suggests that the indole side chain makes van der Waals contact (cation-π interactions) with the quaternary ammonium group of acetylcholine. Although recent structural studies (Celie et al., 2004) do not favor this interpretation (see Chapter 7), it was supported by three lines of observation: first, incorporation of a tyrosine graft bearing a quaternary ammonium [Tyr-O-$(CH_2)_3$-$N(CH_3)_3^+$] group produces some constitutive activity, thus plausibly mimicking a bound agonist close to tryptophan 149 (Zhong et al., 1998); second, the apparent affinity of acetylcholine for the α7-receptor is particularly sensitive to mutation at this position, with a 100-fold increase in EC_{50} following replacement of the tryptophan by phenylalanine, compared with a 10-fold increase for the same replacement at tyrosines 93 or 190 (Galzi et al., 1991b); and third, analysis of protein structures indicates that tryptophan presents the most potent cation-π-binding site, especially in the case of acetylcholine esterase, within which the quaternary group of acetylcholine makes van der Waals contact with tryptophan 84 in the X-ray crystallographic structure (Sussman et al., 1991).

Mutations of the homologous aspartic acids γ174 and δ180 to asparagine were also found to decrease the affinities for acetylcholine and tetramethylammonium (Martin et al., 1996). It is noteworthy that two other aspartates from the principal binding component, α152 from loop B (Sugiyama et al., 1996) and α200 from loop C (O'Leary and White, 1992), have been shown to decrease the acetylcholine-binding affinity when mutated to asparagine. This indicates that aspartates may further help to stabilize the ammonium ion, possibly through long-range electrostatic interactions. In the case of α-conotoxin M1, which exhibits a 10,000-fold higher affinity for the α/δ-binding sites of mouse muscle-type receptor than for the α/γ sites—in contrast to d-tubocurarine, which displays a 100-fold preference for the α/γ site (Sine, 1993; Sine et al., 1995a)—mutations such as Tyr → Arg at position γ111 and Arg → Tyr at position δ113 (within the highly variable loop E of the complementary binding component) alter primarily the apparent affinities for d-tubocurarine and α-conotoxin M1, but not for acetylcholine—evidence of the specific contribution made by these residues to the binding of these large antagonists (Chiara et al., 1999). Taken together, these data establish that acetylcholine interacts with a cluster of electron-rich or charged aromatic and acidic amino acid side chains within the nicotinic site that primarily stabilizes the ammonium portion of the molecule.

CHAPTER 4

Pharmacological and Physiological Diversity of Nicotinic Receptor Sites

The unconventional diversity of receptor subunits, and their interpretation in terms of a variety of receptor oligomers, confer novel physiological and pharmacological properties upon both muscle and brain receptors. Before it was established that the acetylcholine-binding sites are located at the boundary of subunits, preliminary work had revealed the nonequivalence of the two binding sites present on *Torpedo* receptor. Initial evidence was supplied by the two-step kinetics of α-toxin binding and dissociation (Weber and Changeux, 1974a, 1974b, 1974c; Maelicke and Reich, 1976; Maelicke et al., 1977). Furthermore, the equilibrium binding curve disclosed two distinct affinities for the antagonist [^{3}H]*d*-tubocurarine, but not for the physiological neurotransmitter acetylcholine, nor for carbamylcholine (Neubig and Cohen, 1979, 1980). Additionally, the affinity reagents MBTA and bromoacetylcholine were reported to attach to only one-half the α-toxin- and acetylcholine-binding sites on the reduced receptor (or to both, but with a different affinity) (Karlin et al., 1976; Damle and Karlin, 1978; Deleglane and McNamee, 1980; Ratnam et al., 1986). Similar data were reported with the diterpenoid coral toxin lophotoxin (Culver et al., 1984). Some antibodies (monoclonal or from myasthenic sera) were also found to block one of the two binding sites for α-bungarotoxin (Watters and Maelicke, 1983; Mihovilovic and Richman, 1984; Gu et al., 1985; Whiting et al., 1985). Among twelve antibodies that block the binding of α-bungarotoxin to the receptor, six block the binding to one site, four block the binding to the other site, and two block the binding to both sites (Dowding and Hall, 1987).

Finally, structural differences in the two agonist-binding sites of the *Torpedo* nicotinic acetylcholine receptor were directly revealed by time-resolved fluorescence spectroscopy involving the partial nicotinic agonist dansyl-C6-choline: 1-(5-dimethylamino naphthalene-1-sulfuramide) propane-3-trimethyl ammonium iodide, a compound introduced by Weber et al. (1971). Although dansyl-C6-choline binds with comparable affinity to the two sites, the microenvironment of the probe differs in each case, with dansyl-C6-choline at the α/δ site being associated principally with a 8.7 ns lifetime, and at the α/γ site with a 20.2 ns lifetime (Martinez et al., 2000). Distinct chemical interactions thus contribute to the interaction of pharmacological agents with the acetylcholine-binding site.

The pharmacological properties of nicotinic receptors vary with species, and in a manner that does not necessarily mirror the interaction with the physiological neurotransmitter. For example, receptors reconstituted in *Xenopus* oocytes from rat α4β2-subunits exhibit a preferential order of affinity: acetylcholine ∼ nicotine > DMPP > cytisine; for rat α4β4, by contrast, the order is cytisine > nicotine > acetylcholine > DMPP; and for rat α7, the order is nicotine > cytisine > DMPP > acetylcholine (Role, 1992). This issue is discussed in further detail in Chapter 11.

The diversity of binding properties results from differences in the structure of the agonist-binding site, which is itself determined by the subunit composition of the different receptor oligomers. In electric organ and muscle receptors, as we noted earlier, the two binding sites are located at the α/γ and α/δ interfaces. Neuronal heteropentameric receptors composed of one type of α- and one type of β-subunit— e.g., $(α4)_2(β2)_3$—possess two equivalent agonist-binding sites (Buisson et al., 1996). One would expect to find nonequivalent binding sites in receptors having a more complex subunit composition. Moreover, α5 has been observed only in receptors containing another α-subunit and may be incapable of forming a functional receptor only with β-subunits (Ramirez-Latorre et al., 1996; Wang et al., 1996; Nelson and Lindstrom, 1999). Similarly, β3 appears to require the presence of other β-subunits (Groot-Kormelink et al., 1998).

Overall, the multiplicity of chemically distinct subunits in various oligomeric combinations has a number of interesting consequences. First, distinct pharmacologies occur as a function of the subunit components at a particular interface. Second, physiology does not automatically follow pharmacology: low affinity for acetylcholine is observed for homooligomers, but both high and low affinities are observed for heterooligomers. Third, some subunits are pharmacologically neutral, such as α6, but may nonetheless be important for targeting particular receptor oligomers to axonal (versus somatodendritic) compartments (Champtiaux et al., 2002, 2003). Fourth, it would appear that "canonical" amino acid side chains in the binding site determine the specificity for acetylcholine, whereas "neighboring" amino acid side chains account for the observed pharmacological diversity.

With respect to the final point, for example, the loop-C region was found to play a major role, with the 180–208 segment of α7/α8 contributing to relative acetylcholine and DMPP (1,1-dimethyl-4-phenylpiperazinium) affinities, and the 195–215 segment of α2/α3 and the 183–191 segment of α7/α4 contributing to relative acetylcholine and nicotine affinities. By contrast, the 152–155 loop-B segment in α7/α4 chimeras was shown to alter the pharmacology of all agonists, independently of their chemical structure (Corringer et al., 1998). Amino acids involved in toxin binding were found near loop C as well. A glycosylation at this level was shown to interfere with α-bungarotoxin-binding, thus rendering cobra and mongoose resistant to α-toxin (Kreienkamp et al., 1994). Additionally, mutations at positions valine 188, tyrosine 190, proline 197, and aspartic acid 200 were found to decrease the affinity of the short-chain toxin from *Naja mossambica mossambica* by 60- to 400-fold (Ackermann and Taylor, 1997).

On the complementary side of the site, amino acids contributing to pharmacological diversity were found in most cases at the level of loops D, E, and F. Mapping of the amino acids involved in the different affinities of *d*-tubocurarine, α-conotoxin M1, and carbamylcholine for the α/γ-, α/δ-, and α/ε-binding sites of the mouse muscle-type receptor highlighted the role played by variable residues at position $i + 2, 3$, or 4 from the tryptophan at position δ55, $i + 4$ or 6 from the tyrosine at position δ113, and $i - 2$ and $i + 1$ or 2 from the aspartic acid at position δ180. However, two other amino acids outside these loops—serine at position δ36 and lysine at position δ163—were shown to account for some of the differences in affinity (Sine, 1993; Sine et al., 1995b; Prince and Sine, 1996; Bren and Sine, 1997; Osaka et al., 1999). A residue from loop D and several residues from loop E determine, respectively, the differing affinities of DHβE/α-conotoxin M2 (Harvey and Luetje, 1996) and the differing sensitivities of cytisine for α3β2 and α3β4 (Figl et al., 1992).

Taken together, these studies support the view that variable residues located in the vicinity of affinity-labeled amino acids are the major determinants of the specificity of the receptor site for pharmacological ligands. Yet differences in the binding properties of the acetylcholine neurotransmitter do not seem to correspond with the variability of pharmacological properties. In the picture that is now beginning to emerge, the binding site consists of a conserved core of aromatic residues that interact with the neurotransmitter, with variable amino acids adjacent to these positions (as well as several amino acids from the nonconserved loops E and F) endowing each receptor subtype with its individual pharmacological properties. In other words, pharmacological diversity may not exactly coincide with the variety of physiological properties enjoyed by endogenous neurotransmitters (Le Novère et al., 2002a).

Conclusion

The results presented in this chapter reveal diversity in both the manner of binding and the pharmacological properties of acetylcholine-binding sites. This structural diversity results from the capacity of nicotinic receptor subunits to form oligomers with well-defined subunit stoichiometries: from pentameric homooligomers (α7–α10) to multiple heteropentamers associating up to four different subunits, $(\alpha 1)_2\beta\gamma\delta$ in muscle, $\alpha 4\alpha 6\alpha 5(\beta 2)_2$ in the substantia nigra (Klink et al., 2001; see Chapter 12). The multiple interfaces between pairs of distinct or identical subunits create a broad spectrum of pharmacological properties that have not yet been fully explored. In particular, it is not known how many interfaces accommodate neurotransmitters in neuronal receptors, and whether some interfaces do not bind neurotransmitters but instead interact specifically with pharmacological agents, such as benzodiazepines in the case of the GABA receptors (see Galzi and Changeux, 1995; Mohler et al., 2001). It is nonetheless anticipated that receptor heterooligomers will be found that carry several sites with different pharmacological specificities. This diversity is, of course, a matter of crucial importance for the successful design of pharmaceutical remedies, particularly with regard to drugs that act on brain receptors.

These various populations of receptor heterooligomers are not distributed at random in the nervous system. On the contrary, they exhibit well-defined patterns that imply the existence of exquisitely precise mechanisms of differential gene expression, supported by complex processes of assembly, transport, and targeting of specific oligomers to the neuronal surface. A better understanding of the differential distribution of pharmacologically distinct oligomers holds out the prospect that a finely grained and focused pharmacology can be developed at the synaptic and molecular levels. Future research along these lines may be expected to illuminate not only the effects of nicotine as a drug of abuse, but also the therapeutic properties of nicotinic agents and their differential action with respect to pain, anxiety, attention, and working memory—ultimately, with the aim of alleviating severe pathologies such as Tourette's syndrome and Parkinson's and Alzheimer's diseases (see Chapter 12).

CHAPTER 5

Identification and Properties of the Nicotinic Receptor Ion Channel

Prior to the isolation of the nicotinic receptor, several possibilities were considered with regard to the nature of the coupling between the ligand-binding entity and the ion channel, for example a direct linkage within the same molecule or a mechanism with separate ligand-binding, transducing, and response-generating components (Birnbaumer et al., 1970; see also Rodbell, 1995). Voltage-gated ion channels were thought to represent a form of "independent" ion channels. An initial hypothesis arguing for the presence of both the binding and channel components within the same molecule was inspired by the theory of allosteric proteins (Changeux, 1969). Two elements of protein structure were distinguished: the *receptor* moiety that carries the acetylcholine-binding site (the regulatory "allosteric site") and the *ionophore*, or *ion channel* (the biological "active site") involved in the selective transfer of ions through the membrane. According to the model, these two elements are coupled by an allosteric transition occurring within the same macromolecular complex. Experiments undertaken to purify the acetylcholine-α-toxin-binding component failed to indicate, however, whether or not the ligand-binding component actually served as the ion channel. The possibility was not thereby discounted that the recently identified 43K-protein (Sobel et al., 1977b) (renamed "rapsyn" by Frail et al. [1988]; see Chapter 10), which is abundant in nicotinic receptor-rich membranes, might constitute part of the ion channel. Even when it was demonstrated that the 43K-rapsyn can be removed from the membrane without loss of ionic response (Neubig et al., 1979), uncertainty persisted until early reconstitution experiments of acetylcholine-regulated ion fluxes unambiguously demonstrated that the $α_2βγδ$ oligomer contains the ion channel gated by acetylcholine (see Chapter 2).

In spite of this advance, however, more than a decade of additional work was required to understand the general principles of organization of the ion channel and to identify some of the amino acids that line the ion path. Pharmacological agents known as noncompetitive inhibitors (or "channel blockers") played a critical role in this identification, as well as the application of recombinant DNA technology. Here again success came from the encounter of two cultures, in this case biochemistry and physiology, which led to the identification of the membrane channel (or "pore"), something that scientists thought would forever escape biochemical detection.

Characterization of the First Channel Blockers of the Nicotinic Receptor

Early studies with *E. electricus* electroplaque by members of David Nachmansohn's laboratory (Podleski and Bartels, 1963) revealed that local anesthetics, such as tetracaine, block the electrical response to the agonist carbamylcholine at the postsynaptic level, although in a manner different from curare. They decrease the amplitude of the permeability response to nicotinic agonists without markedly altering the apparent affinity of these agonists. In other words, they behave as *noncompetitive* blockers. At the motor endplate, local anesthetics change the shape of the postsynaptic potential, which becomes bi- or triphasic, with one component acting more rapidly and another more slowly than the single component seen in the absence of local anesthetics (Furukawa, 1957; Maeno, 1966; Steinbach, 1968; reviewed in Colquhoun, 1979). Moreover, because their action at the neuromuscular junction does not correlate with their local anesthetic activity on nerve action potentials, it was interpreted in terms of a blocking action at the level of the postsynaptic acetylcholine receptor, though by means of a mechanism *different from that of curare* (Steinbach, 1968). This analysis led to the conclusion that most local anesthetics (though not all) exert little action unless the ion channel has been opened by acetylcholine (see review by Changeux, 1990).

It was also proposed, more specifically, that these noncompetitive blockers sterically inhibit the transport of ions by entering the open channel and binding to a site located *within the channel*. This interpretation, initially suggested to account for the effect of tetraethylammonium ions on the transport of K^+ by the K^+ channel (Armstrong, 1966, 1969), was extended to apply to the effect of some noncompetitive blockers on the acetylcholine-elicited single channel activity of nicotinic receptors recorded using patch-clamp techniques (Neher and Steinbach, 1978; Neher, 1983). Low-affinity agents such as the quaternary lidocaine derivative QX-222 chop up the single current square pulses into bursts of much shorter pulses, the duration of which decreases with increasing concentrations of effector (*fast channel block*). The observed flickering was viewed as resulting from the repetitive binding of the blocking molecule to a site located inside the ion channel. Other agents of higher affinity, such as QX-134, cause an irregular sequence of very short pulses, which was interpreted as a consequence of the blocker's longer residence time within the nicotinic receptor channel, itself due to a slow rate of unblocking (*slow channel block*) (Neher and Steinbach, 1978; Neher, 1983).

Toxic compounds extracted from the skin of South American frogs, such as histrionicotoxin and its perhydro derivatives (Daly et al., 1971; Albuquerque et al., 1973, 1974; Lapa et al., 1975; Kato and Changeux, 1976; Burgermeister et al., 1977; Dolly et al., 1977; Eldefrawi et al., 1977, 1980), block endplate currents in a manner similar to that of the local anesthetics. Their blocking action has been found in most cases to be sensitive to voltage, which is consistent with the hypothesis that noncompetitive blockers enter the ion channel and bind to a site located *within the membrane* (see Ikeda et al., 1984, for references and discussion). Chlorpromazine, the well-known neuroleptic, behaves as a noncompetitive inhibitor of the peripheral nicotinic receptor, for example. A low-voltage sensitivity had been reported for chlorpromazine in some studies (Koblin and Lester, 1979; Albuquerque et al., 1980). In patch-clamp recordings on a mouse-muscle cell line, two effects of chlorpromazine were observed: it was shown, first, to decrease the channel-opening frequency and thus to act as a closed-channel blocker (Changeux et al., 1986); and subsequently it was also shown to decrease the mean channel-open time in a concentration- and voltage-dependent manner (Benoit and Changeux, 1993). The second effect is commonly associated with open-channel blockers, supporting the notion that chlorpromazine binds to a site within the nicotinic receptor ion channel. The first effect suggested a possible interaction of ligands such as chlorpromazine with *secondary allosteric sites* located at the lipid-receptor interface and/or a "sensing" of the physical state of the lipids by these sites (see Weber and Changeux, 1974a, 1974b, 1974c; Neher and Steinbach, 1978; Changeux, 1981; and Heidmann et al., 1983b, for discussion). Success in identifying the site where chlorpromazine "plugs" the ion channel paved the way ultimately for the identification of the ion channel itself. In the following sections we shall describe the successive steps that led, over more than a decade, to its unambiguous characterization.

Characterization of the Sites of Action for Noncompetitive Blockers

A crucial aspect of the attempt to elucidate the mechanism responsible for the action of noncompetitive blockers—in particular, chlorpromazine—involved combining *in vivo* physiological experiments with *in vitro* biochemical studies (reviewed in Changeux, 1981, 1990; Changeux et al., 1984). The first step in that direction was the demonstration that the above-mentioned *in vivo* effect of local anesthetics could be reproduced *in vitro* by examining the ion-flux response of microsacs purified from *E. electricus*. The local anesthetic tetracaine was indeed found to block the ion-flux response to nicotine agonists in a noncompetitive manner (Kasai and Changeux, 1971). Since several of these channel blockers interact at relatively high concentrations with the acetylcholine-binding site, the possibility had to be entertained that the blocking action of noncompetitive blockers was due to a direct interaction with agonists at the acetylcholine-binding site. To test this possibility, the effects of a series of local anesthetics (dimethisoquin, dibucaine, tetracaine, procaine, prilocaine) on both the electrical response of *E. electricus* electroplaque (or of excitable microsacs) to carbamylcholine and the initial rates of [^{3}H]-α-toxin binding to the membrane-bound acetylcholine receptor from the same electric organ were quantitatively compared (Figure 5.1A, B). A direct effect of the local anesthetics on α-toxin binding was noted. This effect occurred at concentrations *one or two orders of magnitude higher* than those that inhibited the permeability response, however, whereas the effects of authentic competitive antagonists on binding and response were produced in the same range of concentrations (Weber et al., 1972; Weber and Changeux, 1974b). It was concluded that the sites for the pharmacological action of local anesthetics are "distinct from the cholinergic receptor site but nevertheless are located on or near the receptor protein" (Weber and Changeux, 1974c).

The next step consisted in measuring the equilibrium binding of [^{3}H]-acetylcholine to receptor-rich membranes from *T. marmorata*. The local anesthetic prilocaine was found not to displace, but rather to enhance, [^{3}H]-acetylcholine binding, suggesting that prilocaine binds to a distinct allosteric site (Cohen et al., 1974). The first *direct* evidence for the presence of a distinct class of sites for local anesthetics in receptor-rich membranes from *T. marmorata* was provided by spectroscopic studies using the cholinergic fluorescent probe dansyl-C_6-choline, mentioned earlier in Chapter 4. In the presence of agonists that activate the channel while also blocking dansyl-C_6-choline binding to the acetylcholine-binding site, this compound was associated with distinct

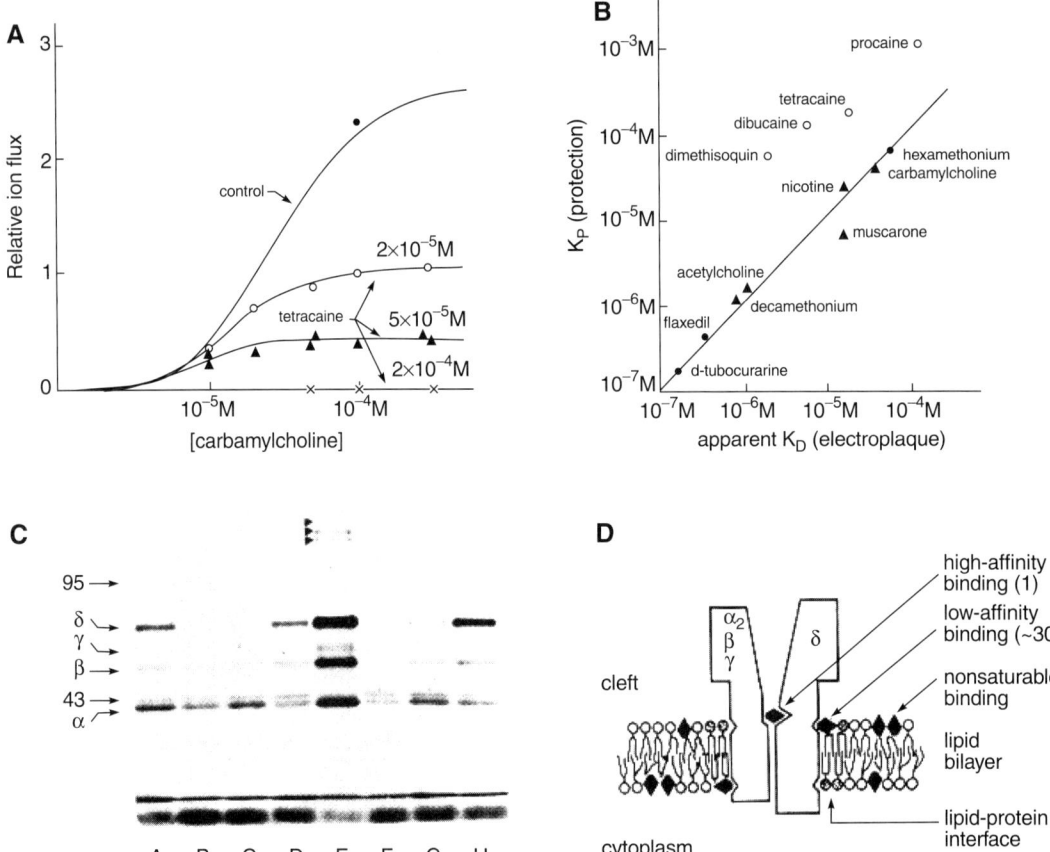

FIGURE 5.1 Experimental strategy for chemically identifying the ion channel. (A, B) *In vitro* effects of noncompetitive blockers on the properties of the membrane-bound receptor. (A) Noncompetitive inhibition of ^{22}Na$^+$ ion-flux response of *E. electricus* excitable microsacs to carbamylcholine. (From Kasai and Changeux, 1971.) (B) Comparison of the apparent dissociation constant (K_d), determined by following the depolarization of a single electroplaque from *E. electricus* by agonists or its blocking by the tested compounds, and the protection constant (K_p), determined by measuring the decrease of the initial rates of ^{3}[H]-α-toxin binding. The noncompetitive blockers fall outside the line along which the agonists and competitive antagonists are distributed, and so their pharmacological action is exerted at a site or sites distinct from the acetylcholine-binding site. (From Weber and Changeux, 1974c.) (C) Interaction of noncompetitive blockers with membrane-bound receptors from *T. marmorata*. Photolabeling of the α-, β-, γ-, and δ-subunits of the membrane-bound receptors by ^{3}H-chlorpromazine and its enhancement by the agonist carbamylcholine (Carb): ***A***, no Carb, no inhibitor; ***B***, no Carb, 50 μM histrionicotoxin (HTX); ***C***, no Carb, 1μM HTX; ***D***, no Carb, 10 μM erabutoxin b; ***E***, 0.1 mM Carb, no inhibitors; ***F***, 0.1 mM Carb, 50 μM HTX; ***G***, 0.1mM Carb, 1μM HTX; ***H***, 0.1 mM Carb, 10 μM erabutoxin b. (From Oswald and Changeux, 1981a). (D) Model of the localization of the high-affinity site for noncompetitive blockers on the quasisymmetry axis of the receptor oligomer. (From Heidmann et al., 1983b.)

fluorescence properties after binding (Cohen and Changeux, 1973a, 1973b) that were abolished by nonfluorescent local anesthetics (Cohen et al., 1974). Additional evidence was supplied by spectroscopic investigations using the simpler fluorescent ligand quinacrine, a compound that does not interact significantly with the acetylcholine-binding site (Grunhagen and Changeux, 1975, 1976) but whose binding sites were revealed in the presence of agonists (Grunhagen and Changeux, 1975, 1976) *and* blocked by prilocaine and histrionicotoxin (Sobel et al., 1978).

Finally, direct binding of the radioactive local anesthetics [^{14}C]-trimethisoquin (Cohen, 1978; Sobel et al., 1980) and [^{14}C]-meproadifen (Krodel et al., 1979) to *Torpedo* receptor-rich membranes was demonstrated. Analysis of the binding curves disclosed a population of agonist-dependent, high-affinity sites from which the local anesthetics were displaced by histrionicotoxin and by other noncompetitive blockers. Detailed equilibrium binding studies with *T. marmorata* receptor-rich membranes and tritiated derivatives of various noncompetitive blockers led to a clear-cut distinction between two main categories of sites for noncompetitive blockers, both of which differ from the acetylcholine-binding site (Heidmann and Changeux, 1981; Heidmann et al., 1983b): high-affinity and low-affinity sites. The high-affinity sites bind noncompetitive blockers whose interactions are blocked by the frog toxin histrionicotoxin. Binding obeys a simple hyperbolic law with an equilibrium dissociation constant in the micromolar range and a stoichiometry of one site per two α-toxin sites (i.e., per $\alpha_2\beta\gamma\delta$-oligomer) (Heidmann et al., 1983b). Interestingly, positive allosteric interactions take place between these high-affinity sites and nicotinic agonists. Carbamylcholine, when bound to the receptor site, enhances the affinity of the noncompetitive blockers for the high-affinity sites, but to a different extent in each case: 1.5-fold for perhydrohistrionicotoxin, 5-fold for phencyclidine, and up to 10-fold for chlorpromazine. Similar equilibrium binding properties of noncompetitive blockers for their high-affinity site were observed with purified receptor reconstituted into lipid vesicles (Sobel et al., 1980; Heidmann et al., 1983b). The reconstituted vesicles also displayed carbamylcholine-stimulated ion fluxes with a sensitivity to channel blockers similar to that of native membranes (Popot et al., 1981). By contrast, a large number of low-affinity sites are present on each receptor; exactly how many probably depends on the lipid-protein interactions in the membrane (see Changeux, 1990).

The High-Affinity Site for Noncompetitive Blockers on the Quasisymmetry Axis of the Receptor Oligomer

In order to chemically identify the high-affinity site for noncompetitive blockers, labeling studies were carried out with a radioactive photoaffinity derivative of the potent local anesthetic trimethisoquin [^{3}H]-5-azido trimethisoquin (5A[^{3}H]-T), synthesized by Waksman et al. (1980b). This compound behaves as an irreversible blocker of the response of *E. electricus* electroplaque to carbamylcholine (Waksman et al., 1980b) and labels almost exclusively the β-subunit of *T. marmorata* receptor under conditions that limit proteolysis of the receptor protein (Oswald et al., 1980; Saitoh et al., 1980). Similar results were obtained with [^{3}H]-perhydrohistrionicotoxin, [^{3}H]-phencyclidine, [^{3}H]-trimethisoquin, and [^{3}H]-chlorpromazine by simple ultraviolet irradiation of their complex with the membrane-bound receptor from *T. marmorata* (Oswald and Changeux, 1981a). It is interesting that in the case of [^{3}H]-chlorpromazine (Figure 5.1C), *all four chains* were labeled, and the labeling was strikingly enhanced by carbamylcholine and decreased by histrionicotoxin (Oswald et al., 1980; Oswald and Changeux, 1981b; Heidmann et al., 1983b).

These data were readily interpreted by a model in which the four chains jointly constitute a *single common high-affinity binding site* for noncompetitive blockers (Oswald and Changeux, 1981b). This site, analogous to the diphosphoglycerate site in hemoglobin, was thought to be located within a crevice in the central part of the 267-kDa oligomer (Arnone, 1972). Accordingly, the agonists were presumed to regulate the interaction of [^{3}H]-chlorpromazine with this site in an allosteric manner, with the ion channel corresponding to the hydrophilic pit observed in the center of multisubunit receptor "rosettes" (Oswald and Changeux, 1981a; Heidmann et al., 1983b; Figure 5.1D). Consistent with this hypothesis, [^{3}H]-chlorpromazine was found to interact with a *unique* high-affinity site per receptor oligomer. Moreover, carbamylcholine *quantitatively* and *simultaneously* enhanced the incorporation of the labeling compound into the four subunits, and this incorporation was reduced by preincubation in the presence of histrionicotoxin, with the *same apparent dissociation constant for the four subunits* (Heidmann et al., 1983b). It was therefore concluded that the high-affinity site is situated in the central cavity, at a location from which the "distance to all five subunits of the receptor is minimum . . . on the *symmetry axis* of the molecule"—a view supported by "the high degree of homology between the four chains of the receptor" (Heidmann et al., 1983b).

Channel-Blocker Binding in Relation to the Functional State of the Receptor

During the course of investigation into channel blockers, an important issue arose concerning whether the high-affinity site labeled by noncompetitive blockers, in particular by [^{3}H]-chlorpromazine, is part of the ion channel and whether access to it depends on the ion channel being open. Since under *physiological conditions* such transient openings occur on a timescale of milliseconds, the interaction of noncompetitive blockers had to be explored with rapid-mixing techniques using stopped-flow equipment (Barrantes, 1976; Grunhagen et al., 1976, 1977; Heidmann et al., 1977; Cohen, 1978). A first step in this direction was taken by experimenting with *T. marmorata* receptor-rich membranes equilibrated with quinacrine (Grunhagen et al., 1976, 1977). Although a relatively fast fluorescence signal (in the 10–20 msec time range) was recorded upon mixing with agonists (acetylcholine, phenyltrimethylammonium), it was *never* observed with competitive antagonists (*d*-tubocurarine, flaxedil, hexamethonium). By using rapid-mixing photolabeling equipment (Heidmann and Changeux, 1984, 1986; Cox et al., 1984; Muhn et al., 1984; Fahr et al., 1985), it became possible to monitor the kinetics involved in the incorporation of radioactive noncompetitive blockers (time resolution of about 20 msec) in experiments carried out on receptor-rich membranes from *T. marmorata* (Heidmann and Changeux, 1984, 1986). Under these conditions, [^{3}H]-chlorpromazine was incorporated into the α-, β-, γ-, and δ-subunits with similar kinetics, although the rate was consistently maximal for the γ-subunit and minimal for the δ-subunit.

The time course of [^{3}H]-chlorpromazine incorporation was adequately fitted by the equation for a simple bimolecular binding reaction. Rapid association of [^{3}H]-chlorpromazine was not observed in the presence of the competitive antagonist *d*-tubocurarine, flaxedil, or snake venom α-toxins. The initial rate of radiolabel incorporation increased with agonist concentration to a *maximum* of 0.6 for carbamylcholine, 0.2 for phenyltrimethylammonium (when normalized to the value for acetylcholine), with apparent K_d values of 30 μM, 400 μM, and 300 μM for acetylcholine, carbamylcholine, and phenyltrimethylammonium, respectively. Finally, the shape of the dose-response curve was slightly sigmoid, with a Hill coefficient of 1.1–1.3 (Heidmann and Changeux, 1984).

All of these features closely parallel those found *in vitro* following an initial ion flux (Neubig and Cohen, 1980; Dunn and Raftery, 1982a, 1982b; Heidmann et al., 1983a), activating the ion channel. The rapid association of [^{3}H]-chlorpromazine disappeared on prolonged exposure of the receptor to acetylcholine (Heidmann and Changeux, 1986), presumably because of the rapid phase of desensitization (Hess et al., 1982; Walker et al., 1982). Analysis of the concentration-effect curve for the inactivation of [^{3}H]-chlorpromazine labeling yielded apparent K_d values of 40 μM for acetylcholine and 0.4 μM for carbamylcholine (Heidmann and Changeux, 1986), coinciding with the expected values for the rapid desensitization of the permeability response (Hess et al.,

1982; Walker et al., 1982). These data thus supported the conclusion that [^{3}H]-chlorpromazine rapidly labels the active conformation of the receptor in which the ion channel is open. In this view, chlorpromazine binds in a diffusion-controlled manner to a site common to all subunits that lies within the ion channel and becomes accessible when the channel is opened by acetylcholine (Heidmann and Changeux, 1984, 1986).

Kinetic measurements were also reported for other noncompetitive blockers triphenylmethylphosphonium (TPMP) (Fahr et al., 1985) and quinacrine azide (Kaldany and Karlin, 1983; Karlin et al., 1983; Cox et al., 1985), although here the evidence for labeling of the open state was more ambiguous than for chlorpromazine. These *in vitro* data are thus consistent with the observation mentioned earlier that chlorpromazine decreases the mean channel-open time of the receptor in a concentration- and voltage-dependent manner and behaves as an open-channel blocker. The bulk of the data supports the notion that chlorpromazine binds to a site located within the open ion channel of the receptor (Heidmann et al., 1983b).

Identification of the M2 Segment as a Component of the Ion Channel

The first attempt to identify the amino acid(s) labeled by [^{3}H]-chlorpromazine was undertaken using peptide fractionation and sequencing techniques (Giraudat et al., 1986; see Figure 5.2). The δ-subunit was selected because it is the most heavily labeled subunit in *T. marmorata* (Oswald et al., 1980; Oswald and Changeux, 1981b). Photolabeling of the membrane-bound receptor with [^{3}H]-chlorpromazine was carried out under equilibrium conditions in the presence of carbamylcholine, and both in the presence and absence of phencyclidine, a selective ligand for the high-affinity site for noncompetitive blockers. The δ-subunit was purified, digested with trypsin, and the resulting fragments fractionated by reversed-phase high-performance liquid chromatography (HPLC). The labeled peptide(s) could not be purified to homogeneity, however, because of their marked hydrophobic character. This intrinsic difficulty had somehow to be circumvented. A strategy was developed that took advantage of the presence of a methionine residue at the amino terminus of one of the tryptic fragments (Met-Ser-Thr-Ala-Ile-Ser . . .) (Giraudet et al., 1986). Treatment of the mixture of tryptic peptides with cyanogen bromide (CNBr) was predicted to shorten this peptide by one residue, thus shifting by one cycle the radioactivity released on peptide sequencing by the labeled amino acid in the event it belonged to this particular peptide.

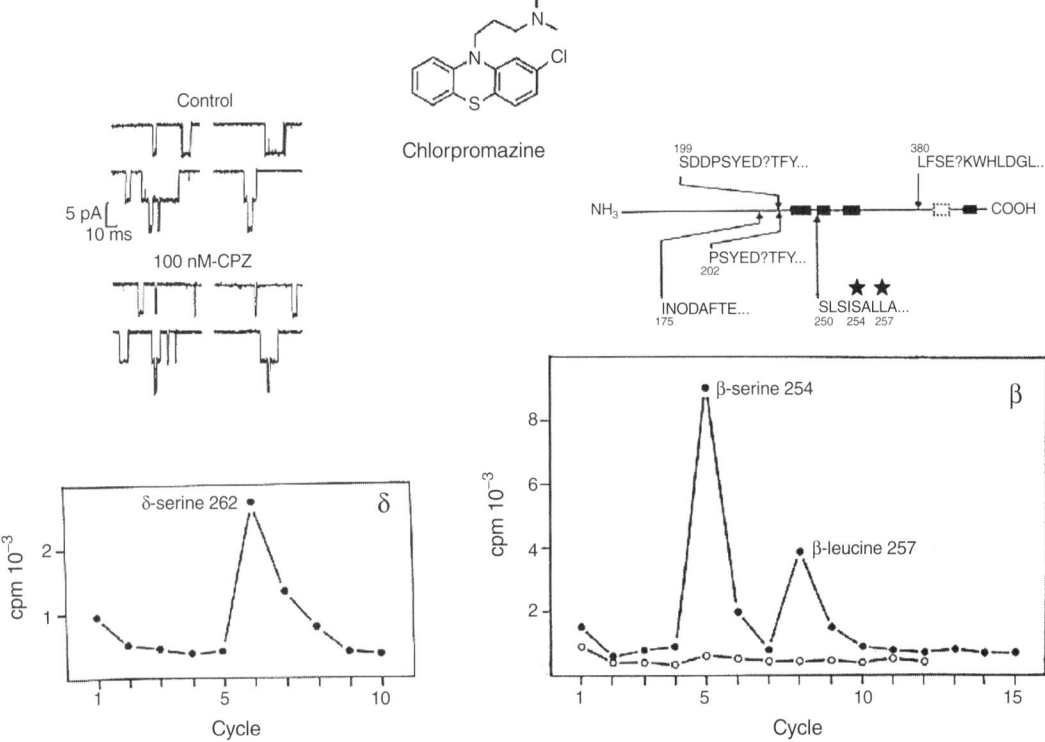

FIGURE 5.2 Identification of the amino acids labeled by chlorpromazine for the receptor β- and γ-subunits, showing detailed chromatographic patterns for γ and β, with (open circles) or without (closed circles) phencyclidine. All the labeled amino acids belong to the transmembrane segment M2. (From Giraudat et al., 1986, 1987.)

Such a shift was in fact observed, and the amino acid labeled by [^{3}H]-chlorpromazine was identified as serine δ262 (Giraudat et al., 1986).

Sequence analysis of analogous material derived from control batches further showed that labeling of the serine residue δ262 was prevented by phencyclidine. Similar experiments were repeated with the β-chain and sequence analysis resulted in the identification of serine 254 and leucine 257 as residues labeled by [^{3}H]-chlorpromazine in a phencyclidine-protectable manner (Giraudat et al., 1987). These residues are both located in the hydrophobic and potentially transmembrane segment M2 of the β-subunit, a region homologous to the one containing chlorpromazine-labeled serine 262 in the δ-subunit.

Oberthur et al. (1986), using ^{3}H-triphenylmethylphosphonium (^{3}H-TPMP) and the same strategy for cleaving the δ-subunit as Giraudat et al. (1986), were also able to identify serine δ262 as an amino acid labeled on the δ-subunit. In addition, Hucho's group provided suggestive evidence for labeling on the α- and β-subunits at homologous positions (Hucho et al., 1986). Homologous regions of different receptor subunits thus constitute the unique high-affinity site for noncompetitive blockers within the ion channel (Hucho et al., 1986; Giraudat et al., 1987), a finding consistent with the proposal by Heidmann et al. (1983b) that this site is located on the quasisymmetry axis of the receptor molecule. Moreover, the M2 segments are not homologous with the regions labeled by affinity reagents of the acetylcholine-binding site on the α-subunits, a result consonant with the view that the acetylcholine-binding site and the high-affinity site for channel blockers are structurally distinct binding sites (Figure 5.3).

The contribution of segment M2 to the ion channel has subsequently received support from further labeling experiments with meproadifen mustard (Pedersen et al., 1992), 3-(trifluoromethyl)-3-(m-[^{125}I]iodophenyl)diazirine (White and Cohen, 1992), as well as its benzoic acid ester and phospholipid analogues (Blanton et al., 1998b), 2-[^{3}H]diazofluorene (Blanton et al., 1998a) and tetracain (Middleton et al., 1999). Electrophysiological recordings from recombinant receptor confirmed and extended this conclusion.

Site-Directed Mutagenesis in the Characterization of the Channel

In the early 1980s a novel encounter between scientific cultures, in this instance recombinant DNA technologies and electrophysiology, created an experimental paradigm that facilitated the investigation of the relationship between the linear amino acid sequence of the receptor and ion channel functions. The method was successfully adopted to study receptor function, and even to isolate receptors—such as the glutamate-NMDA receptor (Moriyoshi et al., 1991)—despite the fact that it overlooks the three-dimensional structure of the receptor. Sumikawa et al. (1981) and Barnard et al. (1982) injected exogenous mRNA coding for nicotinic acetylcholine receptor subunits from *T. marmorata* into *Xenopus* oocytes and were able to record acetylcholine-responsive channels. The same experiment was repeated with mRNA from cloned subunits of *T. californica* and calf receptor protein. Differences in intrinsic ion channel conductances were recorded by patch-clamp measurements at low divalent cation concentrations (Mishina et al., 1984; Imoto et al., 1986). The domain of the receptor δ-subunit responsible for the observed difference in conductance was located by replacing the δ-subunit mRNA by various *Torpedo*/calf chimeric δ-subunit mRNAs. The data suggest that M2 and the segment located between M2 and M3 are involved in determining the conductance difference noted between *Torpedo* and calf channels (Imoto et al., 1986). Yet in this early work, which happened to be consistent with prior labeling studies using chlorpromazine and ^{3}H-TPMP, no single amino acid was identified as critical for the ion-transport mechanism.

In later studies (Imoto et al., 1988; Kienker et al., 1994), particular amino acids involved in channel conductance were identified by in-

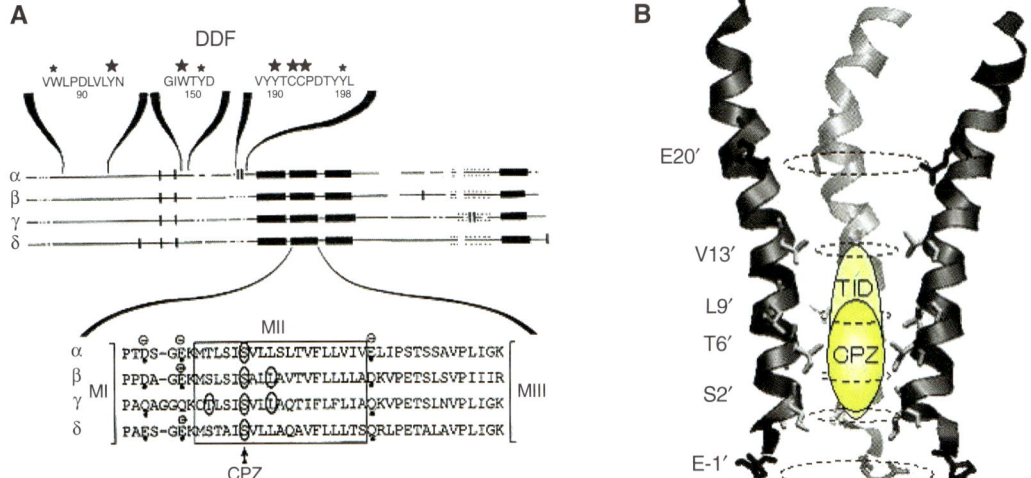

troducing point mutations into the *Torpedo* receptor subunit cDNAs to alter the net charge of the negatively charged or glutamine residues around the M2 segment. Under conditions of low divalent ion concentration, mutations in three rings of negatively charged residues (and glutamine residues) neighboring the hydrophobic M2 segment were accompanied by changes of channel conductance for monovalent cations, namely the residues Asp-α238, Glu-α241, and Glu-α262, as well as the homologous positions on the other subunits (see Figure 5.3). On the basis of the sidedness of Mg^{2+} effects, the anionic ring between M2 and M3 (Glu-α262 and its homologues) was located on the extracellular side, with the two other positions between M1 and M2 (Asp-α238 and its homologues, and Glu-α241 and its homologues) being situated on the cytoplasmic side. These rings were referred to as "inner," "intermediate," and "outer" negative rings. Thus the M2 segment spans the membrane, and the site for [^{3}H]-chlorpromazine and TPMP binding in M2 is "framed" by the anionic rings *within* the membrane. The site-directed mutagenesis data strikingly confirm the models of Heidmann et al. (1983b), Hucho et al. (1986), and Giraudat et al. (1987).

The physiological importance of the ring of serines homologous to Ser-δ262 (Giraudat et al., 1986) was first explored by site-directed mutagenesis and expression in *Xenopus* oocytes of mouse receptor subunit mRNAs (Leonard et al., 1988). In the injected mRNAs (Boulter et al., 1986b), these serines were replaced by alanines (α- and δ-subunits) or by phenylalanine (β-subunit). As the number of serine residues decreased, the residence time (and thus affinity) of the channel blocker QX-222 diminished (Leonard et al., 1988). Receptors with Ser → Ala mutations also displayed a selective decrease in outward single channel potassium currents. Furthermore, in agreement with

FIGURE 5.3 Structure of the ion channel and of the binding site for noncompetitive channel blockers. (A) Summary of amino acids labeled by p-N,N-dimethyl-aminobenzenediazonium fluoroborate (DDF) at the acetylcholine-binding site and by the noncompetitive blocker chlorpromazine (CPZ) in the transmembrane region M2. (From Changeux, 1990.) (B) Three-dimensional model of the ion channel for nicotinic receptor α7. The walls of the ion pore are composed of five α-helices (two have been removed for the sake of clarity). The amino acids S2', T6', and L9' are marked by chlorpromazine (CPZ), whereas the probe trifluoromethyl-3-(m-iodophenyl)diazirine (TID) reacts with the same residues as CPZ in addition to a ring of Val residues (V13'). The labeled residues constitute the narrowest portion of the pore. (Courtesy of Thomas Grutter.)

the models of Hucho et al. (1986) and Giraudat et al. (1987), the data concerning voltage dependence and rectification suggested that the mutated residues were located within the ion channel, near its cytoplasmic end. Additional studies on the role of M2 residues involved the polar residues Thr-244 and Ser-248 of the α-subunit (Leonard et al., 1988; Villarroel and Sakmann, 1992), as well as hydrophobic amino acids (α7 residues Leu-247 and Val-251, corresponding to α-subunit positions 251 and 255, respectively) (Révah et al., 1991; Bertrand et al., 1993). The latter residues alter channel conductance and, in some cases, the selectivity among monovalent (Konno et al., 1991; Cohen et al., 1992) or divalent cations (Bertrand et al., 1993).

Toward a Stereochemical Model of the Acetylcholine-Gated Ion Channel

All available data are consistent with the notion that the ion channel is located along the quasisymmetry axis of the receptor oligomer, at the fivefold interface of the subunits, with each subunit contributing residues from its M2 segment. The pattern of labeling—i.e., residues homologous to Thr-244, Ile-247/Ser-248, Leu-251/Ser-252, Thr-254/Val-255, Leu-259, and Glu-262 of the α7 receptor—strongly supports the view that M2 folds into an α-helix, in agreement with secondary structure predictions (Le Novère et al., 1999). Electron microscopy images of the channel in its open conformation identified five rods bordering the ion channel, attributed without biochemical evidence to the M2 segment (Toyoshima and Unwin, 1988; Unwin, 1995). The rods lie ∼18 and ∼11.5 Å from the axis of the pore, at the upper and lower faces, respectively—a measurement that is consistent with a funnel-shaped pore having a minimal diameter of ∼10 Å. The α-helical section would appear to merge with a nonhelical loop at the cytoplasmic end, however. Mutations within the loop component have been found to alter various aspects of the channel's ion selectivity.

Mutations in the rings composed of α-subunit residues Ser-244 and Glu-241 (and their homologues in other subunits) progressively decrease the conductance of large cations when the bulk of the side chain increases, which suggests that these residues are involved in cation selection in proportion to their size (Cohen et al., 1992; Villarroel and Sakmann, 1992; Wang and Imoto, 1992). Furthermore, decreasing the net charge of the ring corresponding to Glu-241 (and, to a lesser extent, to Asp-238) results in a proportional decrease in potassium unitary conductance, in agreement with their direct or indirect (electrostatic) interaction with cations (Imoto et al., 1988). The mutation Glu → Ala at position 237 in α7 (241 in α1) abolishes the permeability of the α7-receptor to calcium while preserving permeability to monovalent cations (Bertrand et al., 1993).

A major insight into the structural determinants of the ion selectivity of the channel came from the introduction of a new technique referred to as "distributed chimera." The question was simple: if M2 and neighboring sequences determine the ion selectivity of the channel, could a cationic channel be transformed into an anionic one? To answer this question, chimeras were constructed between the cationic α7-nicotinic acetylcholine receptor and the anionic α1-glycine receptor. When all or major portions of the glycine receptor M2 segment were used to replace *en bloc* the corresponding portions in α7, no current was detected in response to acetylcholine application in the oocyte system. Instead, a distributed set of amino acids thought to face the lumen of the ion pore in one receptor was exchanged within the channel sequence of the other receptor. Interestingly, when six individual residues from the chloride channel at the same meridian in the glycine receptor were introduced into the cationic α7-channel sequence (plus a proline residue in the M1-M2 loop), robust acetylcholine-activated currents were observed that were almost entirely carried by chloride ions. In order to identify the critical residues necessary for converting ion selectivity from cationic to anionic, the number of amino acid residues introduced was then progressively reduced, leading to the conclusion that the introduction of only three amino acids—insertion of a proline residue between positions G236 and E237, along with mutations of Glu → Ala and Val → Thr at positions 237 and 251, respectively—suffices to convert the selectivity from cationic to anionic (Galzi et al., 1992). For the first time, a cationic excitatory receptor was transformed into an anionic inhibitory one.

Further insight into the role of the M1-M2 loop was provided by the systematic variation of residues in this region. The addition of a proline residue yields an anionic channel only when inserted between positions 234 and 237; insertion before 234 gives a cationic channel and insertion after 238 alters receptor surface expression. These results have been interpreted in terms of a coiled loop that directly contributes to the charge-selectivity filter of the α7-channel (Corringer et al., 1999). Moreover, in spite of anion-anion repulsion, mutation V251D also yields an anionic channel, indicating that the mechanism is not related solely to the nature of the incorporated side chain (Corringer et al., 1999).

The converse mechanism—the transformation of an anionic channel into a cationic one—has been shown to occur in the case of the *Drosophila* anionic GABA-gated Cl^- channel, for which introduction of a positively charged Lys or Arg in place of the naturally occurring asparagine at position 319 (corresponding to α7 Ala-257) results in significant cationic permeability (Wang et al., 1999a). Similar effects were observed in the case of glycine receptor, where selectivity was converted to cationic from anionic by introducing three mutations (Keramidas et al., 2000) that were the "inverse" of those used by Galzi et al. (1992). In addition, selectivity was switched from cationic to anionic for the 5-HT(3a) receptor by introducing three mutations (Gunthorpe and Lummis, 2001) that correspond to those used by Galzi et al., 1992.

The 10 Å diameter of the α-helical component noted above is consistent with the notion that ions cross the membrane at this level in a fully hydrated state, whereas the narrower diameter of the loop component is thought to accommodate only partially dehydrated ions. The critical role of the loop component in both cation discrimination and charge selectivity further supports its postulated contribution to the selectivity filter of the channel by specific ion dehydration. It is therefore reasonable to suppose that the α-helical component selects on the basis of stabilization of hydrated ions within the membrane. Preliminary explanations of several phenotypes observed for mutation-altered residues in this region have been formulated in an initial attempt to analyze this portion of the channel (Corringer et al., 1999).

In this view, the structure of the nicotinic receptor ion channel calls to mind that of the tetrameric voltage-gated Na^+, Ca^{2+}, and K^+ channels, which are composed of similar elements, only with an inverted arrangement: the loop component of these channels (called P-loop) is located on the extracellular side (Armstrong et al., 1998), a result established by X-ray crystallography in the case of a phylogenetically related bacterial potassium channel, KcsA (Doyle et al., 1998; Zhou et al., 2001). Furthermore, the P-loops of the Na^+ and Ca^{2+} channels display two similarities with the loop component of nicotinic receptors: first, they consist of two rings of negatively charged residues separated by two amino acids analogous to α7D234/α7E237; second, mutation of the inner ring of the Na^+ channel to Glu, as in the Ca^+ channel, confers Ca^{2+} permeability (Heinemann et al., 1992), and interestingly, a change in the opposite direction for the mutation α7E237A abolishes Ca^{2+} permeability (Bertrand et al., 1993). One may thus tentatively suggest that similar—though not necessarily identical—mechanisms of cation permeation selectivity operate in these evolutionarily distant channels, in which no homology of primary structure is observed (Corringer et al., 1999, 2000).

It is generally accepted by researchers in this field that two distinct structural entities control the ion selectivity of the channel (the selectivity filter) and the opening and closing of the channel (the gate) (see Hille, 1992). Various attempts have therefore been made to identify the so-called nicotinic receptor channel gate, and to establish its relationship with the set of amino acids known to contribute to ion selectivity.

In the case of muscle nicotinic receptor, amino acids located within M2 or in its vicinity were systematically substituted for by cysteines (a procedure known as the substituted cysteine accessibility method, or SCAM) to the α-subunit. At the level of the loop component, cysteine substitution indicated that residues corresponding to α7E237, K238, and I239 act as barriers to the permeant methanethiosulfonate ethylammonium when applied either extracellularly or intracellularly (Wilson and Karlin, 1998). These positions, the authors suggest, correspond to the location of the gate. In keeping with this hypothesis, the coapplication of methanethiosulfonate ethylammonium with acetylcholine has been found to result in the removal of this barrier. Furthermore, the insertion of a proline residue between positions 233 and 238 in the α7-receptor (along with E237A and V251T mutations) causes an increased acetylcholine EC_{50} and high levels of spontaneous activity (Corringer et al., 1999). These observations strongly suggest that the loop component may serve the dual function of selectivity filter and physical gate in the channel's activation transition. Alternatively, the gate may involve a set of different residues, analogous to those proposed for the K^+ channels (see above) and possibly closer to the equatorial region of the channel. Such an interpretation might superficially accord with the constriction observed by electron microscopy near the conserved Leu (Unwin, 2000; Miyazawa et al., 2003). This point is discussed more fully in Chapter 7.

The Distance between the Binding Site and the Channel

Affinity reagents specific to the agonist-binding site and to the high-affinity site for channel blockers unambiguously label distinct domains of the primary sequence of the α and other subunits of the receptor protein: the large NH_2-terminal hydrophilic domain, on the one hand, and the membrane-spanning segment M2 on the other. However, the distances between these sites may be considerably modified by the tertiary folding and quaternary assembly of the subunits—a question to which site-directed mutagenesis does not provide a simple answer. The distance between these sites on the folded protein was estimated directly by Taylor and his coworkers using fluorescence spectroscopy (Herz et al., 1987, 1989; Johnson et al., 1984, 1987). They proceeded initially by measuring energy transfer between two fluorescent-labeled α-toxins, and estimated a distance of about 67 Å between toxin molecules bound to the two acetylcholine-binding sites. These data indicated that a portion of the toxin resides near the outer perimeter of the receptor molecule (Johnson et al., 1984). Then, using the same technique (Johnson et al., 1987), they found a distance of at least 50 Å between a fluorescent lysine 23-labeled α-toxin and decidium (a small fluorescent competitive antagonist) bound to the two acetylcholine sites. Measuring energy transfer between several fluorescent toxin derivatives and eosine alkylamide incorporated into the membrane, they also estimated that the distance between the tip of the toxic loop of the α-toxin (and, therefore, the agonist-binding site) and the bilayer surface was larger than 20 Å.

To tag the high-affinity site for channel blockers, Herz et al. (1987, 1989) used ethidium bromide and found an ethidium stoichiometry of one site per $\alpha_2\beta\gamma\delta$-oligomer, in agreement with the results of Heidmann et al. (1983b). Steady-state polarization of ethidium fluorescence further indicated that bound ethidium is strongly immobilized. Moreover, the magnitude of quantum-yield enhancement and the shifts of excitation and emission maxima of bound ethidium suggest, again in agreement with the structural data mentioned above, that its binding site lies within a *hydrophobic* domain (Herz et al., 1987).

The spatial relationship between the agonist-binding sites and the high-affinity site for channel blockers was further explored using the fluorescent agonist dansyl-C_6-choline (Waksman et al., 1980a) and the competitive antagonist bis(choline)-N-(4-nitrobenzo-2-oxa-1,3-diazol–7-yl)-iminodipropionate (BCNI) (Bolger et al., 1984) as energy donors when bound to the agonist site, and ethidium as energy acceptor from both donors. Fluorescence lifetime and energy transfer measurements revealed similar efficiencies of transferred energy between the donors (bound to the two agonist sites) and ethidium (bound to the high-affinity site for channel blockers). The distance between agonist and high-affinity sites was estimated to be between 21 and 35 Å for the BCNI-ethidium pair and 22 to 40 Å for the dansyl-C_6-choline-ethidium pair.

On the basis of these measurements and of the known molecular dimensions of the receptor protein, Herz et al. (1989) suggested locating the high-affinity site for noncompetitive blockers along the quasi-symmetry axis of the molecule, but on the *extracellular* surface of the membrane bilayer. This raises a problem, however. The ring of serines homologous to Ser-δ262 had been located in the NH_2-terminal third of M2 (Giraudat et al., 1986; Oberthur et al., 1986), which, according to site-directed mutagenesis experiments (Imoto et al., 1988) and in keeping with the original four-membrane-spanning segment models, is oriented toward the cytoplasmic face of the membrane. Accordingly, the site for channel blockers should be closer to the intracellular than the extracellular surface of the membrane bilayer. In this connection it should be noted that the binding site for the noncompetitive allosteric blocker quinacrine has recently been localized to the inner half of M2 for mouse muscle nicotinic receptors in the open state (Yu et al., 2003). Overall, these studies clearly demonstrate that interaction within the ion channel between the acetylcholine-binding sites and the high-affinity site for noncompetitive blockers involves *topographically distinct (and distant)* sites.

Conclusion

Success in identifying the ion channel carried by the acetylcholine receptor protein—the first ion channel to be chemically understood—was the result of employing a family of pharmacological agents referred to as noncompetitive, or "open channel," blockers. These compounds, which include local anesthetics, covalently attach to a unique high-affinity site, distinct from the acetylcholine-binding site, and to which all five subunits contribute. This site was therefore presumed to be located within the axial crevice of the receptor oligomer. Because the access of noncompetitive blockers such as [^{3}H]-chlorpromazine to the site as a consequence of rapid mixing with acetylcholine or other nicotinic agonists bears striking parallels with the physiological conditions for channel opening, it was postulated that the site is located within the ion channel.

Chemical identification of the amino acids labeled by [^{3}H]-chlorpromazine (as well as by other noncompetitive blockers) provided the first evidence of a symmetrical contribution from each of the five subunits contributing their transmembrane segment M2 to the wall of the ion channel, and also of the α-helical coiling of this segment. Subsequent site-directed mutagenesis experiments first pointed to a functional role for the identified amino acids, but later revealed the existence of negatively charged inner, intermediate, and outer rings that "frame" the high-affinity site for [^{3}H]-chlorpromazine nearer the equatorial region of the receptor channel.

Combinations of at least three mutations within or in the neighborhood of M2 may result in a change of the channel's ion selectivity from cationic to anionic in the α7 neuronal receptor. A converse change of selectivity has also been achieved in the case of the anionic glycine receptor channel. Amino acids from the loop that links M1 and M2 play a critical role in this change of ion selectivity. The structural elements of the channel's physiological gate—whether local structures or structures responsible for the *en bloc* movement of part or all of the α-helices—remain to be identified. In the meantime, the distance between the acetylcholine-binding site and the ion channel has been estimated by energy transfer experiments to be between 22 and 40 Å, which is to say within the range of measured distances between hemes in hemoglobin (Perutz, 1989). The physiological activation of the ion channel by acetylcholine may thus be said to result from an interaction between topologically distinct sites. In other words, it is in fact a typical *allosteric* interaction (Monod et al., 1963).

CHAPTER 6

Activation and Desensitization at the Structural Level

In the decades that followed the isolation of the nicotinic receptor (Changeux et al., 1970b), multidisciplinary investigations revealed structural properties that must be taken into account by any satisfactory description of the molecular mechanisms of signal transduction that mediate the ionic response to neurotransmitters. As described in Chapter 4, specific sites for the binding of nicotinic agonists and antagonists were identified and localized to subunit interfaces. The ion channel was found to be constructed from elements within the quasi-symmetry axis of the transmembrane protein oligomer, with contributions from all of the subunits, as we saw in the previous chapter. Measurements of the distance between the binding sites and the ion channel indicated a separation of 22–40 Å, therefore requiring a protein-mediated coupling between these two essential elements of receptor function. These aspects of nicotinic receptor structure are consistent with the notion that an allosteric mechanism analogous to the mechanism linking nonadjacent regulatory and catalytic sites in feedback-inhibited enzymes (Monod et al., 1965; Changeux, 1969; Edelstein, 1972) mediates the signal transduction process elicited by acetylcholine.

However, an adequate mechanistic description of receptor physiology must go beyond the allosteric enzyme model in order to take into account the distinctive features of neurotransmitter receptors, in particular the slow process (or processes) of desensitization generally observed in addition to the fast activation of the ion channel triggered by the neurotransmitter. The early studies of activation described in Chapter 1, which involved recordings of the postsynaptic potential at the neuromuscular junction with intracellular electrodes and stimulation by the iontophoretic application of acetylcholine onto the subsynaptic membrane, set the stage for characterization of the electrical parameters of channel conductance (Del Castillo and Katz, 1956; Takeuchi and Takeuchi, 1960). These early studies also revealed that the response curve for acetylcholine was sigmoidal rather than hyperbolic (Katz and Thesleff, 1957; Changeux and Podleski, 1968), pointing to a cooperative interaction between ligand-binding sites.

A striking complementary aspect of the electrophysiological investigations was the discovery of a reversible decline in the conductance response to cholinergic agonists brought about within seconds or minutes by steady exposure to exogenous acetylcholine (Fatt, 1950; Thesleff, 1955; Katz and Thesleff, 1957). This desensitization phenomenon had been recognized by Langley (1905) and extensively studied by both pharmacologists (see review in Triggle, 1980) and physiologists (reviews in Hamill and Sakmann, 1981; Magleby and Pallotta, 1981). The advent of single channel recordings (Neher and Sakmann, 1976) provided additional detailed information on the kinetics of channel-opening and -closing transitions and revealed the occurrence of channel openings in bursts separated by silent periods, indicative of a passage to desensitized states (Sakmann et al., 1980).

As studies on the biochemistry and biophysics of the nicotinic receptor protein progressed—establishing the various features of activation and desensitization, identifying the acetylcholine-binding site and the ion channel, and describing their properties—a major challenge presented itself: generating a comprehensive model that would satisfactorily represent these fundamental aspects in terms of the molecular architecture of the receptor and of the dynamics of its structural transitions. As we recalled in Chapter 1, initial attempts to provide a quantitative description of receptor function were inspired by the analogy with enzyme catalysis (Del Castillo and Katz, 1957a), leading to a sequential binding model that is still currently employed by many electrophysiologists. An alternative approach, developed as an extension of principles derived from studies of hemoglobin and regulatory enzymes, was based on the experimentally demonstrated existence of multiple conformational states that interconvert spontaneously, even in the absence of ligand. Although this approach had been suggested in the early stages of receptor research (Changeux et al., 1967b; Karlin, 1967a; Changeux, 1969; Edelstein, 1972), its application to the full range of properties of the receptor occurred only later and required the incorporation of data from single channel recordings, as well as from quantitative measurements of desensitized states (see review in Changeux and Edelstein, 1998).

Activation and Desensitization of Ionic Response *in Vitro*

The first steps toward understanding the molecular mechanism of activation and desensitization were made possible by the use of closed membrane fragments, or "microsacs," purified from electric organs in their native form. As we noted in Chapter 2, these preparations respond to nicotinic agonists by an "activation" of ion fluxes (Kasai and Changeux, 1970, 1971). Similar results were also obtained with purified receptor protein reconstituted into lipid bilayers from detergent-containing solutions (Changeux et al., 1972). These observations led to a critical advance in conceptualizing the link between chemistry and physiology, since ligand binding could now be measured directly under conditions of agonist-dependent ion channel opening—a possibility never realized by conventional electrophysiological recordings. However, the chemical process of ligand binding itself presented perplexing features that complicated preliminary investigations. Early reports (reviewed by Weber et al., 1975; Colquhoun, 1979) on the receptor protein from *Electrophorus*, both in sodium deoxycholate crude extracts (Changeux et al., 1970a, 1971) and in purified membrane fragments (Kasai and Changeux, 1971), yielded low affinities for agonists, but after purification of the same protein on an affinity column in the presence of Triton X-100, affinities were found to be some 15–40 times higher for the same agonists (Meunier and Changeux, 1973).

Clarification was forthcoming from a detailed analysis of the effects of detergents on acetylcholine binding to the well-defined receptor-rich membranes from *T. marmorata* (Sugiyama and Changeux, 1975). At equilibrium, acetylcholine was found to bind to these membranes with a dissociation constant close to 3×10^{-8} M, but dispersal of the membranes by sodium cholate caused a marked decrease in affinity associated with the appearance of medium ($K_D \sim 10^{-7}$ M) and low ($K_D \sim 10^{-6}$ M) affinities, without a significant change in the *total* number of sites. Clearly the same molecules could exist under *different* states of affinity. Two possibilities then came to mind: either isolation and purification of the membranes altered the binding properties of the receptor in a nonphysiological manner, or prolonged contact with the cholinergic ligand during the assay caused an apparent change of the binding constants of the membrane-bound receptor.

Dynamics of Binding and Flux

To distinguish between the two possibilities noted above, the properties of the receptors were compared as a function of time after mixture with the cholinergic ligand using the simplest method then available:

decreasing the initial rates of α-toxin binding (Weber et al., 1975). The data unambiguously showed that in the native membrane the receptor protein is present in a state displaying low affinity for binding cholinergic agonists; prolonged exposure to the agonist then stabilizes the receptor protein, in a slow (i.e., seconds or minutes) and reversible manner, in a high-affinity state for agonists, with little change in the ability to bind α-toxins (Weber et al., 1975; Changeux et al., 1976). Similar observations were subsequently reported with the receptor from muscle cells (Colquhoun and Rang, 1976) and with *T. californica* membrane fragments (Weiland et al., 1976, 1977; Lee et al., 1977; Quast et al., 1978a, 1978b).

This slow transition of the membrane-bound receptor to a high-affinity state was analyzed further by measurements of α-toxin binding (Sine and Taylor, 1979; Blanchard et al., 1979a, 1979b). A significant improvement in time resolution resulted from the application of rapid-mixing techniques (review in Changeux et al., 1984; Ochoa et al., 1989). For example, studies with the fluorescent agonist dansyl-C_6-choline (Heidmann and Changeux, 1979; see Chapter 3 and Figure 6.1) and [^{3}H]-acetylcholine (Boyd and Cohen, 1980) show that 20% of the receptor spontaneously appears in a state of high affinity (3 µM for acetylcholine and 2 µM for dansyl-C_6-choline) in the absence of agonist, with the remaining 80% exhibiting a low affinity for agonists. Beyond the details of the analysis, which are often linked to a particular kinetic model, two major conformational transitions (Heidmann and Changeux, 1979, 1980; Dunn et al., 1980) have been resolved: a slow transition, with an apparent rate constant of about 0.01 s^{-1}, which either increases to a plateau (Heidmann and Changeux, 1979) or decreases after reaching a maximum (Dunn et al., 1980) with agonist concentration; and an intermediate transition, with a rate constant of 2 s^{-1} (Dunn et al., 1980; Heidmann and Changeux, 1980), as presented in Figure 6.1C. The receptor state stabilized after equilibration with agonist binds acetylcholine and dansyl-C_6-choline with respective K_d of 3 µM (Boyd and Cohen, 1980) and 2 µM (Heidmann and Changeux, 1979); in this respect, it does not differ from the state that prevails before agonist binding in the case of 20% of the population in the membrane (see, however, Sigman and Young, 1981, 1983). The state observed following the intermediate transition has an apparent equilibrium constant for agonists in the range of 1 µM.

Finally, techniques were developed that permitted rapid kinetic measurements of agonist binding and permeability response to be carried out *in parallel,* with the same membrane preparation and on the same timescale (Figure 6.1D). This system made it possible to establish a direct correlation between ion fluxes and the binding of the fluorescent agonist dansyl-C_6-choline (Heidmann et al., 1983a). Analy-

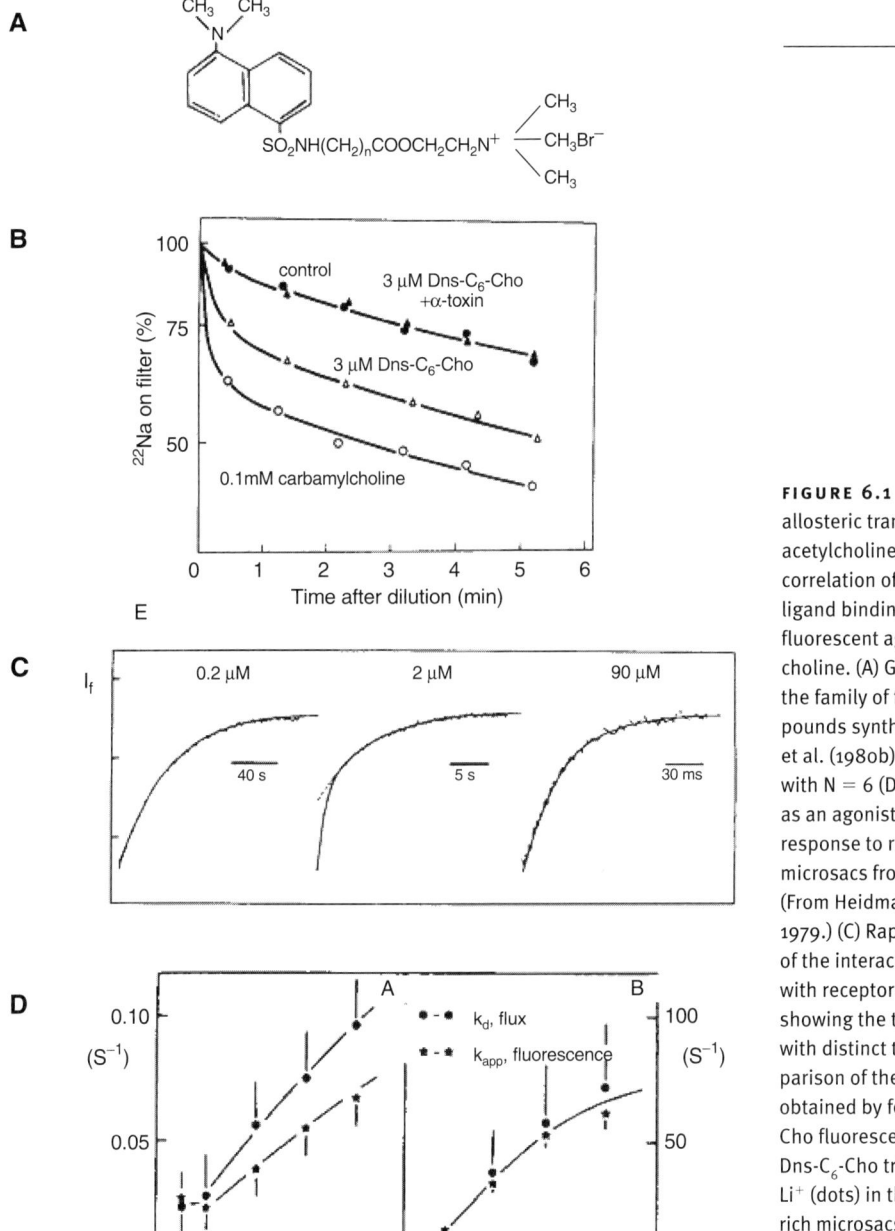

FIGURE 6.1 Analysis of allosteric transitions of the acetylcholine receptor through correlation of ion fluxes with ligand binding using the fluorescent agonist dansyl-C_6-choline. (A) General formula for the family of fluorescent compounds synthesized by Waksman et al. (1980b). (B) The derivative with N = 6 (Dns-C_6-Cho) behaves as an agonist for the $^{22}Na^+$ flux response to receptor-rich microsacs from *T. marmorata*. (From Heidmann and Changeux, 1979.) (C) Rapid kinetic analysis of the interaction of Dns-C_6-Cho with receptor-rich membranes showing the three kinetic steps with distinct timescales. (D) Comparison of the rate constants obtained by following Dns-C_6-Cho fluorescence (stars) and Dns-C_6-Cho triggered efflux of Li^+ (dots) in the same receptor-rich microsacs from *T. marmorata*. *Right:* fast channel closing; *left:* slow channel closing. (From Heidmann et al., 1983a.)

sis of the kinetics of dansyl-C_6-choline binding over a millisecond-to-minute time range led to the identification of a minimum of three conformational states of the acetylcholine receptor: a *low-affinity* state (in the vicinity of 50 μM) that can be interconverted in a fraction of a second to a transient state of *intermediate affinity* (in the vicinity of μM), followed by a final stabilization, in the second-to-minute range, producing a state of *high affinity* (in the vicinity of 3 µM). These experiments revealed that the slow transition coincides exactly with the slow desensitization phase of the ionic response (Neubig et al., 1982; Heidmann et al., 1983a) and that the intermediate transition monitored with dansyl-C_6-choline (Heidmann and Changeux, 1980) fits the kinetics of fast desensitization (Heidmann et al., 1983a).

The experiments further showed that the initial rate of agonist-dependent ion transport and the apparent rate constant of the intermediate relaxation process vary in parallel with agonist concentration (Heidmann et al., 1983a). Since both parameters are directly proportional to the low-affinity activatable conformation of the receptor, it was concluded that *low-affinity binding causes the opening of the ion channel, whereas high-affinity binding is associated with receptor desensitization.*

Contributions from Single Channel Recordings

The link between ligand-receptor interaction kinetics and ion flux was further characterized by patch-clamp recordings of single molecules (Neher and Sakmann, 1976), although still in the absence of direct measurement of agonist (or antagonist) binding. As noted in Chapter 1, a pivotal advance in the electrophysiological study of receptors involved the resolution of earlier noise measurements into discrete pulses through the application of a patch pipette with a high-resistance seal (several GΩ) to a small area of membrane (1–2 µ in diameter). Each of the molecules examined exhibited stochastic passages between closed and open states, with virtually all open states characterized by the same level of intrinsic conductance. The only substantial variations in the recordings concerned the time elapsed between openings and the duration of these openings, with the former diminishing and the latter increasing as a function of the acetylcholine concentration. Accordingly, the trains of registered discrete pulses indicated a rapid, all-or-none transition (< msec) from the closed to the open state. Individual molecules displayed a consistent conductance level. The data were analyzed using a model based on rapid binding (i.e., near the diffusion limit) of acetylcholine to two sites (characterized by values of $k_{on} = 1.5 \times 10^8$ M^{-1} s^{-1}, $k_{off} = 8000$ s^{-1}, leading to an affinity for the basal state of $K_D = k_{off}/k_{on} = 53$ µM), followed by brief channel opening of biliganded molecules (average open time ~1 msec) alternating with rapid channel closing (average closed time ~0.03 msec) (Colquhoun and Sakmann, 1981, 1985). In addition, some openings of putative monoliganded receptors were also detected. At high acetylcholine concentrations, however, bursts of rapid open-closed transitions were interrupted by prolonged silent periods identified with desensitization (Sakmann et al., 1980). In all of these situations, the so-called binding of the agonist was inferred from the channel recordings and never directly measured.

The desensitization phenomenon for muscle receptors is kinetically complex, with somewhat different results obtained by electrical and flux measurements. The process consists of two distinct components: a fast phase with a rate of 2 to 7 s^{-1} as recorded electrophysiologically (Sakmann et al., 1980; Feltz and Trautmann, 1982; Walker et al., 1982) or 75 s^{-1} as recorded by the method noted above following ion fluxes (Heidmann et al., 1983a), and a slow phase with a rate of 0.01 to 0.1 s^{-1} as recorded by both techniques (Heidmann et al., 1983; Neubig et al., 1982; Walker et al., 1982). The rapid phase leads to a 250-fold decrease in initial flux, and the slow phase to undetectable ion transport (Walker et al., 1982). The relative amplitudes of the two phases as observed by recovery measurements depend on the time of exposure to the agonist, an indication that the components are interdependent (Feltz and Trautmann, 1982). The rate of the rapid phase varies in a cooperative manner with agonist concentration (Heidmann et al., 1983a; Pennefather and Quastel, 1982), while that of the slow phase does not (Heidmann et al., 1983; Neubig et al., 1982). At low concentrations of agonist (Neubig et al., 1982), complete desensitization can occur without significant activation. In *Electrophorus,* only the fast component was initially detected (Aoshima et al., 1980; Hess et al., 1983), but a much slower phase (with a rate constant of 0.19 hr^{-1}) was discovered with purified microsacs (Aoshima, 1984). Differences in the quantitative features observed with alternative methods may be related to effects of voltage or to the specific membrane environment. Nevertheless, these measurements show that the receptor is a transmembrane oligomeric protein able to interconvert among multiple conformational states, in contrast to allosteric enzymes, which can be adequately represented in general by *two* conformational states.

Molecular Models of Activation and Desensitization

The Sequential (KNF) Model

Early work on the quantitative modeling of ligand-gated channel opening, as we noted in Chapter 1, relied on analogy with enzymatic catalysis and resulted in the formulation of a "sequential" mechanism (Del Castillo and Katz, 1956). The first attempt to fit the cooperative binding of four molecules of oxygen to hemoglobin was also expressed in terms of a sequential model (Adair, 1925). Subsequently, models were proposed that included ligand-induced conformational changes (Pauling, 1935; Koshland et al., 1966):

$$\bigcirc + S \rightleftharpoons \boxed{S}.$$

In this respect, the formulation proposed by Koshland, Némethy, and Filmer in 1966, known as the KNF model, can be categorized as an *instructive* schema since the ligand is supposed to "instruct" the protein to assume a conformation that is complementary to the ligand. "The conformational change for individual subunits," the authors noted, "occurred *only when* a molecule of ligand was bound" (emphasis added). The KNF model was applied to hemoglobin as well as to regulatory enzymes (Koshland et al., 1966), with the structural implication of some kind of "plastic motion" of functional units as "soft globules" (Changeux and Edelstein, 2001). Although aspects of the sequential mechanism had earlier been employed by Katz and Thesleff (1957) to describe channel activation, the new model was not strictly sequential, since it implied a preexisting equilibrium between resting and desensitized states. At that time, such models were presented in phenomenological terms, however, without specific reference to protein structure.

The Concerted (MWC) Model

The schema proposed by Monod, Wyman, and Changeux in 1965, known as the MWC model, represented a radical departure from the customary description of ligand interactions. It assumed the quaternary organization of the protein into symmetrical oligomers and further postulated its spontaneous interconversion among a limited number of discrete, preexisting quaternary conformations, with strict symmetry relations governing the organization of "protomers" within the oligomers. In this view, the ligand preferentially binds to one of the preexisting conformations, in effect "selecting" the one with the highest affinity for the ligand. This *selective* schema, as it may therefore be called, implies mechanisms involving interactions of "rigid bodies" (Changeux and Edelstein, 2001), and so stands in clear contrast to the instructive schema, according to which the conformation associated with bound ligand cannot occur in the absence of ligand. As Shulman (2001) remarked, the choice between the KNF and MWC models is not a "conflict of abstractions," but rather a question of experimentally distinguishing between alternative mechanisms: under both models affinities depend solely on the *number* of ligands already bound, but in KNF they are linked to the protomer's tertiary structure and in MWC they are linked to the oligomer's quaternary structure. In the case of allosteric enzymes, only two states (T and R) were needed to describe virtually all relevant properties. In the case of receptors, however, a "four-state" model is required (see below) to accommodate measurements over a broad timescale (Heidmann and Changeux, 1980; Neubig and Cohen, 1980), although more states are suggested by various experimental observations, particularly for receptors bearing pleiotropic mutations (Bertrand et al., 1992; Galzi et al., 1996a) or under conditions of "up-regulation" (Buisson and Bertrand, 2001).

Generalized Allosteric Models

It was recognized soon after the presentation of the MWC and KNF models that certain principles of each could be combined into generalized models (see Figure 6.2A). On the one hand, the symmetry requirement could be relaxed so as to permit all combinations of oligomers containing protomers to exist in one of two conformational states (Eigen, 1967). On the other hand, the idea that conformational changes are induced by ligand binding (or by voltage changes in the case of voltage-gated receptors) could be regarded as compatible with local changes at the level of individual protomers, as in certain proposals formulated to describe particular features of oxygen binding to hemoglobin (Henry et al., 2002) and properties of voltage-gated channels (Cox et al., 1997). The generalized models take on particular significance in the analysis of subconductance states, as we shall have occasion to observe in connection with cGMP receptors and voltage-gated channels.

Applications to the Nicotinic Receptor

As we noted earlier, the first single channel measurements were interpreted in terms of a sequential model involving the binding of two acetylcholine molecules to the receptor, followed by a conformational change from closed- to open-channel form. This change is represented by the first line of the reaction diagram shown below. Moreover, ever since the benchmark studies of Katz and Thesleff (1957), it has generally been confirmed that recovery occurs spontaneously in the aftermath of desensitization upon removal of the agonist, and "silently" as well, which is to say with no observed opening of the channel during the recovery period. This cyclic reaction pattern, which was subsequently expanded to incorporate the two agonist-binding sites, remains the basis for most analyses of kinetic data (Colquhoun and Sakmann, 1985; Franke et al., 1993):

FIGURE 6.2 General model of possible conformational transitions mediated by an oligomeric receptor protein. (A) Schematic description of the conformational states for a protein having four subunits and two tertiary states (□ and ○), along the lines first proposed by Eigen (1967); ligand binding (or voltage activation) is indicated by the filled symbols. The two-state concerted (MWC) model corresponds to the vertical columns at the left and right extremes; the sequential (KNF) model corresponds to the descending diagonal. (Changeux and Edelstein, 1998.) (B) Free energy diagram for the four-state allosteric model, including all liganded and unliganded allosteric states, as well as their respective transition states from the analysis of Edelstein et al., 1996. The B, A, I, and D states are each represented by a free energy ladder with step sizes for the B, A, I, and D states that increase with affinity according to the series of dissociation constants $K_B > K_A > K_I > K_D$. The vertical alignment of the ladder for each state is set by the values assigned to the relative concentrations of B_0, A_0, I_0, and D_0. The transition-state heights are determined by the rates of interconversion between conformational states. The progression of doubly liganded states $B_2 \rightarrow A_2 \rightarrow I_2 \rightarrow D_2$ represents the allosteric cascade for the conformational changes elicited by application of a strong and prolonged pulse of agonist, with the time of passage through A_2 corresponding to the average open time in single channel measurements. Following termination of the pulse, agonist dissociation drives the system to the unliganded states and the initial distribution is reestablished relatively rapidly along the pathway $D_2 \rightarrow D_1 \rightarrow D_0 \rightarrow I_0 \rightarrow A_0 \rightarrow B_0$. (continues)

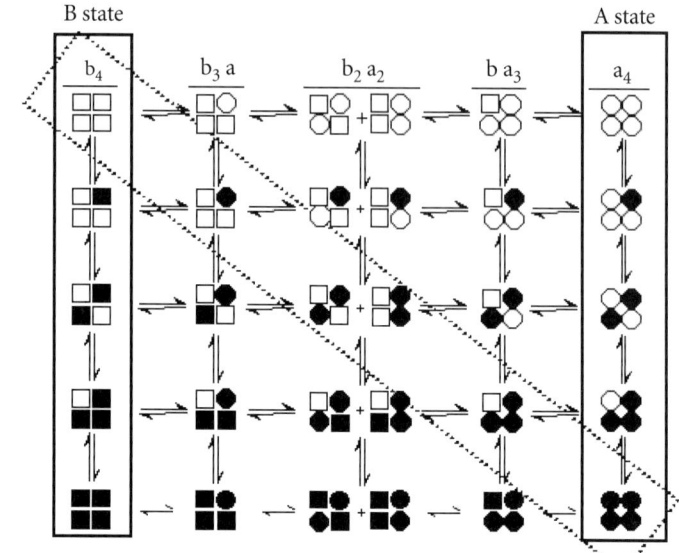

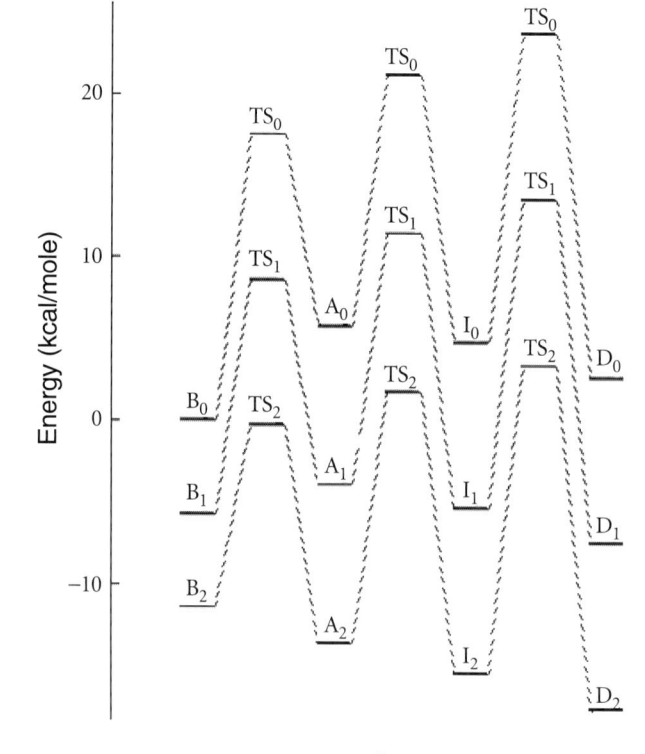

$$B + 2X \underset{k_{-1}}{\overset{2k_1}{\rightleftharpoons}} BX + X \underset{2k_{-1}}{\overset{k_1}{\rightleftharpoons}} BX_2 \underset{\alpha}{\overset{\beta}{\rightleftharpoons}} AX_2$$

$$d_{-2} \updownarrow d_2 \qquad\qquad\qquad\qquad d_{-1} \updownarrow d_1$$

$$D + 2X \underset{k_{-3}}{\overset{2k_3}{\rightleftharpoons}} DX + X \underset{2k_{-3}}{\overset{k_3}{\rightleftharpoons}} DX_2.$$

In this description, due to Katz and Thesleff, the B and D states are assumed to be present in equilibrium prior to the addition of ligand, in a manner that anticipated the formulation of preexisting states in the MWC model. However, since return to the resting state from DX_2 via AX_2 after a long agonist pulse would imply channel-opening events, it has been argued that a distinct "recovery pathway" via DX and D must exist (Franke et al., 1993), although how the conformational transitions of the receptor could be constrained to follow this sequence has yet to be explained.

The 1957 schema dealt only with the slow process of desensitization, whereas subsequent *in vivo* and *in vitro* data showed both that desensitization of the acetylcholine receptor includes a fast process and that modulating effects produced by competitive antagonists must also be considered (Rang and Ritter, 1970). A minimal model based on the tetrahedral arrangement of conformational states (Heidmann and Changeux, 1980; Neubig and Cohen, 1980) that is compatible with all these results was proposed as an alternative to both the MWC concerted model for allosteric transitions (Monod et al., 1965) and the earlier Katz and Thesleff desensitization model:

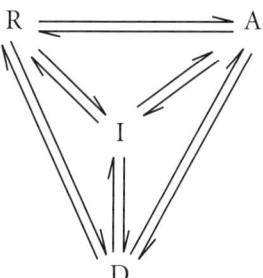

Here A corresponds to the active, open-channel state, and I and D correspond to rapidly and slowly desensitized states, respectively. The same two primary acetylcholine sites are involved in each state, but their affinity increases from the resting state R to D via A and I. All these states are discrete, interconvertible, and, in some cases, present before ligand binding. Their respective K_d values are 50 to 100 μM (R), less than 1 μM (I) and 3 to 5 μM (D) for acetylcholine (Boyd and Cohen, 1980) and dansyl-C_6-choline (Heidmann and Changeux, 1979, 1980). During synaptic transmission, the local concentration of acetylcholine is in the range of 10^{-3} to 10^{-4} M, which is to say about 10 times

Activation and Desensitization

FIGURE 6.2 (*continued*) Since $^{01}k_0 >> {}^{80}k_0$, passage through the A_0 state occurs so rapidly during recovery along the pathway $I_0 \rightarrow A_0 \rightarrow B_0$ that channel opening is negligible. The four-state model therefore predicts a negligible channel opening during the recovery period, but without the necessity of invoking a separate recovery pathway from I (or D) to B that arbitrarily disallows passage through A, as does the "cyclic" mechanism of the sequential model. (From Edelstein et al., 1996.)

the K_d value for the R state, resulting in near saturation for the low-affinity state (Katz and Miledi, 1973c; Kuffler and Yoshikami, 1975). It should also be noted that the K_d value for the desensitized state D fits with the nonquantal "leak" concentration of acetylcholine measured for the postsynaptic membrane in the presence of esterase inhibitor (Katz and Miledi, 1977).

In order to arrive at a quantitative formulation of the tetrahedral model, the two additional steps described below were introduced (Edelstein, 1996; Edelstein and Changeux, 1998), and the designation "R" was replaced by "B" (for Basal) to avoid confusion with the high-affinity "R" state in the original MWC model.

First, for the interconverting B, A, I, and D states, the time ranges observed for the various transitions (~msec for B → A; ~100 msec for A → I; ~10 sec for I → D) lead to selection of a predominant kinetic pathway, B → A → I → D that corresponds to the passage between states over the lowest transition state barriers. Since the secondary pathways contribute less than 1% of each state to the various rates observed for the nicotinic receptor (Edelstein et al., 1996), the tetrahedral arrangement reduces to the linear cascade:

$$B_i \rightleftharpoons A_i \rightleftharpoons I_i \rightleftharpoons D_i.$$

For variant forms of the receptor that may involve more similar reaction rates between each pair of interstate interconversion reactions, however, it is possible that this simplification would no longer be appropriate. Indeed, for mutant forms of the receptor observed in congenital myasthenic syndromes, complicated models with branching patterns and multiple open states have been presented (Sine et al., 1995c; Ohno et al., 1996; Milone et al., 1998). But it is not yet clear whether these models provide useful insights into the dynamics of the structural transitions of the receptor molecule.

Second, a number of individual rate constants must be assigned within the linear cascade, since interconversion rates for the transitions between each pair of states vary with the number of ligands bound. However, in accordance with the linear free-energy relations described by Edelstein et al. (1996), the variations in these sets of rate constants obey a scaling factor that may be fixed by a transition state position parameter (see the Appendix). In this way, the six rate constants for each pair of states in the linear schema (in the case of a receptor having 0, 1, or 2 molecules of bound ligand) are reduced to two rate constants and one positional parameter.

The complete set of rate constants obtained using linear free-energy relations yields an adequate fit for the various kinetic properties of activation and desensitization for muscle receptors (Edelstein et al., 1996). Linear free-energy relations were also subsequently used to analyze the functional properties of a series of mutations at different locations of the nicotinic receptor by estimating the position of their transition state along a reaction pathway between the open and closed states (Grosman et al., 2000).

Experimental Evidence Supporting the Allosteric Schema

Single Channel Openings in the Absence of Ligand

Single channel measurements can be used to distinguish between alternative models, but only under certain conditions. For example, at high ligand concentrations the doubly liganded open state (A_2) predominates, and the results predicted by the sequential and concerted models are effectively indistinguishable (Edelstein et al., 1997b). If experiments are appropriately designed, however, the channel opening of nonliganded receptors predicted by the MWC model can be observed. Such "spontaneous" all-or-none openings have been recorded at a low frequency (Jackson, 1986). They constitute decisive evidence in favor of the MWC model, since the sequential model excludes channel opening in the absence of agonist binding. Moreover, the frequency of channel openings in the absence of ligand dramatically increases for certain combinations of subunits and mutant phenotypes (Jackson et al., 1990; Ohno et al., 1995; Bertrand et al., 1997; Grosman and Auerbach, 2000), with a particularly dramatic increase in spontaneous openings observed for the myasthenic mutant εT264P (Ohno et al., 1995), analyzed in detail in Chapter 8.

Spontaneous channel openings may plausibly be asserted to contribute to the pathological properties of mutant human nicotinic receptors producing slow-channel congenital myasthenic syndromes (Engel et al., 1999), although the *in vivo* effects may also be exacerbated by activation due to endogenous choline (Zhou et al., 1999). Additionally, the presence of spontaneously desensitized states in the absence of acetylcholine (Heidmann and Changeux, 1979; Boyd and Cohen, 1980) has already been noted. Spontaneous openings have also been observed for cGMP and NMDA receptors (Picones and Korenbrot, 1995; Zagotta and Siegelbaum, 1996; Tibbs et al., 1997; Turecek et al., 1997), reinforcing the view that the allosteric model applies to other ligand-gated ion channels, in addition to the acetylcholine receptor.

Partial Agonists

The fact that different agonists activate receptors to different extents is readily explained by the concept of nonexclusive binding (Rubin and Changeux, 1966), whereby the extent of channel opening by the partial agonist is determined by the ratio of the affinities to the closed and opened conformational states. However, alternative interpretations are possible with respect to the observed restriction of full response to agonists, such as the noncompetitive blocking of the ion channel by high agonist concentrations.

Cooperativity

A primary characteristic of allosteric proteins is the cooperative binding of ligands, readily explained by the MWC model for homooligomers (Monod et al., 1965). In the case of heterooligomers such as the nicotinic receptor, however, a question arises as to the extent to which two nonidentical sites may interact. The two main acetylcholine-binding sites on each electric organ or muscle $\alpha_2\beta\gamma\delta$-oligomer are located at the interface between α-γ- and α-δ-subunits (see Chapter 4). Yet the two sites nonetheless interact in a *positively cooperative manner*. The Hill coefficient of the dose-response curve recorded as a function of acetylcholine concentration ranges between 1.5 and 2.0 at the neuromuscular junction (Katz and Thesleff, 1957), for *E. electricus* electroplaque (Higman et al., 1963), and on excitable microsacs (Kasai and Changeux, 1971). Moreover, as in the case of many allosteric proteins, the equilibrium binding curve of acetylcholine is clearly sigmoid, displaying a Hill coefficient of 1.3 to 1.4 (Weber and Changeux, 1974a, 1974b, 1974c; Cohen et al., 1974). Such cooperative binding at equilibrium could in principle be accommodated by either the concerted or the sequential model. Nonetheless, the degree of cooperativity measured by the Hill coefficient of the dose-response curves can be used as a critical test of the two models in the case of certain pleiotropic mutations, since each model predicts distinctly different linkage relationships between changes in affinity and changes in cooperativity. In a number of cases described below, the MWC model readily provides explanations of mutant phenotypes that predict changes in cooperativity in good agreement with observed values (Edelstein and Changeux, 1998). For certain mutants, very large changes in cooperativity have been reported compared to wild-type, as in the case of $\alpha 7$ L254T, for which the Hill coefficient is near 5 (Bertrand et al., 1993). Systematic modeling has not been carried out for all of the reported mutations, however.

Pleiotropic Mutations

Dramatic new phenotypes of the nicotinic receptor have been discovered by mutating amino acid residues in the M2 channel that were labeled with the noncompetitive blocker chlorpromazine, as described in Chapter 5. The first site to be mutagenized was the Leu at position 247 in α7 (Révah et al., 1991), a residue that is conserved at the corresponding position throughout the acetylcholine receptor superfamily. The site-directed mutant genes were engineered, introduced into vectors, and examined in the *Xenopus* oocyte expression system with respect to their physiological responses. The recordings revealed surprisingly "complex" and pleiotropic phenotypes (Révah et al., 1991; Bertrand et al., 1992; Devillers-Thiéry et al., 1992; Yakel et al., 1993; Langosh et al., 1994; Rajendra et al., 1994). For example, a single mutation of α7 nicotinic receptor, Leu-247 to Thr (Révah et al., 1991; Bertrand et al., 1992; Devillers-Thiéry et al., 1992), yields a receptor that is insensitive to the channel blocker QX-222, does not desensitize, and displays an apparent affinity for acetylcholine about 200-fold higher than the wild-type. In addition, the mutant receptor exhibits two conducting states activated by high (the 40 pS state) or low (the 80 pS state) concentrations of acetylcholine. A competitive antagonist of the wild-type receptor, dihydro-β-erythroidine (DHβE), acts on this mutant as a full agonist (with 10-fold higher apparent affinity than acetylcholine) and exclusively activates the high conductance state. Moreover, mutations at several nearby positions facing the channel along the M2 helix produce similar phenotypes.

Such properties are readily accommodated by the MWC model, since effects can be produced in several distinct ways (Galzi et al., 1996a), for instance by changing independently the following three parameters: the interconversion equilibrium between different conformations (L phenotype); the intrinsic affinity for agonist of one or more specific conformational states (K phenotype); and the conductance properties of individual states (γ phenotype). Various combinations of the three phenotypes may also occur. The sequential model, on the other hand, is unable to describe all of these phenotypes since, as we have already seen, it does not account for the occurrence of spontaneous opening, nor for specific relationships between cooperativity and affinity. Let us consider each of these phenotypes in turn.

The L phenotype is assumed to result from mutations that selectively alter the equilibrium constant between two given, interconvertible conformations. As L increases, agonists may progressively become partial agonists and competitive antagonists (see Appendix). The model furthermore predicts that, for very low values of L, significant spontaneous stabilization in the active state may occur, yielding constitutively active mutants (spontaneous channel opening). This phenomenon is readily accounted for by the MWC model. Plausible examples of the L phenotype occur for the α7-nicotinic receptor upon substitution of M2 residues at positions identified on the basis of chemical labeling of *Torpedo* nicotinic receptor with noncompetitive blockers

(Révah et al., 1990), such as the ring of Val-251 to Thr or of Thr-244 to Gln. Indeed, acetylcholine dose-response curves can be simulated for wild-type and mutant receptors (Galzi et al., 1996a), on the sole assumption that L values are high for the wild-type ($\sim 8 \times 10^5$) and low for the mutants (~ 20). The fit of such simulations lies within the limits of precision of the experimental data. Furthermore, the competitive antagonist of the wild-type receptor, DHβE, is predicted to become a partial agonist when the L value corresponds to those of the V251T or T244Q mutant receptors. Similar interpretations based on an L phenotype can be applied to mutations involved in human diseases, particularly congenital myasthenic syndromes and hyperekplexia, as well as to the Lurcher mutation in mice. These examples will be discussed in Chapter 8, when we turn to the subject of receptor pathology.

The K phenotype is assumed to result from mutations that selectively alter the intrinsic binding affinities of individual conformational states, as in the case of mutations of the amino acids Tyr-93, Trp-148, Tyr-190, and Tyr-198, which were tagged by affinity and photoaffinity labeling of the acetylcholine-binding site in electric organ nicotinic receptor (Révah et al., 1990). Substitution of their homologues to phenylalanine on the corresponding chick neuronal α7-residues Tyr-92, Trp-148, and Tyr-187 (Galzi et al., 1991b), or on the mouse muscle α1-subunit (Tomaselli et al., 1991; Aylwin and White, 1994), yields functional receptors with reduced sensitivity to acetylcholine, but unchanged Hill coefficients and current amplitudes (Galzi et al., 1996a).

The γ phenotype is assumed to result from changes in the state of activity of the ion channel (from nonconducting to conducting, for example) in one (or possibly more) conformations, with no alterations of the intrinsic binding parameters of each state—that is, of its pharmacological specificity—nor of the equilibria (and kinetics) of interconversions. Analogies exist between the phenotypes of the M2 mutants L247T and T244Q or V251T. If the L247T mutant receptor were to correspond to an L phenotype, however, a single change in L value would not fully account for the experimentally determined acetylcholine and DHβE dose-response curves (Révah et al., 1991; Bertrand et al., 1992; Devillers-Thiéry et al., 1992). Indeed, at the L value yielding an appropriate EC_{50} for acetylcholine, DHβE will not behave as a full agonist, but rather as a partial agonist, as in the case of the T244Q or V251T mutants. Moreover, under no circumstances will the apparent affinity for DHβE, as observed experimentally for L247T, be higher than for acetylcholine. The occurrence in L247T of two conducting states with distinct pharmacological profiles (Révah et al., 1991; Bertrand et al., 1992; Devillers-Thiéry et al., 1992) favors an interpretation in terms of the γ-phenotype schema. In such a case, simulated dose-response curves suggested by the MWC model satisfactorily fit the experimental data (see Appendix, Figure A.4) for both acetylcholine (ligand 2) and DHβE (ligand 1), assuming that one of the conducting states is identical to the wild-type conducting state (not stabilized by DHβE), while the other (assumed to correspond to a desensitized conformation of the wild-type) binds DHβE with affinity higher than for acetylcholine (Galzi et al., 1996).

In conclusion, the MWC-type model successfully accounts for the various new phenotypes observed by site-directed mutagenesis. Changes in allosteric equilibrium readily explain L phenotypes that display spontaneous opening and facilitate description of K and γ phenotypes by delimiting distinct conformational states, including A and D in unliganded forms, whereas in the sequential model these states are present only in liganded forms. Moreover, the MWC-type model can accommodate changes in channel residues that lead to effects at the ligand-binding site, whereas such effects lie outside the scope of models based on an induced-fit schema such as KNF. For these reasons, the MWC-type model can readily be generalized to cover all the known phenotypes of nicotinic receptors. However, as we shall see in the following section in connection with ligand-gated receptors of the glutamate and cGMP families, the presence of multiple subconductance states that vary with the extent of ligand binding cannot be accommodated by either a pure MWC or sequential model, and requires a generalized allosteric model involving intermediate states that may not exhibit complete symmetry.

Desensitization

SPONTANEOUS DESENSITIZATION   As we mentioned earlier, binding kinetics of nicotinic agonists to membrane-bound *Torpedo* receptor reveal that, under resting conditions, about 20% of the population is present in a desensitized state (Heidmann and Changeux, 1979; Boyd and Cohen, 1980). This finding is in agreement with the presence of spontaneous equilibria between conformational states, as postulated by the MWC-type model. In this case, however, a fraction of the receptor population would be in the desensitized state in the absence of ligand. In this respect, spontaneous desensitization closely parallels spontaneous opening, as described above, particularly for the pleiotropic mutants whose L phenotype involves exceptionally small values of L (Galzi et al., 1996a).

RECOVERY FROM DESENSITIZATION   The fact that receptors are "silent" during recovery from desensitization requires, as we noted earlier in this chapter, the incorporation in the sequential model of a special "recovery pathway" that is an essential feature of the "cyclic" mechanism (Katz and Thesleff, 1957; Franke et al., 1993). Silent recovery is readily explained for nicotinic receptors by the MWC model, however, using the same parameters and values that satisfactorily represent activation. In effect, simulations of recovery after a strong, desensitizing pulse indicate that the kinetic pathway for return to B_0 can pass sufficiently rapidly through A_0 (Edelstein et al., 1996) that the silent channels observed during recovery may be accounted for without postulating a separate recovery pathway. These properties of the allosteric schema are readily visualized by a free-energy profile (Figure 6.2B). The allosteric schema therefore explains silent recovery as a simple consequence of the basic principles of the model—freely interconverting conformational states, even in the absence of ligand.

Mutant Phenotypes from Human Congenital Diseases

Additional evidence from mutant phenotypes associated with congenital myasthenic syndromes and frontal lobe epilepsy is presented in Chapter 8.

Difficulties and Unresolved Issues

Multiple Conformations and Hybrid States

The physiological properties of the nicotinic receptor are well represented by the MWC model under the conditions of a single discrete conductance state that does not vary with the nature or concentration of the agonist. However, a few exceptional cases have been reported for the nicotinic receptor—for example, the two conductance states of human α4β2 (Kuryatov et al., 1997)—and for several nonnicotinic ligand-gated channels (as well as a number of voltage-gated channels) having multiple conductance states. Their incorporation into mechanistic models raises a number of new issues. Two categories of results have been reported that depend on whether or not the distribution of subconductance states varies with agonist concentration.

In the first category, typified by glycine receptors, multiple conductance states are prevalent (Bormann et al., 1993), but their distribution does not depend on agonist concentration (Twyman and Macdonald, 1991). The simplest interpretation of such deviations is that discrete fluctuations occur in the steric organization of a few amino acid residues within the open-channel state, independent of binding-site occupancy and subunit interactions. Only a limited number of side chain conformations (rotamers) are in fact accessible to residues belonging to α-helices (Dunbrack and Karplus, 1994), some combinations of which may be "frozen" in the channel-lining pore during the transition to the open state. In other words, these fluctuations are postulated to arise from local, sequence-dependent variations in conformation within the channel for certain receptor subunits (such as glycine receptor α1), without in any basic way departing from the all-or-none openings mediated by changes in the quaternary organization of the molecule as predicted by the MWC model.

A second category of receptors displaying multiple subconductance states that vary with the nature of the agonist is exemplified by some voltage-gated receptors and ligand-gated receptors outside the acetylcholine superfamily. These observations raise a more serious challenge to the fully concerted allosteric schema. For example, such multiple conductance states (with no observed variations dependent on agonist concentration) have been reported for AMPA-type glutamate receptors expressed in HEK cells (Swanson et al., 1997) or studied *in vivo* (Cull-Candy and Usowicz, 1987; Jahr and Stevens, 1987; Jonas and Sakmann, 1992; Wyllie et al., 1993), as well as for changes that accompany the induction of LTP (Benke et al., 1998). Subconductance states have also been reported for GluR6/GluR3 chimeric glutamate receptors in connection with changes following from slow dissociation of the competitive antagonist 6-nitro-7-sulphamoyl-benzo(F)quinoxalinedion. These states were presumed to correspond to partially liganded receptors (Rosenmund et al., 1998). Subconductance states related to ligand binding were reported for retinal rod cGMP-gated channels with covalently attached ligands (Ruiz and Karpen, 1999), but the covalent attachment may complicate interpretation. On balance, the agonist-dependent subconductance states of GluR resemble the patterns reported for drk1 voltage-gated receptors (Chapman et al., 1997). It is possible that they reflect common channel-gating mechanisms, distinct from those found within the nicotinic receptor family (MacKinnon, 1995; Zagotta and Siegelbaum, 1996; see, however, Corringer et al., 2000).

A first approach to interpreting the ligand-dependent subconductance states is offered by the extended allosteric model suggested by Eigen (1967) and subsequently developed by Cox et al. (1997), which assumes, in addition to fully symmetric states, the presence of hybrid states containing subunits in different conformations. The assumption of perfect symmetry by the MWC model represents an extreme case that adequately describes certain systems; but in other systems deviations from the symmetry principle may occur, leading to the nonsymmetrical "intermediate states" shown in Figure 6.2A. In cases such as these, the observed discrete subconductances may be assumed to correspond to the intermediate states as a consequence of local changes in subunits that reside within a common allosteric quaternary state (the A state, for instance, distributed in the vertical column at the right in Figure 6.2A). These substates would then be analogous to the tertiary-level contributions to the H+ Bohr effect in hemoglobin (Edelstein, 1975; Henry et al., 2002). Each agonist molecule bound to these tetrameric receptors (as for each individual Bohr proton for hemoglobin) would then produce *local* conformational changes that modulate activity without causing a major transition to another quaternary conformational state. Interestingly, the equivalent of a pH effect on current-voltage relationships has recently been produced for *Shaker* channels by introducing a histidine into the S4 segment responsible for voltage sensitivity (Starace et al., 1997).

While the presence of subconductance states is clearly detected in single channel experiments, the high cooperativity (Hill coefficient ~3) of the cGMP receptors (Ruiz and Karpen, 1999) indicates that the intermediate states are relatively unfavorable by comparison with the symmetric states. For example, a quantitative analysis of oxygen binding by tetrameric hemoglobin revealed that, for a degree of cooperativity corresponding to a Hill coefficient of 3, intermediate states cannot exceed a few percent of the molecular population (Edelstein, 1996). Indeed, in the original presentation of the MWC model, the principle of conservation of symmetry was not imposed as an indispensable condition for allosteric transitions, but rather on the ground that asymmetric intermediate states were expected to be significantly less stable than the symmetric T and R states and, if abundant, to diminish cooperativity.

Single Channel Recordings and the Nonequivalence of Ligand-Binding Sites

Earlier in this chapter it was mentioned that acetylcholine-binding sites formed by receptor oligomers at different subunit interfaces may possess different intrinsic properties. As we have seen, analysis has been complicated by uncertainties concerning the degree of equivalence of the two bound ligands, notably in single channel recordings of muscle nicotinic receptors (Edmonds et al., 1995). Depending on whether or not it is assumed that the sites are equivalent, quite discrepant values for key parameters may be assigned, leading to significantly different mechanistic interpretations (Edelstein et al., 1997b). For example, single channel recordings from the same laboratory on what should be identical wild-type human muscle receptors expressed in HEK cells have been interpreted in terms of acetylcholine-binding affinities varying from a 350-fold difference for the affinity of the two sites in the B state in one study (Sine et al., 1995c) to identical affinities for the two sites in another study (Wang et al., 1997). Simulations based on the allosteric model (Edelstein et al., 1997b) have shown that the same single channel data can in fact be satisfactorily represented by mechanistic schemas incorporating either equivalent or nonequivalent binding affinities at the two sites. In principle, technical advances on the horizon that would permit "single binding events" (see below) to be measured simultaneously with channel opening might lead to an experimental distinction between equivalent and nonequivalent schemas.

Single Binding Events and Single Channel Openings

Among the new approaches that hold out the prospect of yielding deeper insight into ligand-gated channels, methods permitting single binding events to be monitored in parallel with single channel events may turn out to be very powerful. Although single channel measurements of muscle receptors contributed valuable information about ion channel properties and their ligand gating (Sakmann et al., 1980; Colquhoun and Sakmann, 1985), the linked events of single ligand binding have so far only been *indirectly* computed from channel recordings. As a result, these studies do not give the exact relationship between binding events and channel opening events. The MWC formalism, on the other hand, makes specific experimental predictions for these relationships. They include binding without opening, opening without binding, and binding events of longer duration than opening events (Edelstein et al., 1997b). A particularly instructive example of these distinctions is presented in Chapter 8 in connection with the myasthenic mutant εT264P (see Figure 8.2). Such distinctions at the single molecule level are analogous to the distinction between the binding ($\bar{Y}$) and state functions ($\bar{R}$) of the MWC model discussed in Chapter 1. Measurements of single binding events represent an experimental challenge that so far remains unmet. However, recent developments in the field of fluorescence correlation spectroscopy (Eigen and Rigler, 1994; Edman et al., 1996; Rauer et al., 1996; Schwille et al., 1997) have placed such measurements within the realm of possibility. These measurements may also be expected to provide critical information for distinguishing between equivalent and nonequivalent sites (Edelstein et al., 1997b).

CHAPTER 6

Modulation of Conformational Transitions by Multiple Allosteric Effectors

A characteristic feature of allosteric proteins is that their activity and relevant conformational transition may be regulated by ligands binding to sites distinct from the biological active site. On the nicotinic receptor, several categories of functional allosteric sites have been mapped that are distinct from the acetylcholine-binding site (Figure 6.3). A variety of ligands control the functional properties of receptors when they bind to sites distributed throughout the protein molecule, at the N-terminal extracellular domain, the transmembrane segment, and the cytoplasmic loop. With respect to the N-terminal domain, bound Ca^{2+} ions behave as positive effectors of neuronal (but not muscular) nicotinic receptors, with a major Ca^{2+}-binding site involving part of loop F of the acetylcholine-binding site ($\alpha 7$ residues 161–172) (Galzi et al., 1996b; see Chapter 7). Moreover, the data are consistent with a mechanism whereby Ca^{2+} ions primarily affect the isomerization constant, L, governing the interconversion between the resting and open channel conformations of the receptor protein (Galzi et al., 1996b). Variations in the external Ca^{2+} concentration in the millimolar range have been noted in several brain regions and in ganglionic sympathetic neurons during periods of high synaptic activity (Amador and Dani, 1995). Specifically, potentiation in rat central neurons by calcium results in a shift of the dose-response curve to the left and an augmented maximal response accompanying an increase (~threefold) in the frequency of channel opening (Mulle et al., 1992b).

These observations make it plausible to postulate a mechanism for efficient short-term regulation of nicotinic transmission by the action of Ca^{2+} ions on the receptor involving both external Ca^{2+}-binding sites and Ca^{2+} influx through its own ion channel. Moreover, allosteric modulation by Ca^{2+} could provide a mechanism of coincidence detection (Heidmann and Changeux, 1982; Changeux and Heidmann, 1987) supplementing the NMDA Ca^{2+}/Mg^{2+} coincidence detection system (Wigstrom and Gustafsson, 1985), possibly in conjunction with phosphorylation effects (Edelstein and Changeux, 1998).

Other allosteric binding sites occur on the external surface of nicotinic receptors for substances of endogenous origin such as choline (Alkondon et al., 1997), the tryptophan metabolite kynurenic acid (Hilmas et al., 2001), the β-amyloid peptide (Liu et al., 2001; Pettit et al., 2001), steroids (Valera et al., 1992; Buisson and Bertrand, 1998), substance P (Stafford et al., 1993), and ATP (Nakazawa, 1994); and also for pharmacological agents such as local anesthetics and other noncompetitive blockers (Maelicke et al., 1997; Arias and Blanton, 2002).

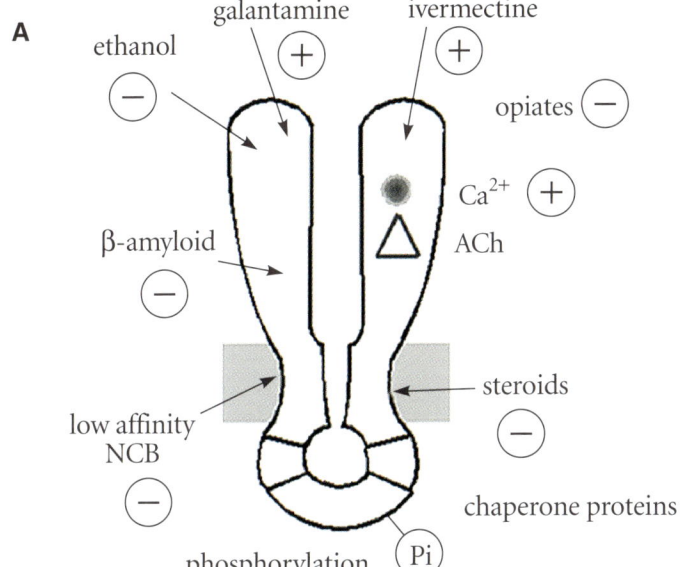 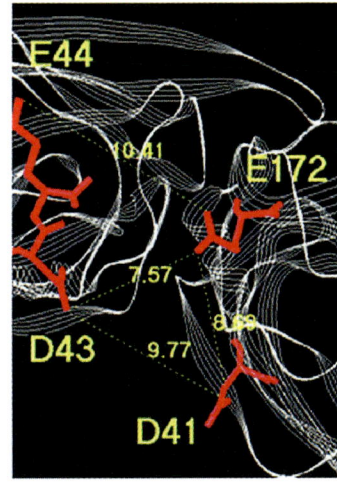

In general, the inhibitory action of local anesthetics is ascribed to one of two possible mechanisms: an open-channel blocking mechanism or an allosteric process whereby the drug binds either to the closed channel or to other nonluminal sites (Arias and Blanton, 2002). For several such compounds it has been possible to identify the site of action on *Torpedo* nicotinic receptors by photolabeling (see Ziebell et al., 2004 and references therein). Neuronal nicotinic acetylcholine receptors are more sensitive to anesthetics than their muscle counterparts, with the exception of the α7-receptor (Tassonyi et al., 2002). Ethanol, an important psychoactive molecule, can interact directly with neuronal nicotinic receptors (Narahashi et al., 1999); but it can also modulate cholinergic effects through interactions with nicotine responses in the mesocorticolimbic dopamine system (Larsson and Engel, 2004). Ivermectin, a powerful antihelminthic drug, behaves as a strong positive allosteric effector of the neuronal α7-receptor by interacting at a site whose precise location remains unidentified (Krause et al., 1998).

The nicotinic receptor is an integral membrane protein. Mutagenesis and photolabeling experiments have shown that some transmembrane segments interact with the membrane bilayer, thus forming a "lipid belt" (Giraudat et al., 1985; Blanton and Cohen, 1994), with specific regulatory roles noted for phosphatidic acid and for cholesterol (Bhushan and McNamee, 1993). General anesthetics constitute an interesting class of allosteric effectors in this respect (Tassonyi et al., 2002). In the case of the GABA receptor, binding pockets for general anesthetics have been identified at the level of the M2 and M3 transmembrane segments (Husain et al., 2003; Siegwart et al., 2003; Bali and Akabas, 2004; Rudolph and Mohler, 2004).

FIGURE 6.3 Allosteric sites. (A) Diverse sites for allosteric effectors. (B) The calcium-binding site model. The principal residues are indicated, along with the distance (in Å) between critical atoms. (From Le Novère et al., 2002b.)

Finally, the cytoplasmic domain is potentially a target for a distinct group of modulators, notably protein kinases (Huganir and Greengard, 1990; Swope et al., 1995). The phosphorylation reactions of specific sites may selectively affect interactions with the cytoskeleton (Sheng and Wysznyski, 1997) and also play a role in synaptic plasticity (Raymond et al., 1993; Levitan, 1994; Kirsch and Betz, 1998), as has been suggested for glutamate receptors in postsynaptic aspects of LTP (McHugh et al., 1996; Rotenberg et al., 1996; Barria et al., 1997) and LTD (Nakazawa et al., 1995). Several specific phosphorylation processes have also been implicated in the mechanism of receptor upregulation (Paterson and Nordberg, 2000).

Pharmacological agents that act upon the nicotinic receptor may occupy binding sites for physiological ligands. But they also interact with protein domains, interfaces, and crevices that do not match any recognized regulatory signal. The benzodiazepines and cyclothiazides, for example, are active at subunit interfaces in $GABA_A$ and glutamate receptors, respectively.

Regulatory effects may also be extended from either the synaptic or the cytoplasmic side through protein interactions that lead to clustering or immobilization. Examples include 43K-rapsyn, which anchors nicotinic receptors at the neuromuscular junction (see references in Duclert and Changeux, 1995); gephyrin, which stabilizes glycine receptors at neuronal postsynaptic sites (Kirsch and Betz, 1998); and various PDZ-domain proteins in the PSD-95/SAP90 family that interact with glutamate receptors (Kornau et al., 1995; Kennedy, 1997; Sheng, 1997) and other PDZ-binding proteins (Kim et al., 1998; Niethammer et al., 1998, and references therein). These interactions serve to stabilize the receptor proteins at a cellular location where they remain directly accessible to the high local concentrations of neurotransmitter released by the nerve ending (see Chapter 10).

General Conclusions and Future Directions

This analysis shows that accounting for high-resolution electrophysiological data hinges upon an understanding of the three-dimensional molecular organization and conformational dynamics of regulatory membrane proteins. It also illustrates that interpreting the basic signal transduction mechanism of neurotransmitters in molecular terms requires both the systematic application of formal models and the continual refinement of such models through comparison with experimental data. The various attempts to model the conformational transitions of receptor molecules emphasize the distinction between major discrete *quaternary* transitions, asserted by the MWC model to be common to both pentameric and tetrameric receptors (e.g., B and A states), and *local* reorganizations at the subunit (tertiary) level directly linked to ligand (or voltage) gating, asserted by the KNF model to occur more probably in tetrameric receptors. Strong experimental and theoretical evidence supports the view that quaternary changes contribute to signal transduction in membrane receptors and classical allosteric proteins alike. The extent to which fractional ligand binding may exert partial effects on the quaternary organization of the molecule is still undetermined at this stage, however, and indeed may vary among different species of receptor. Ultimately, structural and dynamic studies at atomic resolution scale may be expected to lead directly to an objective description of the conductance states, and possibly to a distinction among subconductance states in terms of local flexibility within a common quaternary "open" state, as in the case of hemoglobin (Baldwin and Chothia, 1979; Shulman, 2001; Henry et al., 2002). In particular, it is important to keep in mind that small structural changes produced by ligand binding at the tertiary level may have only minor consequences for the signal transduction process by comparison with those arising from changes in the quaternary structure of the receptor molecule.

In conclusion, experimental data—particularly the data concerning pleiotropic pathological mutations of the nicotinic receptor reported in Chapter 8—that are readily accounted for by the allosteric schema cannot be satisfactorily represented by the sequential schema. To cite only one example, strong experimental evidence points to the existence of freely interconverting conformational states that undergo transitions in the absence of ligand, and therefore are not "induced" by ligand. There is also an important body of experimental data and theoretical analysis that is consistent with the fully concerted transition predicted by the MWC model, although intermediate conformation states appear to be present in the case of some ligand-gated and voltage-gated channels. Determining whether they represent distinct quaternary conformations or tertiary variations within the same quaternary states will require further investigation.

CHAPTER 7

Three-Dimensional Structure at the Amino Acid Level

Elucidating the full three-dimensional structure of the nicotinic receptor at atomic resolution is a long-standing goal that has been pursued ever since the isolation of the nicotinic receptor as a discrete protein in 1970. In the years that followed, the general topology of the individual polypeptide chains was established, and key residues at the acetylcholine-binding site and in the ion channel were identified by chemical labeling and site-directed mutagenesis, as we saw in Chapters 4 and 5, respectively. Concomitant studies, conducted mainly by electron microscopy (EM) and providing preliminary structural information about the overall shape of the receptor (Cartaud et al., 1973, 1978; Nickel and Potter, 1973; Kistler and Stroud, 1981; Bon et al., 1982, 1984), are now able to exploit the advantages of atomic resolution (Miyazawa et al., 2003). In addition, detailed structural predictions have been generated by advanced computational approaches (Le Novère et al., 1999).

A pivotal moment occurred in 2001 with the publication of the high-resolution structure of a soluble pentameric molluscan acetylcholine-binding protein (AChBP) produced by glial cells and secreted into the synaptic cleft, where its "buffering" effect on acetylcholine regulates synaptic transmission (Smit et al., 2001). Surprisingly, this protein turns out to be homologous to the N-terminal domain of the nicotinic receptor, particularly α7 with which it shares 24% sequence identity (Brejc et al., 2001). This result has been a unique source of insights into the structure of the agonist-binding site and other features of the N-terminal domain that we will examine later in this chapter. Many structural questions concerning the transmembrane and intracellular portions of the receptor nevertheless remain unanswered, along with structural details of the allosteric transitions that link the acetylcholine-binding sites to the ion channel and thus govern receptor function (see discussion in Grutter and Changeux, 2001). We begin by reviewing the current understanding of receptor structure.

Overall Shape of the Integral Membrane Receptor Protein

EM projection images of the receptor protein from the electric organ of *Electrophorus electricus* and *Torpedo marmorata* were described in Chapter 2 as revealing ringlike particles 80–90 Å in diameter (Cartaud et al., 1973; Nickel and Potter, 1973)—a picture that subsequently was resolved into a five-subunit rosette with two nonadjacent α-toxin sites (Bon et al., 1982, 1984; Brisson and Unwin, 1984, 1985). Viewed from the side (Cartaud et al., 1978; Stroud, 1983; Mitra et al., 1989), the receptor appears as a 110-Å-long cylinder with a 70–80 Å diameter and an axial well. Similar dimensions were obtained by neutron scattering (Wise et al., 1981). Images reconstructed from crystallized postsynaptic membranes of *Torpedo* at 16–18 Å resolution (Toyoshima and Unwin, 1988; Unwin et al., 1988) provided additional information, although certain points were later reinterpreted (Miyazawa et al., 1999). First of all, in agreement with earlier results, the molecule has been shown to traverse the lipid bilayer. Its inner and outer leaflets, however, are separated by only 30 Å, a value 10–15 Å smaller than anticipated on the basis of profile X-ray diffraction measurements from typical biological membranes, possibly as a consequence of a relatively high cholesterol content equal to 50 mole percent (Popot et al., 1978). The receptor projects 65 Å toward the synaptic cleft from the outer leaflet and approximately 15–20 Å toward the cytoplasm from the inner leaflet (Toyoshima and Unwin, 1988). An additional mass (eliminated at pH 11), centered upon the molecular axis and extending an additional 30 Å from the cytoplasmic extremity, was attributed to the cytoskeletal 43K-rapsyn protein (Toyoshima and Unwin, 1988; see Chapter 10). Subsequent results at 4.6 Å resolution (Figure 7.1A) now attribute a portion of this mass to the five receptor subunits, however, thereby extending the cytoplasmic boundary to ∼40 Å below the membrane (Miyazawa et al., 1999). Additionally, the 43K-rapsyn protein was reinterpreted as being limited to ∼10 Å of the cytoplasmic extremity, in accord with earlier structural observations by electron microscopy (Sealock, 1982; Nghiem et al., 1983; Bridgman et al., 1987).

With respect to the internal organization of the molecule, the averaged structure reveals an overall cylindrical arrangement with an axial

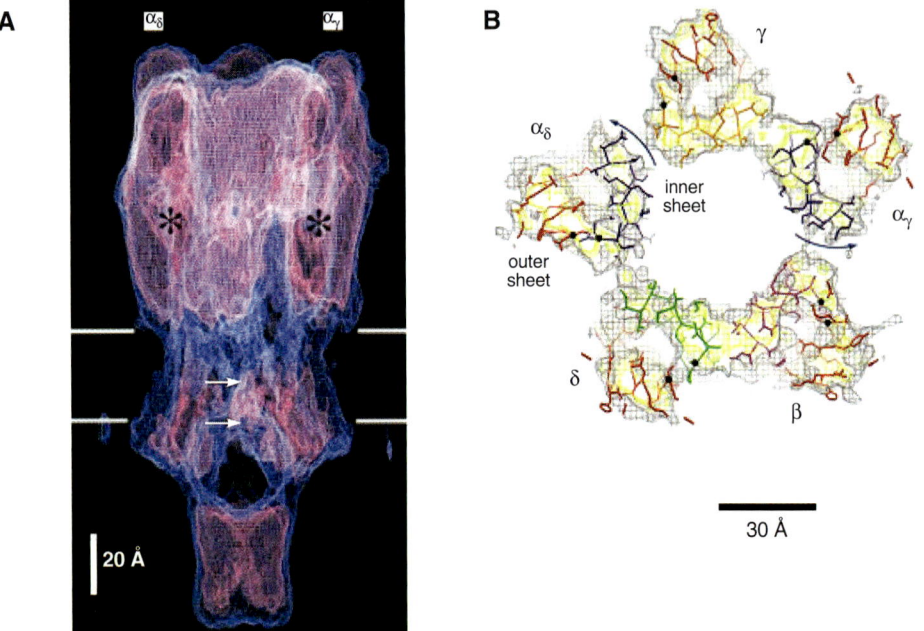

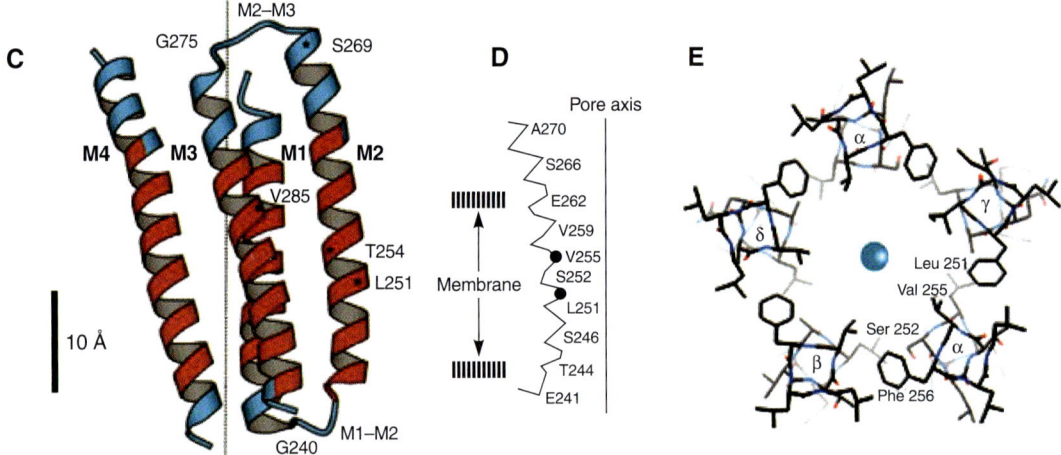

FIGURE 7.1 Structure of the acetylcholine receptor as revealed by electron microscopy, together with the cytoplasmic vestibule. (A) Overall architecture of the receptor. The suggested binding pockets are marked by asterisks. The upper arrow indicates the constricted portion of the closed channel; the lower arrow the narrowest portion of the open channel. (From Unwin, 2000.) (B) Cross section of the extracellular portion of the nicotinic receptor, with superimposed *Torpedo* polypeptide chains predicted on the basis of the AChBP. The pairs of black dots identify the C^α positions of the cysteine residues, which form the disulfide bonds linking the inner to the outer sheets. The inner sheets of the α-subunits are rotated 15°–16° anticlockwise (curved arrows) relative to the inner sheets of the non-α-subunits. (From Unwin et al., 2002.) (C–E) Structure of the transmembrane region. (From Miyazawa et al., 2003.)

well about 25 Å wide at the synaptic end, and little variation in cross section over the next 60 Å downward toward the bilayer. No cavity was resolved across the interior of the bilayer, but an *axial* widening (~20 Å in diameter) appears again on the cytoplasmic side beyond the inner leaflet of the lipid bilayer (Toyoshima and Unwin, 1988). In the more recent studies summarized in Figure 7.1A, a barrier in the axial channel appears limited to a portion of the receptor extending from about the middle of the plane of the bilayer to a region approaching the cytoplasmic limit of the bilayer (Miyazawa et al., 1999). Roughly speaking, the molecule appears as a cylindrical bundle with transverse polarity and subunits arranged like staves around an axis perpendicular to the plane of the membrane. Features of the ligand-binding site are presented in Figure 7.1B, which we will discuss in detail later in this chapter.

Finally, the high-resolution EM data have revealed a surprising feature of the cytoplasmic domain, namely relatively wide openings ("slits" or "windows") between subunits that may be the route by which ions enter or leave the channel (Miyazawa et al., 1999; Unwin, 2005). It has been suggested that the long rods surrounding the slits are amphipathic α-helices, whose existence had been predicted by several studies (Finer-Moore and Stroud, 1984; Le Novère et al., 1999). In the case of the nicotinic receptor, this region includes four heptad repeats of Glu residues (positions 377, 384, 391, and 398 in α-subunits). Since the corresponding positions in the glycine receptor are positively charged (K371, K378, K385, R392), it was proposed that the observed differences in cationic and anionic selectivity between the nicotinic receptor and glycine receptor at least in part might be explained by these charge differences (Miyazawa et al., 1999). But the predominant effects on ion selectivity have been shown to take place in the vicinity of or within the M2 channel domain (Corringer et al., 1999); and studies with the related 5-HT$_3$ receptors in which key arginine residues of the slit region were mutagenized revealed changes in ion conductance, though not in selectivity (Kelley et al., 2003).

Transmembrane Organization

As soon as the full sequences of each of the subunits had been obtained (see Chapter 3), the data were immediately exploited to investigate the transmembrane structure of the receptor subunits. The classical experiments of Del Castillo and Katz (1957a) had shown that acetylcholine applied to the *interior* of the cell had *no* effect on the permeability response. Similarly, antireceptor antibodies block the response to acetylcholine when applied from the cell exterior (Patrick and Lindstrom, 1973; Sugiyama et al., 1973; Blatt et al., 1986). The acetylcholine receptor site is also accessible from the outside to snake venom α-toxins, which do not penetrate the membrane. In addition, the α-, β-, γ-, and δ-subunits all carry carbohydrate residues (Meunier et al., 1974; Vandlen et al., 1979; Nomoto et al., 1986) that, as with all known proteins, are systematically found to be exposed to the exterior of the cell. The α-, β-, and γ-subunits are labeled by lactoperoxidase-catalyzed iodination (Hartig and Raftery, 1977). It therefore followed that the α-, β-, γ-, and δ-subunits are all exposed to the membrane exterior.

Receptor-rich membranes reseal *in vitro* with the right side out. All the receptor polypeptides resist proteases added to the suspension, for reasons that are still unknown, but they become vulnerable to proteolytic cleavage when the vesicles are opened by sonication, or treated at pH 11 or by phospholipases. Proteases thus gain access to the *interior* face of the membrane (Strader and Raftery, 1980; Wennogle and Changeux, 1980). The γ- and δ-subunits can be phosphorylated *in vitro* (Gordon et al., 1977; Teichberg and Changeux, 1977; Saitoh and Changeux, 1980), a process that involves portions of the molecule on the cytoplasmic side of the membrane (Wennogle et al., 1981). More recent studies of other subunits have confirmed the presence of phosphorylation sites on the M3-M4 cytoplasmic domain (see, for example, Guo and Wecker, 2002). Furthermore, iodinating agents acting from both the external and the internal side of the membrane label all four subunits (St. John et al., 1982). Antisubunit antibodies may similarly bind to both sides of the membrane (Strader et al., 1979; Tarrab-Hazdai et al., 1980; Froehner, 1981; Anholt et al., 1985; Lindstrom, 1986). Such transverse polarity is, of course, consistent with the images obtained by electron microscopy and expected for a ligand-gated transmembrane channel. Finally, a combination of selective proteolysis and amino acid-sequencing experiments led to the demonstration that the N-terminal domain of the α- (and δ-) subunits faces the synaptic cleft (Wennogle and Changeux, 1980; Wennogle et al., 1981), making it possible to assert a correlation between primary structure and transmembrane function.

On the basis of these primary sequence data, various models of the transmembrane organization of the four homologous subunits were proposed, as we noted in Chapter 3. These models shared three features in common: orientation of the large hydrophilic domain toward the synaptic cleft; orientation of the small hydrophilic domain toward the cytoplasm; and assignment of the four hydrophobic segments M1–M4 (only M1–M3 for Ratnam et al., 1986b) to transmembrane α-helices by analogy with known membrane proteins such as bacteriorhodopsin and glycophorin A. All of the models assumed that the acetylcholine-binding site is located in the large hydrophilic domain on the α-subunit (subsequently recognized as occurring in association with neighboring subunits: see Chapter 4); and that the walls of the ion channel lie on the quasisymmetry axis of the receptor molecule and are delineated by homologous portions of each subunit. The models differed, however, with respect to the number, orientation, and identity of the transmembrane segments, leading to divergent predictions as to whether the C-terminal segment is intra- or extracellular (see discussion in Popot and Changeux, 1984; Changeux, 1990).

Tests for transmembrane organization were developed in several laboratories using different experimental approaches. Covalent labeling of native acetylcholine receptors with photoactivatable arylazidophospholipids revealed a striking difference in α-subunit labeling between *T. marmorata* and *T. californica*. The fivefold higher labeling of the *T. marmorata* α-subunit was ascribed to the substitution of α-Ser-424 from *T. californica* by a more reactive cysteine at the same position (Giraudat et al., 1985). Since α-Cys-424 belongs to the hydrophobic segment M4, the labeling data supported the view that M4 is exposed to the lipid bilayer. Studies carried out to assess the orientation of the disulfide bond linking δ-subunits in *Torpedo* dimers employing impermeable reducing agents and reconstituted vesicles (McCrea et al., 1987) or native receptor-rich microsacs (DiPaola et al., 1989) demonstrated its extracellular localization, in agreement with the data previously obtained by Giraudat et al. (1985). In the light of evidence that the dimer-forming disulfide bond involves the penultimate Cys residues of the δ-subunits, this finding also indicated that the carboxyl terminus faces the synaptic cleft (Popot and Changeux, 1984). The results of these early protein chemical studies, confirmed by numerous more recent reports (Hucho et al., 1996; Arias, 2000), clearly support the original four-helix model (Claudio et al., 1983; Devillers-Thiéry et al., 1983; Noda et al., 1983a).

Structural Predictions

Indirect approaches to establishing the three-dimensional structure of the nicotinic receptor involved either *de novo* predictions or searches for homologous structures in the three-dimensional database to be used as models. In the latter case, a hidden Markov model approach (sequence-sequence comparisons) identified a local resemblance between the N-terminal domain of nicotinic receptor subunits and some members of the cupredoxin superfamily (Tsigelny et al., 1997). In parallel with this, a threading approach (sequence-structure comparisons) was used to find possible templates for the glycine receptor α1 subunit, prompting one team of researchers to suggest that the extracellular region is homologous to SH2 and SH3 domains (Gready et al., 1997). These two possible homology models have not proven to be useful, however, in further elucidating the three-dimensional structure. By contrast, *de novo* algorithm-based predictions produced a mainly β-sheet structure for the N-terminal region (Le Novère et al., 1999), suggestive of an immunoglobulin fold (Corringer et al., 2000), which subsequently was in fact confirmed by the X-ray structure of AChBP (Brejc et al., 2001).

With respect to the transmembrane domains, attempts have been made to obtain structural models using a number of different approaches. Several models for M2 based on an α-helical structure were proposed (Sansom et al., 1998; Tikhonov and Zhorov, 1998; Opella et al., 1999). The entire transmembrane portion of the receptor was also modeled on the basis of analogy with the heat-labile enterotoxin β-subunits (Ortells et al., 1997). In view of the fact that there exist only a few resolved transmembrane structures of ion channels, the usefulness of computer-based searches for possible templates for the nicotinic receptor transmembrane domain is limited. Nevertheless, two recently resolved ion channels looked to be promising candidates: the tetrameric potassium channel from *Streptomyces lividans* (Doyle et al., 1998) and the pentameric mechanosensitive receptor from *Mycobacterium tuberculosis* (Chang et al., 1998). Both channels are formed by tilted bundles of α-helices, which could conceivably share certain features with the α-helical component of the nicotinic receptor formed by M2. Moreover, the external opening of the channel is lined by a characteristic P-loop structure. A hypothetical model of nicotinic receptors based on the structure of the potassium channel would require that it be transformed into a pentamer, however. Given that the nicotinic M1-M2 loop has been identified as a site of charge selectivity (Corringer et al., 1999)—suggesting a possible analogy (mentioned in Chapter 5) with the P-loop component of the potassium channel, which is also known to contribute to ion selectivity—a plausible template would require the transmembrane organization of the potassium channel to be inverted, so that its P-loop would be located instead at the cytoplasmic border of the pore. Such a template may be helpful in guiding the design of preliminary models, but it needs to be refined further by integrating the structural and functional data so far accumulated concerning the nicotinic receptor channel.

Other studies of the transmembrane domain have suggested a mixture of α-helical and β-sheet structures (Gorne-Tschelnokow et al., 1994; Corbin et al., 1998a; Methot and Baenziger, 1998; Le Novère et al., 1999). The distribution of these elements among the four transmembrane domains remains unclear, however. Recent results disagree with regard to the predominance of α-helical or β-sheet structures, the former possibility being favored by infrared spectroscopy of residual membrane-embedded peptides following extensive proteolytic digestion (Methot et al., 2001), and the latter by spectra obtained from reconstituted α_1-GlyR by circular dichroism (Cascio et al., 2001). M4 displays the largest contact with lipids, and its pattern of labeling by 3-trifluoromethyl-3-(m-[^{125}I]-iodophenyl)diazirine ([^{125}I]TID) is consistent with an α-helical structure (Blanton and Cohen, 1994). Labeling patterns and mutagenesis data concerning M2 are generally interpreted in terms of an α-helical structure (Corringer et al., 2000), a view that has been confirmed by recent EM studies (Miyazawa et al., 2003), which we will consider shortly. Both experimental and theoretical evidence therefore indicate that the transmembrane domain consists primarily of α-helical structures.

Crystal Structure of the Molluscan Acetylcholine-Binding Protein

The discovery of a soluble acetylcholine-binding protein at molluscan cholinergic synapses homologous to the N-terminal portion of the vertebrate nicotinic receptor (Smit et al., 2001), and the determination of the crystal structure of this protein at 2.7 Å resolution, revealing a characteristic nicotinic ligand-binding domain (Brejc et al., 2001), marked a major step forward in the study of nicotinic receptors. The molluscan protein is a homopentamer, each subunit displaying a modified immunoglobulin (Ig) topology that includes an extra strand (b′) and a β-hairpin (Figure 7.2A, B). By comparison with the classical Ig-fold, the AChBP strands are twisted extensively. This structure strikingly confirms the folding predictions of Le Novère and Changeux (1999), which assigned the nicotinic receptor to the Ig family (Corringer et al., 2000). The N-terminal extremity is located at the "top" of the structure (Figure 7.2B) in a region homologous to the main immunogenic region (MIR) implicated in the autoimmune disease myasthenic gravis (see Chapter 8). Although the MIR-related region in AChBP (residues 65–72) has little sequence homology with the α1-subunit, its location in a highly accessible position in loop L3 at the top of the pentamer confirms the assumption of accessibility with respect to this region.

The overall pentameric structure resembles a pinwheel when viewed along the fivefold axis shown in Figure 7.2B, with a diameter of 80 Å. Each protomer participates in two distinct subunit interactions through surfaces that include key residues corresponding to the *principal* and *complementary* components of the ligand-binding sites, respectively (Corringer et al., 1995), identified from labeling and mutagenesis

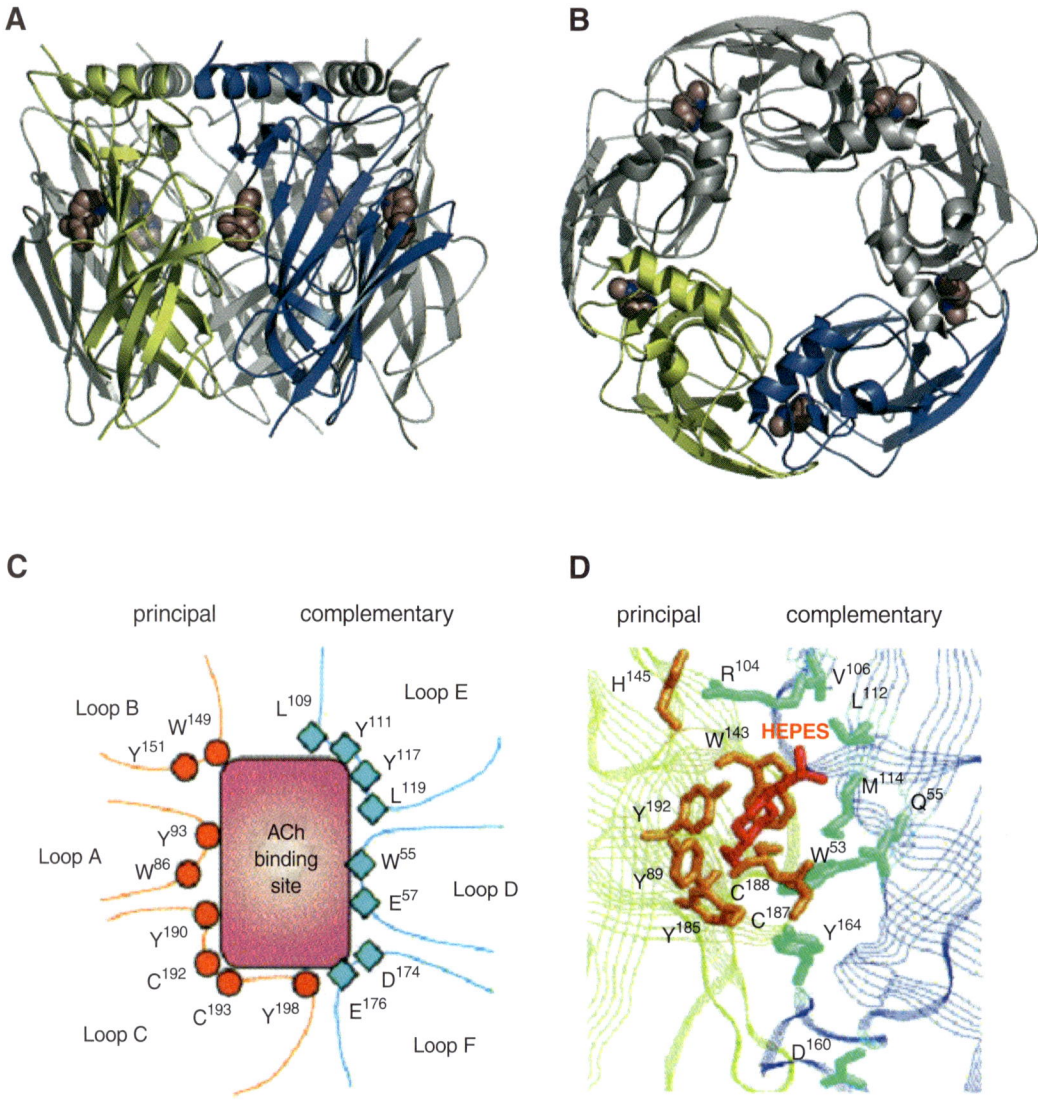

FIGURE 7.2 Three-dimensional structure of the molluscan acetylcholine-binding protein (AChBP) and models of the acetylcholine/nicotine binding site. (A) AChBP pentamer with nicotine in the binding pocket, viewed from the side. (B) The same pentamer shown in (A), viewed from the top. (From Celie et al., 2004.) (C) Model of the ligand-binding site based on chemical labeling and mutagenesis of receptor from *Torpedo*. (From Corringer et al., 2000, courtesy of Annual Reviews.) (D) Detailed view of the binding site modeled in C, based on X-ray data showing the chelating agent HEPES in the binding pocket. (From Brecj et al., 2001.)

studies (Figure 7.2C, D). Although AChBP was initially crystallized in the absence of nicotinic ligands, a 4-(2-hydroxyethyl)-1-piperazine-ethanesulfonic acid (HEPES) buffer molecule was resolved in the crystal at the level of the putative acetylcholine-binding site, yielding further information about the positioning of the nicotinic ligands. The structure of the complex crystallized with carbamylcholine or nicotine has recently been described in detail and compared to the structure obtained with the bound HEPES molecule (Celie et al., 2004).

The available data concerning the residues that participate in ligand binding now give a relatively clear picture of the acetylcholine-binding pocket. The principal side of the interface is framed essentially by loops, whereas the complementary side presents mainly secondary structure elements (β-sheets). The principal side is composed of the three loops: A, containing Tyr-93 (Galzi et al., 1990); B, containing Trp-149 (Dennis et al., 1988); and C, containing Tyr-190, Cys-192, Cys-193, and Tyr-198 (Kao et al., 1984; Dennis et al., 1988; Abramson et al., 1989). The complementary (minus) side is composed of three segments: D, containing Trp-55 and Glu-57 (Corringer et al., 1995; Chiara and Cohen, 1997); E, containing Leu-109, Tyr-111, Tyr-117, and Leu-119 (Chiara and Cohen, 1997; Chiara et al., 1999; Wang et al., 2000); and F, containing Asp-174 and Glu-176 (Czajkowski and Karlin, 1995).

Loop C is folded in a β-hairpin connected at the turn by the two vicinal Cys residues (187 and 188) homologous to the Cys pair that had been identified initially (Kao et al., 1984). The unexpected folding of loop C might serve as a gate controlling access to the ligand-binding site (Brejc et al., 2001). It has been suggested that the loop-B amino acid Trp-143 (homologous to Trp-149 in *Torpedo*) exerts cation/π interactions with the quaternary ammonium group of acetylcholine (see, however, below). This interaction had previously been observed for acetylcholinesterase (Sussman et al., 1991) and hypothesized for nicotinic receptors (Zhong et al., 1998). Residues located at the interface (except for certain of those involved in ligand binding) are not well conserved among nicotinic receptors, and probably contribute to the diversity of subunit combinations. Within the postulated ligand-binding site, residues are not conserved in loops B and E—further evidence that these loops contribute to pharmacological diversity among different members of the nicotinic receptor family (Corringer et al., 1995, 2000; Chiara et al., 1999). Moreover, the residue Asp-161 in segment F, homologous to δAsp-180 and γAsp-174 identified by affinity labeling and mutagenesis in *Torpedo* (Czajkowski and Karlin, 1995), does not contribute directly to the acetylcholine-binding site of AChBP. Nevertheless, since loop F is not folded in a periodic structure in AChBP, local conformational changes in the *Torpedo* nicotinic receptor may have the effect of bringing this residue closer to the acetylcholine-binding site. One residue located in loop F in AChBP (Tyr-164) that had not previously been identified by chemical labeling in nicotinic receptors might contribute to the ligand-binding area because of a Ca^{2+} site that orientates its lateral chain toward loop C. The aromatic character of the amino acid (Tyr or Phe) conserved at this position in the superfamily of ligand-gated ion channels suggests that the geometry of the ligand-binding site has a functional role.

The structure of the AChBP complexes crystallized with carbamylcholine and nicotine (Celie et al., 2004) shows that the ligands are completely buried by the protein, displaying more direct contact with the principal component than with the complementary component. On the principal side, the ligands make extensive aromatic contacts with the Trp-143 side chains (Galzi et al., 1991a; Zhong et al., 1998), and some contacts with Tyr-192 (Dennis et al., 1988; Middleton and Cohen, 1991). Tyr-185 (Dennis et al., 1988; Middleton and Cohen, 1991) contributes to the interactions with the choline moiety of carbamylcholine but not with nicotine, which explains why a mutation at this position affects acetylcholine binding more than nicotine binding (Galzi et al., 1991). Structural studies further indicate that the ligands establish close contact not with the aromatic ring of Tyr-89, but rather with its hydroxyl group—a finding consistent with the observation that mutation from Tyr to Phe strongly reduces the affinity for acetylcholine and carbamylcholine (Sine et al., 1994). The vicinal disulfide is localized in contact with carbamylcholine through Cys-187 and with nicotine through Cys-188 (see Damle and Karlin, 1980). On the complementary side of the acetylcholine-binding pocket, Trp-53 was found to make limited aromatic contact with nicotine, whereas Leu-112 and Met-114 contribute to hydrophilic contacts with both carbamylcholine and nicotine.

Crystallographic studies also reveal original features of nicotine binding (Celie et al., 2004). Two hydrogen bonds contribute to its interactions: one between the pyridine N and the main chain of residues Leu-102 and Met-104 through a bridging water molecule; the other between the pyrollidine N2 and the carboxyl group of Trp-143, which is itself stabilized by the conserved Asp-85.

The charged compensation of quaternary nicotinic ligands has been, and remains, a subject of dispute. Structural studies, while they do not exclude cation-π interaction with aromatic side chains as those of Tyr-192 and -185, suggest that interaction with the partially charged carbonyl of Trp-143 is more likely.

Finally, a close comparison of HEPES-bound and carbamylcholine/nicotine-bound structures does not reveal any change in the quaternary structure of the pentamer. Only local changes were noted:

a backbone movement of the C-loop, and of Tyr-89 in particular; and the formation of a hydrogen bond between Cys-139 and Tyr-183 in the liganded structure. The absence of major conformational transitions (Celi et al., 2004) is expected in the case of a protein that does not display conventional allosteric interactions (cooperative interactions, for instance, are notably absent between ligand-binding sites).

Models anticipating these recent structural studies had generated structures for several bound nicotinic agonists (Le Novère et al., 2002b), as well as for α-toxins (Harel et al., 2001; Fruchart-Gaillard et al., 2002). Figure 7.3 illustrates the remarkable agreement between *in silico* computations and actual X-ray data. In addition, modeling of the allosteric site for Ca^{2+} succeeded in identifying a pocket near loop F in AChBP within a region homologous to the site of Ca^{2+} potentiation identified in α7-receptors (Galzi et al., 1996b), but which involves distinct interactions.

Our current understanding of the topology of the acetylcholine-binding site is sufficiently detailed that the design of pharmacological agents that selectively interact with the acetylcholine-binding pocket, and even with particular subunit interfaces among the broad diversity of known receptor oligomers, may now be contemplated.

Crystal Structure and Electron Microscopic Models Compared

The standard methods of electron microscopy rarely match the resolution of X-ray crystallography. Nonetheless, a casual observation concerning receptor-rich membranes isolated from *Torpedo* (Brisson, 1980) was responsible for a significant improvement in the resolution of EM images. A suspension accidentally left to sit on the bench top for a few days was found unexpectedly to have converted membrane fragments containing randomly distributed receptors into long tubes whose receptor molecules were organized in regular arrays (Brisson, 1980; Brisson and Unwin, 1984) and resembled those found *in situ* at the synapse (Cartaud et al., 1973, 1978; Heuser and Salpeter, 1979). Averaging of numerous images of samples embedded in amorphous ice (over 200 images containing about 800,000 molecules) extended the resolution into the 4–5 Å range (Miyazawa et al., 1999; Unwin, 2000; Miyazawa et al., 2003). Large cavities observed in the extracellular domain, most notably in the α-subunits (Unwin, 1993), were tempting locations for the assignment of the ligand-binding site (Unwin, 2000), particularly since their size made them likely candidates to accommodate an acetylcholine molecule. The crystal structure of the molluscan AChBP (Brejc et al., 2001) clearly shows the ligand-binding site to be

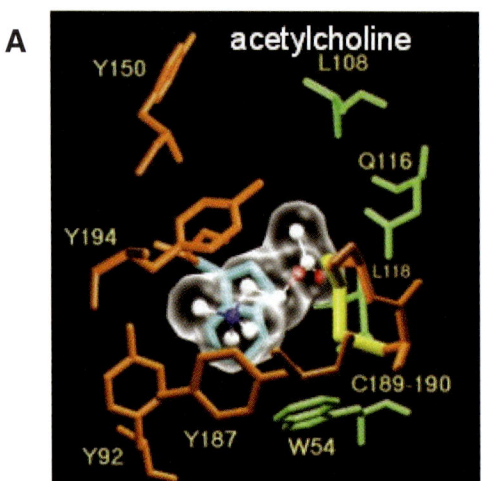

A acetylcholine

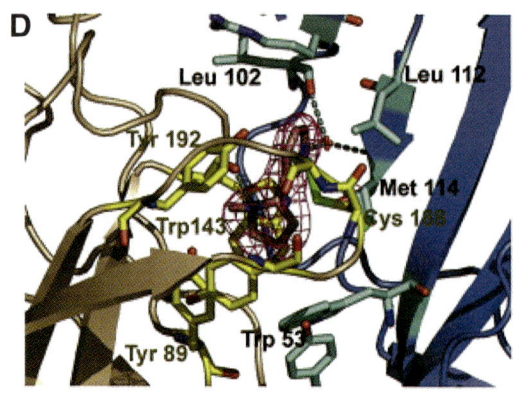

D

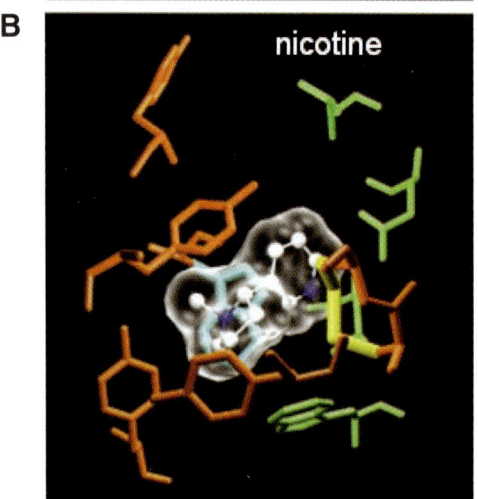

B nicotine

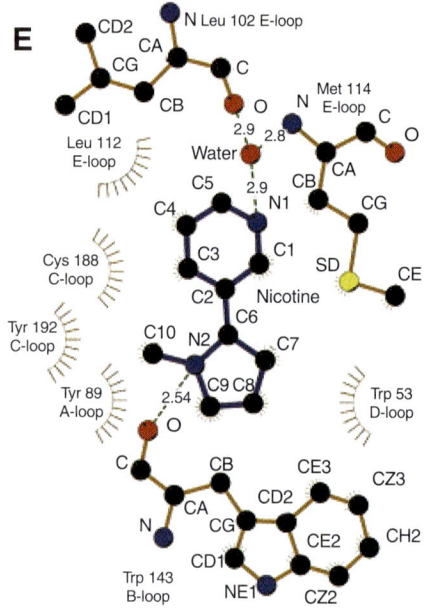

E

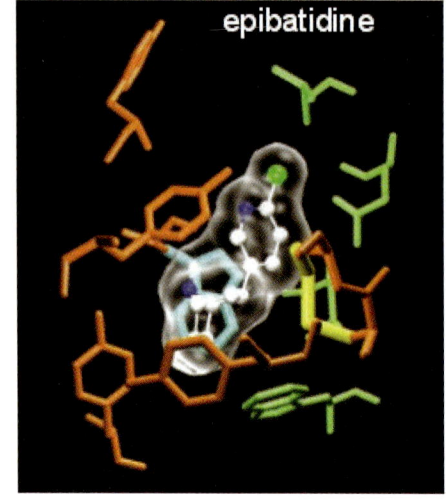

C epibatidine

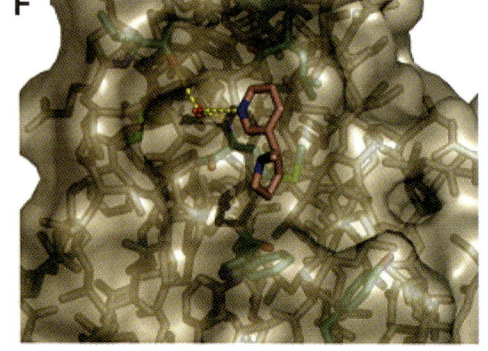

F

situated at the interface between subunits, however, confirming the model suggested by labeling and site-directed mutagenesis (Chapter 4). In the meantime, this issue has been examined in greater detail and a specific structural hypothesis has now been formulated (Unwin et al., 2002). Attempts to place the X-ray structure in correspondence with the EM structure yield a satisfactory match only if portions of the α-subunits are moved from their initial positions in the EM structure of the unliganded receptor, however. This observation has led to the hypothesis that rotations of 15°–16° in the inner pore-facing parts of the α-subunits coincide with the transition from the resting to the active state, the latter more closely resembling the overall structure of AChBP (Figure 7.1B).

With regard to the transmembrane portion of the molecule, until recently only one structural feature had been noted: a kinked, rodlike element that was interpreted as a bent α-helix (Figure 7.1A) lining the ion channel (Unwin, 1993, 2000). Chemical labeling and site-directed mutagenesis experiments had earlier identified a conserved leucine residue near the middle of M2 that plays a key role in determining channel and allosteric properties (Giraudat et al., 1987, 1989; Révah et al., 1991). This led Unwin (1993, 2000) to suggest that the leucine lies in the narrowest portion of the channel, at the level of the kink, which, in his view, corresponds to the residues closest to the axis in Figure 7.1A. Rotations of the subunits accompanying rapid addition of acetylcholine (Unwin, 1995) caused a gating mechanism to be proposed that involves small clockwise rotations of the bent α-helix, so that the constriction rotates out of the channel to allow the passage of ions (see following section). But this interpretation is at odds with other experimental results obtained using cysteine scanning (Wilson and Karlin, 1998) or zinc-histidine mapping (Y. Paas, G. Gibor, R. Grailhe, N. Savatier-Duclert, V. Dufresne, L. Prado de Carvalho, J.-P. Changeux, and B. Attali, unpublished), which have been interpreted as evidence of a contribution to gating by residues at the cytoplasmic extremity.

A more complete picture of the transmembrane region has recently been pieced together from data to 4 Å resolution, revealing a structure composed entirely of α-helices (Miyazawa et al., 2003) that is summarized in Figure 7.1C, D, E.

Structural Basis of the Allosteric Transition

Given the results of these crystallographic and electron microscopy studies, the next major hurdle is to elucidate the allosteric mechanism mediating the activation and desensitization of the nicotinic

FIGURE 7.3 Docking models of several nicotinic receptor ligands and the three-dimensional structure for nicotine bound to the acetylcholine binding site. (A–C) Docking models of acetylcholine, nicotine, and epibatidine bound to the neuronal α7-nicotinic receptor. (From Le Novère et al., 2002b.) (D–F) Several three-dimensional representations of nicotine in the binding pocket of AChBP. (From Celie et al., 2004.)

receptor ion channel, which are enhanced by agonist binding at a site 20–30 Å away. The challenge will be to understand how the extracellular "crown," examined until now as an autonomous structure, comes to be coupled with the transmembrane domain. Various mechanistic possibilities have been considered (Changeux, 1990), among them "twisting" or "blooming" with respect to subunit juxtaposition, as well as more local changes. These alternatives can be interpreted in terms of "rigid body" as opposed to "plastic" motion (Changeux and Edelstein, 2001). Moreover, a critical aspect of any particular structure is its relationship to a specific functional state. For example, the observation that AChBP does not display standard allosteric interactions and exhibits relative high affinities for nicotinic ligands (agonists and antagonists) with Hill coefficients either equal to or less than unity (Smit et al., 2001) has led to the suggestion that AChBP may have been crystallized in a frozen "desensitized" state (Grutter and Changeux, 2001). A hypothetical model along these lines has been proposed for the transition from the resting state (Taly et al., 2005) based on a normal mode analysis that revealed a concerted, symmetric *en bloc* twist motion of the protein (see Figure 7.4).

In general, membrane receptors may conceivably differ from classical allosteric enzymes (Monod et al., 1965) or hemoglobin (Edelstein, 1975), where a higher affinity for the substrate or active site ligand is associated with "relaxed" quaternary interactions. Additionally, the situation is more complicated for binding sites at intersubunit interfaces since ligand binding may help to stabilize the interface, as observed in the case of the catalytic subunit of the allosteric enzyme aspartate transcarbamylase (Yang and Schachman, 1987). Indeed, for nicotinic receptors, some amino acids that selectively alter desensitization are common to the ligand-binding site and intersubunit association surface, in particular Trp-143 in AChBP in loop B (Corringer et al., 1998), which may assist the coupling of ligand binding and conformational change. Valuable indicators of individual conformations have been provided by distinctive fluorescence spectra (Grunhagen and Changeux, 1976; Grunhagen et al., 1976) and by specific chemical labeling with DDF (Galzi et al., 1991a), which revealed a reorganization of the subunit interface during the transition toward the desensitized state.

On the basis of these and other observations it has been hypothesized that the basal (B) activatable conformation of the nicotinic receptor derives from AChBP structure through reorganization of the receptor protein's quaternary structure (Grutter and Changeux, 2001), possibly involving a concerted rotation of all the N-terminal domains. Such movements could lead to a more compact conformation for

CLOSED   OPEN

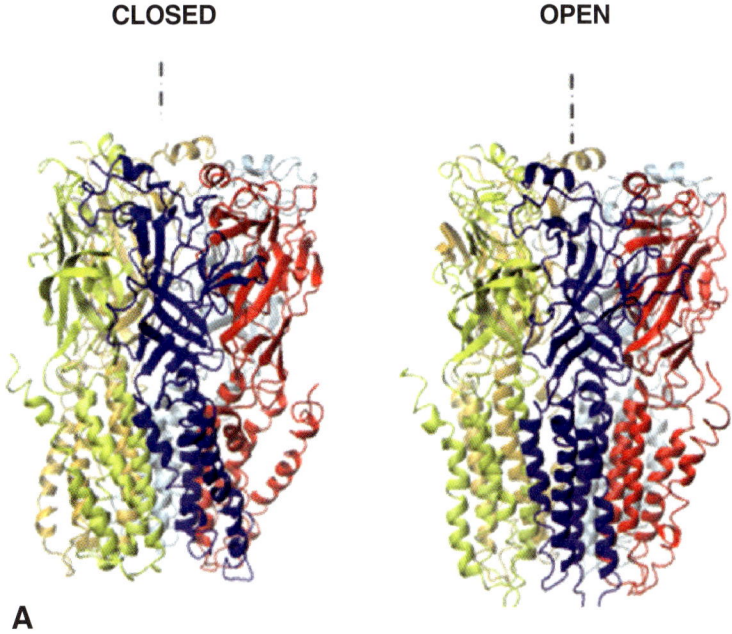

 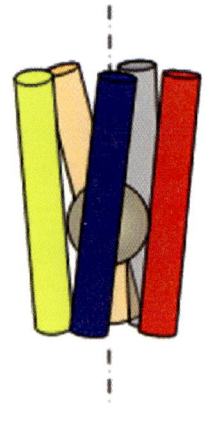

FIGURE 7.4 Model of conformational change based upon normal mode analysis suggesting a quaternary twist model for the α7 homopentameric nicotinic receptor gating mechanism. (A) Models of closed and open states in lateral view. (B) Schematical representation of the *en bloc* quaternary structure conformational changes. Each "subunit" is represented as a cylinder. (From Taly et al., 2005).

the D state. Indeed, when a model of the complex between α7 and α-cobratoxin (Fruchart-Gaillard et al., 2002) is compared to α7 alone for computer-generated docking of agonists (Le Novère et al., 2002b), affinities some 100-fold weaker are estimated for the system with α-cobratoxin, suggesting that the toxin stabilizes the basal rather than the desensitized state (Le Novère et al., 2002b).

The data presently available support the notion that the nicotinic receptor protein undergoes conformational transitions that significantly affect its quaternary structure and, correlatively, its tertiary structure as well. But additional information is needed if we are to be able to make progress in understanding these transitions at the amino acid level.

Conclusion

A closer look at the three-dimensional structure of the acetylcholine receptor was made possible first by electron microscopy. Steadily improved resolution revealed the heteropentameric organization of the receptor molecule, the presence of two nonneighboring α-toxin-binding sites per pentamer, and the protein's transmembrane structure. Further improvements of this technique have recently given rise to an interpretation of the α-subunit synaptic domain in terms of a crescent, with putative β-sheets surrounding a cavity, and an α-helical rod in the transmembrane domain, which may plausibly be supposed to line the ion channel, with a kink in its equatorial region.

The crystal structure of an acetylcholine-binding protein from molluscs, a related homologue of the synaptic domain of the nicotinic receptor, discloses a set of features that strikingly confirm biochemical and structural predictions obtained mainly from *Torpedo* receptor protein (Chapter 4): first, the occurrence of a set of β-pleated sheets, in agreement with the suggestion that the nicotinic receptor belongs from the structural point of view to the immunoglobulin superfamily; second, the location of acetylcholine-binding sites at the interface between subunits; third, the close correspondence between the amino acids forming the ligand-binding pocket and the residues formerly identified by affinity labeling on the *Torpedo* receptor. The available structural data point to the interaction of the α-toxin with the intersubunit clefts on the outside of receptor oligomers.

Convergent biochemical, biophysical, and electron microscopy data convincingly demonstrate the occurrence of conformational transitions in the receptor protein. But they are nonetheless too fragmentary to support firm conclusions of a more detailed nature. A great deal of additional structural data will be required to fully describe the mechanisms responsible for the allosteric transitions of activation and desensitization in the nicotinic receptor.

Even so, the advent of a novel pharmacology may already be envisioned that aims at designing molecules that will differentially match the various conformations of the acetylcholine-binding site, as well as of the other allosteric sites in the different allosteric states of the receptor protein. These molecules would thus behave not only as full agonists or antagonists, but also as partial agonists that bind nonexclusively to several states, in addition to selectively stabilizing desensitized states.

CHAPTER 8

Inherited Pathologies of the Acetylcholine Receptor

Naturally occurring mutations in the human population provide important insight into structure-function relations in proteins, particularly since these "experiments of nature" generate information independent of any conceptual design or any preconceived idea or model. For example, the screening of newborns for sickle cell hemoglobin and detection of novel hemoglobinopathies have led to the identification of more than five hundred mutations in the α- and β-chains of human hemoglobin (Hardison et al., 1998), many with unexpected phenotypes. Although the number of spontaneous or pathological mutations identified in nicotinic receptor subunits remains an order of magnitude lower, unusual phenotypes have been produced by several of the mutations responsible for inherited diseases. Analysis of these phenotypes has yielded a clearer understanding of the pathologies associated with such mutations, the consequences of altering specific amino acid residues, the functional role of transitions between conformational states, and possible pharmacological treatments of these diseases.

To date, the only monogenic inherited diseases directly associated with the nicotinic receptor are a large fraction of the congenital myasthenic syndromes (Engel, 1993; Engel et al., 2003a) and autosomal dominant nocturnal frontal lobe epilepsy (ADNFLE), a rare inherited form of epilepsy (Steinlein et al., 1995; Steinlein, 2000, 2004). Characterization of the genetic mutations that produce these two pathological states, and of their consequences for the receptor phenotype, has given us a new perspective on the relationship between structure and function in the nicotinic receptor, particularly with respect to congenital myasthenic syndromes, for which over fifty distinct mutations have been observed (Engel et al., 1999, 2003a). Interestingly, several of these mutations reproduce the "gain of function" pleiotropic phenotypes initially discovered with the neuronal α7-receptor (Révah et al., 1991; Bertrand et al., 1992; Devillers-Thiéry et al., 1992). In this chapter we shall describe these two genetic diseases and examine the role that allosteric mechanisms play in their development.

Myasthenia Gravis and Congenital Myasthenic Syndromes

The neuromuscular junction is a complex intercellular organelle that mediates the transmission of nerve signals to muscle through the release of acetylcholine, which acts on nicotinic receptors present in the muscle postsynaptic membrane (see Chapter 9). As a result, alterations that substantially increase or decrease activation of the nicotinic receptor, either directly or indirectly, can have serious pathological consequences for muscle contraction, in addition to debilitating secondary effects on the neuromuscular junction.

Interest in pathologies linked to nicotinic receptors initially derived from studies of myasthenia gravis in humans. This autoimmune disease is manifested by flaccid paralysis of skeletal muscles and high levels of antinicotinic receptor antibodies in blood serum. In 1973, James Patrick and Jon Lindstrom recognized that similar symptoms were produced in rabbits following immunization against nicotinic receptors purified from the electric organ of *Electrophorus electricus* or *Torpedo marmorata* (Sugiyama et al., 1973; Heilbronn et al., 1973; Changeux, 1990). Subsequent investigation of myasthenia gravis led to the identification of a particular structure in a part of the synaptic domain of the receptor protein known as the main immunogenic region (MIR), on the α1-subunits (Tzartos and Lindstrom, 1980). Recent discoveries regarding the three-dimensional structure of mollusc AChBP (see Chapter 7) make it plausible to suppose that the corresponding residues lie in a highly accessible loop at the synaptic extremity of the receptor (Brejc et al., 2001), in agreement with suggestions from earlier structural studies at lower resolution (Beroukhim and Unwin, 1995). Some antibodies block acetylcholine-mediated signal transduction. Additionally, the interaction of antibodies with the nicotinic receptor is followed by an internalization of the protein. This internalization leads to a regression of the neuromuscular junction that, together with the blocking of signal transduction, produces the clinical symptoms. Why the mechanism of antibody induction should be operative only in certain individuals is not clear, however. Several different causes have been suggested (involving, for example, rheumatoid arthritis or thymoma), with the triggering immunogen postulated to be either a muscle nicotinic receptor or molecular mimicry arising from a bacterial or viral infection (Lindstron, 2000, 2002).

Myasthenic syndromes are not restricted to autoimmune reactions and can have a genetic basis. There may in fact be six or more distinct origins for such syndromes (see Engel et al., 2003a); at least five have been reported (McConville and Vincent, 2002):

1. Antibodies to, or mutations in, presynaptic voltage-gated channels (Vincent et al., 1989). Autoimmunity to the voltage-gated calcium channel underlies the Lambert-Eaton myasthenic syndrome, for example (Rosenfeld et al., 1993). It is thought that this syndrome may be indirectly related to the nicotinic receptor, since tobacco smoking can cause small cell carcinomas of the lung: these tumors express voltage-dependent calcium channels that elicit the autoantibodies, which in turn impair the release of acetylcholine at neuromuscular junctions (Lang and Vincent, 1996; Benatar et al., 2001). It has also recently come to light that similar symptoms may be produced by mutations (hereditary episodic ataxia type 2) in voltage-gated calcium channels (Jen et al., 2001), as well as by antibodies (Shillito et al., 1995; Hart, 2000) and mutations (hereditary episodic ataxia type 1) that cause related abnormalities in voltage-gated potassium channels (Eunson et al., 2000).
2. Mutations responsible for defects in choline acetyltransferase, the enzyme responsible for acetylcholine synthesis (Ohno et al., 2001). Mutations in the choline acetyltransferase lead to decreased miniature endplate potentials and are associated with frequently fatal episodes of apnea, but can be prevented or mitigated by anticholinesterase drugs.
3. Mutations responsible for a deficiency of acetylcholinesterase at the neuromuscular junction (Donger et al., 1998; Ohno et al., 1998). The absence of junctional acetylcholinesterase causes prolonged postsynaptic currents that produce cationic overloading and degeneration of the junctional folds (Engel et al., 1977).
4. Mutations in nicotinic receptor genes or their promoters. These mutations result in markedly reduced synaptic transmission (Engel et al., 1999).
5. Antibodies against, or mutations in, other postsynaptic proteins (McConville and Vincent, 2002). Antibodies to a synapse-specific tyrosine kinase referred to as MuSK (see Chapters 9–10) are responsible for acquired myasthenia gravis (Hoch et al., 2001). Mutations in the cytoskeletal protein 43K-rapsyn (see Chapters

9–10) cause a congenital myasthenic syndrome (Ohno et al., 2002). MuSK mediates the clustering of nicotinic receptors during synapse formation, while 43K-rapsyn participates directly in receptor clustering and stability (see Chapters 9 and 10).

With regard to the fourth category, about sixty of the mutations of the nicotinic receptor reported to date are in genes for the α1-, β1-, δ-, or ε-subunits. A disproportionately large fraction of these mutations are in the gene for the ε-subunit, presumably because its disabling can partially be compensated by the γ-subunit. The known mutations may be grouped into four main categories. So-called loss-of-function mutations give rise to a first class of receptors displaying diminished intrinsic response to acetylcholine (fast-channel receptors). Mutation may also produce various forms of nicotinic receptors, forming a second class, with reduced or absent expression due to problems in biosynthesis, folding, or assembly (Engel et al., 1999, 2003a), including mutations in the N-box promoter region of the ε-subunit (Ohno et al., 1999; Engel et al., 2003a; see Chapter 9). A third class of mutant phenotypes may also exhibit the effects of compensating expression of the fetal subunit, γ, which results in adult receptors having the lower conductance typically associated with fetal receptors (Milone et al., 1998). Fourth, and finally, a smaller class of ten mutants displays a "gain of function." These phenotypes, which we will examine in detail below, have furnished valuable insight into molecular transitions in nicotinic receptors (Révah et al., 1991; see Chapter 6). They are also known as "slow-channel" mutations because of the distinctive behavior they exhibit (prolonged open times) in single channel recordings.

Slow-Channel Myasthenic Syndrome: A Prototype of "Allosteric Diseases"

One of the main properties of several mutations leading to slow-channel myasthenic syndromes is the occurrence of prolonged channel openings that in some cases can reach on average the 100 msec range (Ohno et al., 1995), compared to the msec range of normal channels. Additionally, in some mutants (see Table 8.1) spontaneous openings have been observed to accompany the slow-channel phenotype. Whether such openings have been found (or not found) in the case of a particular mutation may depend more, however, on the experimental conditions under which observations were made than on the intrinsic properties of the mutant receptor. Of the ten slow-channel mutations so far reported (Table 8.1), six of them occur in the gene for the α-subunit, with two found in β and another two in ε

(Engel et al., 1999). Several of these mutations alter amino acids at positions for which site-directed mutations, in the α7 nicotinic receptor, were found to exert profound phenotypic changes (Table 8.1 and Figure 8.1). For example, six of the mutations change residues in the M2 segment that line the ion channel—including a site on the β-subunit, L262M (Gomez et al., 1996), that initially was found to be labeled by chlorpromazine (Giraudat et al., 1987). The remaining slow-channel mutations are distributed either in the large extracellular N-terminal domain that includes the ligand-binding site, or in the M1 transmembrane segment, or in the M2-M3 loop. No myasthenic mutations have yet been observed in the "anionic ring" residues at the ends of the M2 ion channel segment, which is thought to control ion conductance properties (Imoto et al., 1988).

Table 8.1

Slow-Channel Myasthenic Mutations and Site-Directed Mutations at Corresponding Positions in Chick α7.

Domain	Muscle		Chick Neuronal α7	
	Myasthenic	Properties	Site-directed	Properties
N-terminal:	α1 G153S	Prolonged openings (Sine et al., 1995c)	α7 G152K	Low EC_{50} and IC_{50} (Corringer et al., 1998)
	α1 V156M	Prolonged openings (Croxen et al., 1997)	α7 L155I	Normal IC_{50} (Corringer et al., 1998)
M1:	α1 N217K	Prolonged openings (Wang et al., 1997)	α7 N213	—
M2:	α1 V249F	Spontaneous openings (Milone et al., 1997)	α7 V245	—
	β L262M	Prolonged openings (Gomez et al., 1996)	α7 L247T	Low EC_{50} and IC_{50} spontaneous openings (Révah et al., 1991; Bertrand et al., 1992; Bertrand et al., 1997)
	α1 T254I	Prolonged openings (Croxen et al., 1997)	α7 T250	—
	ε T264P	Spontaneous openings (Ohno et al., 1995)	α7 T250	—
	β V266M	Spontaneous openings (Engel et al., 1996)	α7 V251T	Low EC_{50} (Devillers-Thiéry et al., 1992)
	ε L269F	Spontaneous openings (Gomez and Gammack, 1995)	α7 L255	—
M2–M3 loop:	α1 S269I	Prolonged openings (Croxen et al., 1997)	α7 D265N	High EC_{50} (Campos-Caro et al., 1996)

Adapted from Edelstein et al., 1997a.

A striking feature of these slow-channel mutations is that they alter distinct and sometimes quite distant regions of the receptor molecule (see below), but nevertheless produce similar receptor phenotypes. One possible explanation, which will be developed in the following section, is that these phenotypes result from global changes in quaternary transitions, affecting the relative stability of the open state, for instance, in such a way as to reduce the magnitude of the allosteric equilibrium constant, L, that governs the B $\rightleftharpoons$ A equilibrium (Monod et al., 1965).

Mutations in the N-Terminal Domain

Among the mutations observed in the N-terminal domain (Figure 8.1), αG153S mutant receptors expressed in human embryonic kidney (HEK) cells display longer open-channel times than ones recorded in native receptors (Sine et al., 1995c), a characteristic feature of the gain-of-function phenotype. The unusual "slow" channel properties were attributed by the authors to an increase in affinity (~50-fold) at one of the acetylcholine-binding sites (Sine et al., 1995c). Mechanistic interpretations can be complicated by compensation of parameters, however, with the result that markedly different hypothetical mechanisms can produce similar effects (Edelstein et al., 1997b). For example, differences in equilibrium between conformational states (L phenotype)—which it is thought may account for the mutation at the corresponding position in α7, G152K, in the loop-B region of the binding site—indirectly increase the apparent affinity for ligand (Corringer et al., 1998) through the stabilization of an H-bond with loop C (Grutter et al., 2003). The mutation αV156M in the N-terminal domain also involves prolonged openings (Croxen et al., 1997) and may arise from a similar mechanism. Indeed, examination of the crystal structure of AChBP (Brejc et al., 2001) suggests that the positions of both αG153S and αV156M in the N-terminal domain (corresponding to S147 and I150, respectively, in AChBP), while they are in the vicinity of ligand-binding residue αW149 (W143 in AChBP), do not interact directly with the ligand-binding site. It may in any case be said that the phenotypes of mutations in the N-terminal domain are significantly affected by changes in equilibrium between conformational states in addition to direct changes in the binding site.

Mutations in Transmembrane Segments

The mutation αN217K, located in the M1 transmembrane segment (Figure 8.1), also causes prolonged channel openings. This position does not coincide with any known site-directed mutations of α7. In contrast, two mutations in the M2 transmembrane segment of the β-subunit, βL262M (Gomez et al., 1996) and βV266M (Engel et al., 1996), lead to prolonged openings, with spontaneous openings also reported for the latter mutation. With regard to the homologous sites in α7 site-directed mutants (Table 8.1), a pleiotropic high-affinity phenotype was observed for the L247T mutation (Révah et al., 1991) that included spontaneous opening (Bertrand et al., 1997), and high affinity for acetylcholine was observed in the case of V251T as well (Devillers-Thiéry et al., 1992). For the other M2 myasthenic mutants with slow-channel phenotypes—αV249F (Milone et al., 1997), αT254I (Croxen et al., 1997), εT264P (Ohno et al., 1995), and εL269F (Gomez and Gammack, 1995; Engel et al., 1996)—prolonged or spontaneous openings were observed, but corresponding mutants in α7 are not available. The mutation εT264P has a particularly interesting phenotype that is described in detail below. No slow-channel mutations have been observed in M3 or M4, although a fast-channel mutation has been reported in M3 (Wang et al., 1999b).

Mutation in the M2-M3 Loop

A slow-channel mutation, αS269I, lies within the short extracellular sequence between M2 and M3 (Croxen et al., 1997; Figure 8.1). This residue of the M2-M3 loop corresponds to the position of a site-directed mutation in α7, D265N (Campos-Caro et al., 1996). Diminished affinity for acetylcholine has been observed for this mutation in α7, however, so in this case a parallel between mutant muscle and α7 phenotypes cannot be sustained.

Allosteric Model for a Slow-Channel Myasthenic Mutant

The myasthenic mutant εT264P in M2 (Ohno et al., 1995) is unusual in that three distinct times of channel openings are observed when receptors carrying this mutation are expressed in HEK cells (Ohno et al., 1995). The trimodal distribution is obtained by measuring the length of time of successive openings in single channel recordings. For normal receptors and for most other mutant receptors, histograms of the population of single channels in bins of openings over different time ranges can be fitted to a simple unimodal exponential distribution that exhibits a single peak (occurring when the number of events is plotted against "log time"), with the time at the peak, τ, corresponding to the average open time (about 0.7 ms). For εT264P mutant receptors, however, a single peak for spontaneous openings in the absence of ligand is observed (with a τ value ~ 170 μs), and at a low

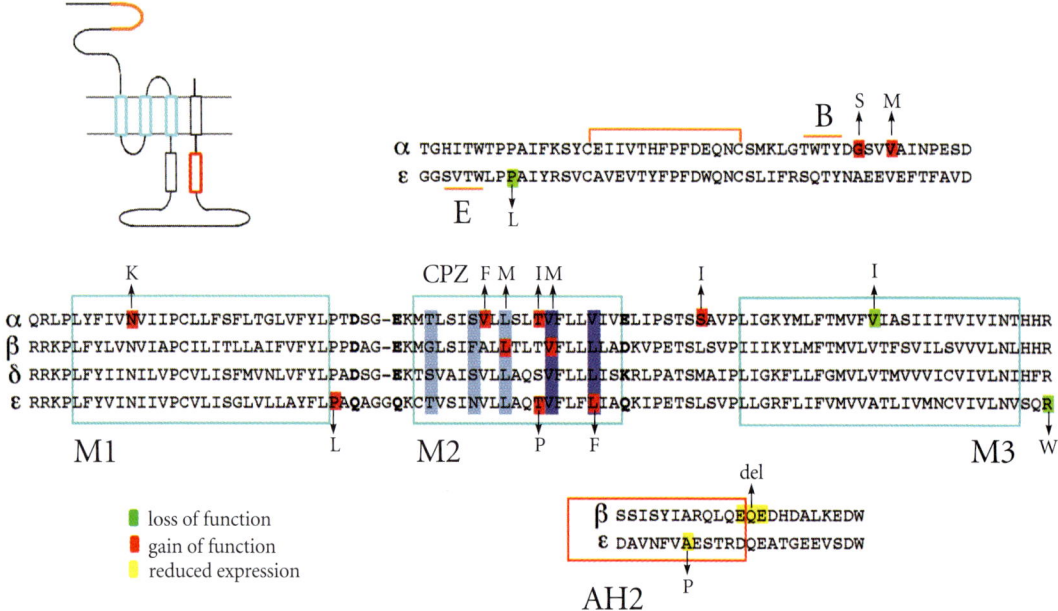

FIGURE 8.1 Mutations responsible for congenital myasthenic syndromes, based on the work of A. Engel, K. Ohno, and S. Sine (1995–2002). The relevant mutations are shown in the appropriate regions of the α-, β-, δ-, and ε-subunits: gain of function (red); loss of function (green); reduced expression (yellow). (Illustration courtesy of P.-J. Corringer.)

concentration of acetylcholine (0.3 μM) three distinct peaks are present (Figure 8.2D), with τ values of 150 μs, 1.8 ms, and 69.5 ms (Ohno et al., 1995). Since no mechanistic interpretations were provided for this mutant by the authors, the data were examined to test whether the allosteric model could provide an explanation of these exceptional properties. Simulations based on the data reported with εT264P were carried out and compared to wild-type human muscle receptors expressed in HEK cells (Edelstein et al., 1997a).

The simulations presented in Figure 8.2 closely fit the properties of εT264P (recorded at 0.3 μM acetylcholine), but differ strikingly from the normal adult human muscle nicotinic receptor. An adequate fit of the mutant data was achieved by modifying the interconversion rates between the basal (B) and open (A) states. The wild-type value of the B $\rightleftharpoons$ A interconversion equilibrium constant ($^{BA}L_0$) was lowered from 9×10^8 for normal receptors to a value of $^{BA}L_0 = 100$ for the εT264P mutant (Edelstein et al., 1997a). In this way the B → A transition was facilitated, yielding a substantial increase in the sensitivity to acetylcholine. For instance, at 0.3 μM acetylcholine the probability of opening increases to $P_{open} = 0.22$ for the εT264P mutant, compared to the wild-type value of $P_{open} = 10^{-4}$ at the same concentrations. As a consequence, three distinct peaks were predicted in the dwell-time profiles of opening events (Figure 8.2D), in contrast to a single peak for wild-type receptors (Figure 8.2C). For the latter, the allosteric model fits the data points (o) obtained from the kinetic constants reported for wild-type receptors expressed in HEK cells (Milone et al., 1997). The simulations demonstrate that for wild-type receptors the

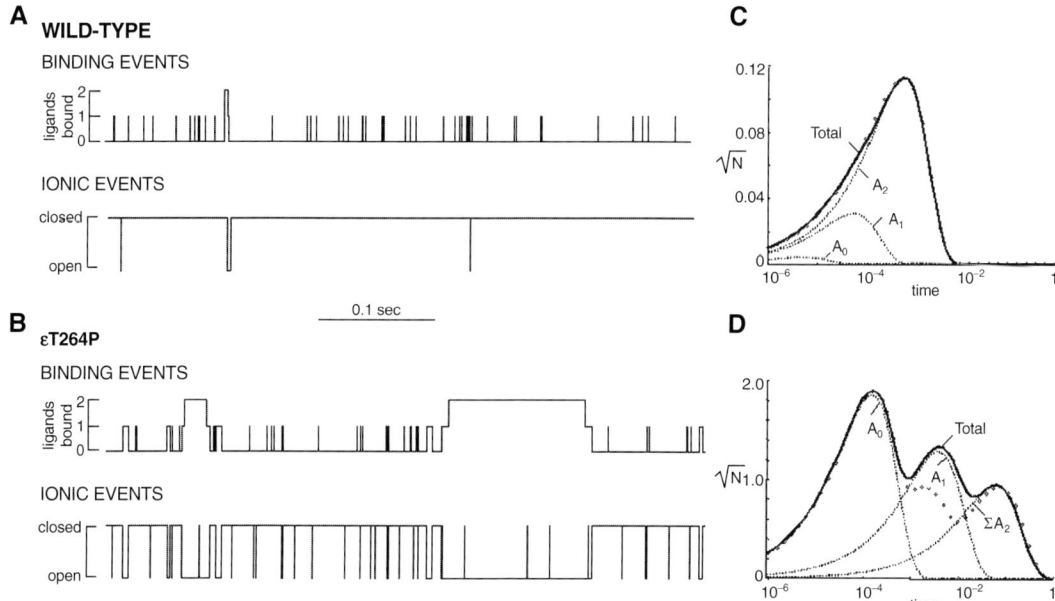

FIGURE 8.2 Allosteric interpretation of the physiological properties of a gain-of-function mutant receptor from a patient with congenital myasthenia. (A) Simulations of binding and ionic events for human wild-type muscle receptors expressed in HEK cells, based on the experimental observations of Ohno et al. (1997). (B) Simulations as in (A), but here for εT264P mutant muscle receptors expressed in HEK cells, based on the experimental observations of Sine et al. (1995c). The binding and ionic events do not coincide, and there are many spontaneous channel openings of unliganded receptors. (C) Dwell-time profiles predicted by the model and experimental data points for wild-type receptors. Individual points (open circles), obtained from the kinetic rate constants specified by Milone et al. (1997), represent activation over a wide range of acetylcholine concentrations. (D) Dwell-time profiles predicted by the concerted (MWC) model and experimental data points for εT264P mutant receptors. Individual points (open circles) are shown for the sum of the three components of the experimentally observed open-channel dwell times corresponding to the published values of Ohno et al. (1995) for 0.3 μM acetylcholine. The data for wild-type and mutant receptors were simulated with values of $^{BA}L_0$ of 9×10^8 and 100, respectively. (From Edelstein et al., 1997a.)

allosteric model can represent patch-clamp data as satisfactorily as sequential models; additionally, however, it predicts single channel openings in the absence of ligand. But since such openings occur in a rare (~1/15 s) and rapid (~5 μs) manner, they escape easy detection. Accordingly, the three peaks of the εT264P mutant receptors are interpreted to represent non-, mono-, and biliganded receptors, respectively. Sequential models, which do not include an open state for nonliganded receptors, cannot adequately represent such data (see Chapter 6).

In conclusion, the simulations and comparison to experimental data based on the MWC model illustrate the key role predicted for the value of $^{BA}L_0$, the conformational equilibrium constant established prior to ligand binding, in determining the relation between binding and ionic events. The altered properties of εT264P receptors thus represent an L phenotype of the sort described in Chapter 5. This interpretation may hold for most slow-channel gain-of-function phenotypes, although alternative explanations—for example, an increase in the intrinsic affinity of the A state for agonist—cannot be excluded on the basis of the existing data for a few of the myasthenic mutants.

Epilepsy

The role of high-affinity neuronal α4β2 nicotinic receptors in brain functions (see Chapter 12) is illustrated by mutations causing a rare form of autosomal dominant congenital epilepsy (Figure 8.3) that affects the frontal lobe and creates crises during sleep that constitute a condition known as autosomal dominant nocturnal frontal lobe epilepsy (ADNFLE) (see reviews by Steinlein, 2000; Bertrand et al., 2002; Raggenbass and Bertrand, 2002). Affected individuals in the first family to be identified (Steinlein et al., 1995; Weiland et al., 1996) were found to carry a mutation in the α4-gene of Ser-248 to Phe, at the precise homologous site of the chlorpromazine-labeled Ser in *Torpedo* receptor M2 (Giraudat et al., 1986; Hucho et al., 1986). This alteration causes a significant enhancement of desensitization and a reduction of maximal response at saturating acetylcholine concentrations. Interpretation of these observations is complicated, however, by the fact that the mutant α4-subunits expressed in oocytes (in the presence of β2) also lead to reduced channel conductances and lifetimes compared to wild-type α4β2-receptors (Kuryatov et al., 1997), an effect described previously when the same mutation was introduced into the muscle receptor α-subunit (Forman et al., 1996). Moreover, substantial reduction of calcium permeability has been reported in the same system for mutant receptors, though the effect is eliminated by the addition of α5-subunits (Kuryatov et al., 1997), which constitute a physiologically significant component of receptors containing α4-subunits (Ramirez-Latorre et al., 1996).

Exactly how the functional changes observed in the mutant α4-subunits expressed in oocytes or cultured cells cause the pathological symptoms is not yet fully understood. The mechanism remains unclear because, in heterozygous patients, the mutated subunit is incorporated in nicotinic receptor oligomers whose subunit composition is unknown. Moreover, in reconstituted receptors, the mutations elicit a broad diversity of functional alterations, as we saw earlier, including kinetics, extent of desensitization, and recovery. No functional models that adequately integrate these findings have so far been proposed.

Distinct mutations responsible for ADNFLE in other families (Figure 8.3), though they have been less extensively studied, display a num-

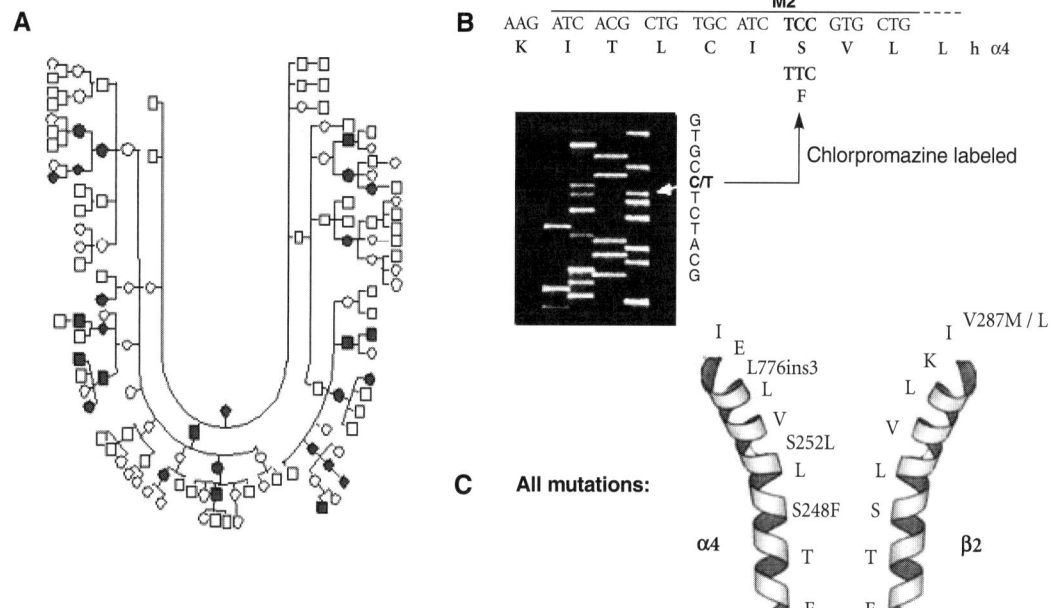

ber of other interesting features. In the second mutation to be discovered, insertion of an additional Leu at the C-terminal end of M2 caused an increase in affinity associated with a lower calcium permeability (Steinlein et al., 1997). V287L, a β2-missense mutation in M2 that leads to a "hyperactive" phenotype manifested by slower desensitization, has also been implicated in ADNFLE (Fusco et al., 2000). In addition, the higher affinity for acetylcholine shown by the V287M mutation in β2 is reminiscent of the Leu insertion mutation (Phillips et al., 2001).

One hypothesis capable of accounting for all these observations might be that the mutations in neuronal nicotinic receptors associated with epilepsy produce somewhat increased sensitivity to acetylcholine without causing significant variations in the maximal acetylcholine-evoked currents (Raggenbass and Bertrand, 2002; Steinlein, 2004). Such a gain of function would provoke the hyperexcitability characteristic of epilepsy if these receptors were, for example, part of a cortical excitatory circuit involved in the presynaptic release of glutamate, since this release could lead indirectly to excessive synaptic activity by *reducing* inhibitory effects. However, other properties of the α4-subunit S248P mutation responsible for ADNFLE have been reported, involving faster desensitization, slower recovery from desensitization, and reduced Ca^{2+} permeability (Kuryatov et al., 1997), which suggests that loss of function could also trigger epilepsy, possibly via receptors that are part of an inhibitory circuit involving GABAergic neurons (Léna et al., 1993). As in the case of the neuromuscular junction, both gain and loss of function of neuronal nicotinic receptors can lead to pathological states of the brain in relation to epilepsy.

FIGURE 8.3 Human autosomal dominant nocturnal frontal lobe epilepsy (ADNFLE) is caused by gain-of-function allosteric mutations in the genes of the high-affinity brain α4β2 nicotinic receptor. (A) Pedigree of a family affected by the first observed mutation in the α4-subunit at the homologue locus of the particular amino acid (serine 262) found labeled by chlorpromazine in *Torpedo* receptor δ-subunit. (B) Details of the base change and amino acid substitution observed in the M2 transmembrane helix for the first ADNFLE mutation. (C) Summary of the five known ADNFLE mutations, three in α4, two in β2. (From Bertrand et al., 2002.)

Related Mutations in Other Receptor Families

Additional insight into these nicotinic receptor pathologies may be obtained from the study of pathologies observed in other families of ligand-gated channels. Rasmussen's encephalitis, a rare autoimmunological syndrome that shares certain features of myasthenia gravis, has been reported in the glutamate receptor family, specifically in connection with GluR3 receptors (Rogers et al., 1994; He et al., 1998; Paas, 1998).

In different families of individuals having a high incidence of generalized epilepsy marked by febrile seizures, two distinct mutations have also been reported in the γ2-subunit of the $GABA_A$ receptor, R43Q (Wallace et al., 2001) and K289M (Baulac et al., 2001). The R43Q mutation occurs in the extracellular N-terminal domain at the benzodiazepine-binding site. Since benzodiazepines are positive effectors of these receptors, used to treat related forms of epilepsy, finding a mutation at a site that eliminates sensitivity to benzodiazepine suggests that a naturally occurring effector may modulate activity through interactions at this site. But this is not the only possible interpretation. Indeed, an earlier hypothesis proposed that the benzodiazepine site is homologous to the agonist-binding site at the interface between other pairs of subunits (Galzi and Changeux, 1994). A mutation at this level may therefore cause allosteric interactions that result in decreased sensitivity for GABA binding at the agonist site.

In addition to mutations in brain nicotinic receptors and $GABA_A$ receptors, loss-of-function mutations in voltage-gated Na^+ channel and K^+ channel genes, as well as gain-of-function mutations in voltage-gated Cl^- channel genes, have been reported to cause inherited epilepsies (see review by Steinlein, 2004).

Another pathology associated with the glycine receptor is the genetic disease hyperekplexia (also known as startle disease), due to a mutation in the Gly α1-gene. Six mutation sites have been identified so far, including several in M2. Afflicted individuals in different families all exhibit a loss of receptor function, in this case reduced sensitivity to glycine (Shiang et al., 1993; Schofield, 2001). Allosteric effects have also been proposed to account for startle mutations outside M2, in the M1-M2 and M2-M3 loops of GlyR α1-subunits (Schofield, 2001). One of them in particular, the K289M mutation, which results in markedly reduced GABA-evoked currents, is located in the M2-M3 loop at a position five residues beyond the mutated position in the Gly α1-receptor that was first identified in individuals with startle disease (Shiang et al., 1993).

With regard to pathologies of the nervous system in general, mutations in genes of sodium, potassium, calcium, and chloride voltage-gated channels have been implicated in over twenty genetic diseases

(Celesia, 2001). Several have gain-of-function phenotypes, as in the case of mutations in the sodium channel responsible for hyperkalemic periodic paralysis (Hayward et al., 1999). Another interesting mutation occurring in the glycine receptor substitutes valine for glycine directly in the channel and leads to a pathological phenotype triggered by a minor drop in temperature (McClatchey et al., 1992). Very few of these channelopathies have been subjected to detailed mechanistic analysis, however, and so no general interpretation is yet possible, whether in allosteric or other terms.

Toward an *à la Carte* Pharmacology of Allosteric Diseases

The therapies currently prescribed for myasthenia gravis aim at counteracting the inactivation of the endplate nicotinic receptor by autoimmune antibodies—for instance, by enhancing the local concentration of acetylcholine through the use of anticholinesterase agents. Mechanistic analyses of the various mutations causing congenital myasthenia (as well as other receptor diseases mentioned above) point to a diversity of receptor phenotypes resulting from the conformational versatility of the nicotinic receptor as an allosteric protein. The distinction between fast-channel and slow-channel mutations, which yield receptors that are respectively *less* or *over* active, leads to alternative courses of rational therapy. Nicotinic receptor inhibitors such as curare are excluded with patients suffering from myasthenia gravis as a consequence of fast-channel mutations. Reducing the excessive receptor activity associated with slow-channel mutations, on the other hand, might be a beneficial form of treatment (Engel et al., 2003b). Quinidine was identified as a long-lived open-channel blocker of wild-type acetylcholine receptor by Sieb et al. (1996). It has been shown to improve muscular endurance, as predicted, and partially to correct the decremental and repetitive compound muscle action potential of patients suffering from slow-channel congenital myasthenia (Harper and Engel, 1998). Treatment with another long-lived open-channel blocker, fluoxetine, was also reported to improve the condition of two patients (Engel et al., 2003b). These medications are only in the first stages of development. In addition to adjusting doses to minimize possible side effects, the concentrations of pharmacological agents administered in clinical practice will need to be adapted to the actual consequences of the particular mutation for the allosteric phenotype of the receptor: certain modifications of the isomerization constant L due to mutation, for example, may quantitatively affect the increase of affinity. Similar caution may be warranted in the case of genetic epilepsies, as well as of other inherited channelopathies that present the additional difficulty of heterooligomeric diversity with respect to brain receptors and channels (see Bertrand et al. [2002] in the case of ADNFLE).

These findings herald the advent of an *à la carte* pharmacology, as it might be called, that would be qualitatively and quantitatively tailored to the specific requirements of individual mutation phenotypes. Such observations clearly demonstrate how insights gained from the understanding of allosteric mechanisms can be translated into clinical practice (Engel et al., 2003b).

Conclusion and Outlook

A specific prediction of the theory of allosteric interactions is that mutations, although they may occur at locations distinct from the ligand-binding site and the biological active site, nonetheless play a pivotal role in the conformational transitions that link these sites. Mutations of the nicotinic receptor in human populations that cause myasthenic syndromes or epilepsy are located in various regions of the structure (Figures 8.1 and 8.4). In hemoglobin, for example, the "Kansas" mutation (βN102T) is located far from the heme pocket and confers a low oxygen affinity and a low Hill coefficient because the R state is destabilized, whereas other mutations such as "Chesapeake" (αR92L) display a high oxygen affinity and a low Hill coefficient because the T conformation is destabilized (Edelstein, 1971). Gain-of-function mutations may apply to the conversion of allosteric inhibi-

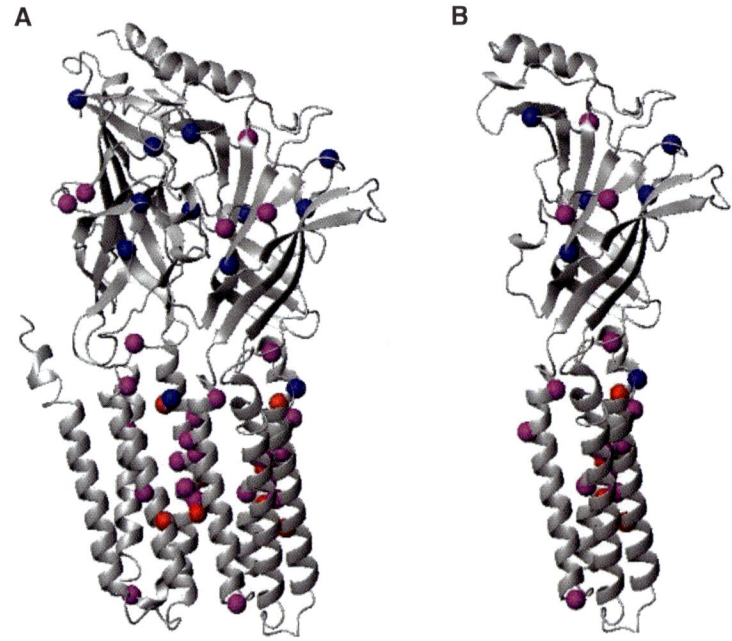

FIGURE 8.4 Location of myasthenic mutants. (A) Two receptor subunits are represented in a ribbon model with positions marked by spheres that correspond to the sites of natural mutations responsible for congenital myasthenic (magenta) or epileptic syndromes (red), as well as loss-of-function mutations (blue). (B) As in (A), but only the subunit on the right is represented. This subunit was arbitrarily divided into inner and outer blocks: most of the mutations are located at the interface between these two blocks or at the interface between subunits. The cysteine and M2-M3 loops, as well as the ion-conduction pathway, are also shown. (Personal communication from T. Grutter and A. Taly.)

tion to activation in regulatory enzymes, such as threonine deaminase (Sanchez and Changeux, 1966) or phosphofructokinase (Lau and Fersht, 1987).

In the case of membrane receptors, fresh insight was provided by analysis of site-directed mutations in the homopentameric α7 neuronal nicotinic receptor that produce dramatic pleiotropic alterations of functional properties, several of them associated with a gain of function (Chapter 6). Such mutations were initially discovered in the channel-lining residues of M2, but they have been found in several other domains as well, as we saw earlier in this chapter. Their mechanistic interpretation poses a serious challenge to KNF-type sequential models of channel activation. The MWC receptor model, on the other hand, provides simple and plausible explanations on the basis of changes in equilibrium between conformations, particularly in the context of an L phenotype (Galzi et al., 1996a). The study of such changes has introduced the concept of a distinctive pathology caused by alteration of the allosteric properties of receptors or ion channels, whose typical disorders may now rightly be referred to as "allosteric diseases" (Changeux and Edelstein, 1998). This interpretation has prompted in its turn a search for pharmacological treatments that can be specifically designed to suit the particular allosteric phenotypes caused by individual mutations. From now on, an understanding of the nature and function of allosteric receptors will be an essential part of standard clinical procedure.

CHAPTER 9

Genesis of the Postsynaptic Membrane by Targeted Receptor Gene Transcription at the Motor Endplate

Understanding the functional organization of muscle and brain nicotinic receptors is not enough to fully explain their role in intercellular communication. The receptor molecule is embedded within a cellular and subcellular context that permits it to operate as an efficient device for the polarized transfer of information between cells. The idea that specialized structures are responsible for establishing contact between motor neurons and skeletal muscle fibers was introduced in the mid-nineteenth century by the French anatomists M. Doyère (1840) and C. Rouget (1862). Soon afterward it was discovered that these structures involve the elaborate juxtaposition of a ramified motor axon terminal and muscle fiber (Kühne, 1862; Ranvier, 1875). As we noted in Chapter 1, electron microscopy of the motor endplate provided the first direct demonstration of Cajal's theory of the autonomy of the neuron, revealing that its two basic components—the membranes of the nerve and of the muscle—are closely aligned with one another, but separated by a "synaptic" cleft of approximately 50–100 nm (Palade and Palay, 1954; Robertson, 1956, 1960; Couteaux, 1972; Heuser and Salpeter, 1979).

On one side of the cleft lies the motor-axon terminal, which is filled with clear vesicles 50 nm in diameter that contain a neurotransmitter, acetylcholine, a few dense core vesicles, and probably also neuropeptides, such as calcitonin gene-related peptide (CGRP). Many of the synaptic vesicles are grouped in the vicinity of regions called "active zones," where they release their contents into the cleft in a highly polarized manner.

On the other side of the cleft, the postsynaptic membrane forms a complex "subneural apparatus" that results from the repeated folding of the membrane, as René Couteaux (1958) was the first to note, with a marked thickening visible opposite the active zones (Figures 1.1, 9.1). A dense electron layer, the basal lamina, covers the cytoplasmic membrane on the cleft side, while on the cytoplasmic side abundant bundles of filaments may be observed. The entire endplate is anchored to an intricate network of fibers and filaments and capped by processes of Schwann cells. Finally, the "granular and nucleated" appearance of the sarcoplasm underlying the terminal arborization is due to the accumulation of mitochondria and of between four and eight large nuclei characterized by clear chromatin and hypertrophied nucleoli. They were recognized to be distinct from the Schwann cell nuclei, and referred to as "fundamental" by Louis-Antoine Ranvier (1875). Couteaux (1978) prophetically likened their differentiation to the biosynthesis of proteins specific to the postsynaptic membrane, in particular the acetylcholine receptor (Figure 1.1).

This complex supramacromolecular architecture channels the transfer of signals in a unidirectional, robust, and reliable manner

Genesis of the Postsynaptic Membrane

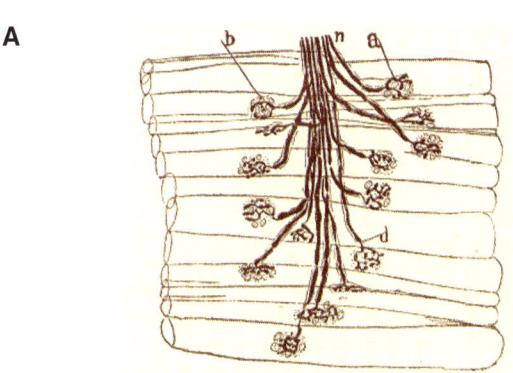
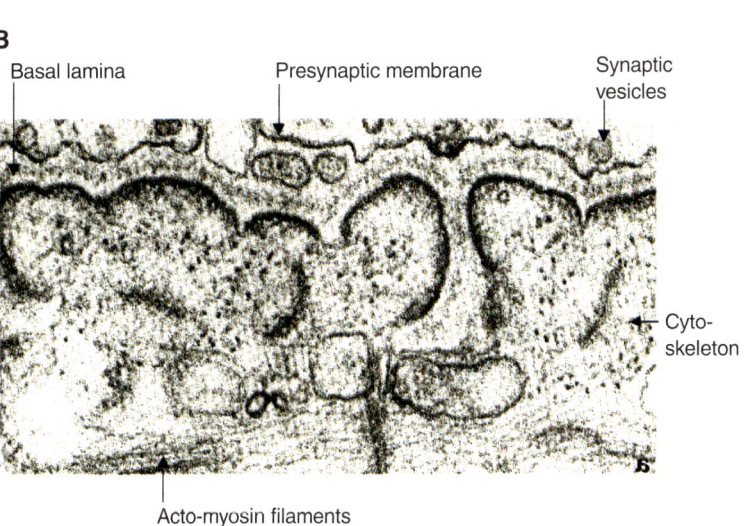

FIGURE 9.1 The neuromuscular junction. (A) Overview of skeletal muscle innervation showing that in fast skeletal muscle there is one motor endplate per muscle fiber. (From Ramón y Cajal, 1909–1911.) (B) Electron microscopy of the contact between nerve ending and muscle showing the synaptic cleft and the complex subneural apparatus, with cytoskeletal arrays and actomyosin contractile filaments below. (From Couteaux, 1978.)

from the nerve to the muscle. Invasion of the nerve ending by an action potential, as we saw in Chapter 1, releases about 300 quanta of acetylcholine, providing a transient rise in acetylcholine concentration to about 3 to 5 $\times$ 10^{-4} M over a background release of 10^{-8} M acetylcholine. Furthermore, as we shall see, because of the distinct distributions of acetylcholinesterase and receptor molecules, acetylcholine rapidly reaches the receptor molecules before being degraded.

Acetylcholine Receptors in the Postsynaptic Membrane

Early pharmacological experiments (Langley, 1905) and electrophysiological recordings (Axelsson, 1959; Miledi, 1960) indicated that in normal muscle the sensitivity to acetylcholine and nicotine is restricted to the endplate. Furthermore, the response to nicotinic agents persists after denervation, suggesting that the putative receptor for acetylcholine is exclusively located on the postsynaptic side of the endplate (the sensitivity to acetylcholine increases in the extrajunctional area as a consequence of denervation as well). Similarly, light microscopy autoradiography using snake venom toxins, which are known to be exquisitely specific labels of muscle nicotinic receptor (as noted in Chapter 1), shows a dense localization of α-toxin sites at the endplate level (Lee and Tseng, 1966; Sato et al., 1970; Barnard et al., 1971).

A preliminary evaluation of the absolute density of α-toxin sites per μm² of subsynaptic membrane was then performed by quantitative analysis of electron microscopic autoradiographs (Bourgeois et al., 1972, 1986) of *E. electricus* electroplaque (Figure 9.2). In this system, α-toxin sites are observed only on the innervated face of the electroplaque, which is to say the side that receives the nerve ending from the electromotor neuron of the electric lobe (Bourgeois et al., 1971). Moreover, the sites are distributed unevenly. They cluster under the nerve ending at densities as high as 30,000–50,000 sites/μm², whereas outside the junctional clusters this figure is reduced by a factor of at least 100. Assuming two toxin sites per α2βγδ-oligomer and a protein molecular weight per site of about 150,000, the receptor molecules are very closely packed. These values are consistent with the density of receptor rosettes (8–12,000/μm²) observed after freeze-etching or negative staining with receptor-rich membranes purified from the *Torpedo* electric organ (Cartaud et al., 1973; Nickel and Potter, 1973; see Chapter 3). They are also consistent with global estimates made at vertebrate motor endplates by optical microscopy (Barnard et al., 1971; Miledi and Potter, 1971; Berg et al., 1972).

High-resolution electron microscopy (EM) autoradiographic studies of the vertebrate neuromuscular junction subsequently revealed that the α-toxin sites are not evenly distributed throughout the postsy-

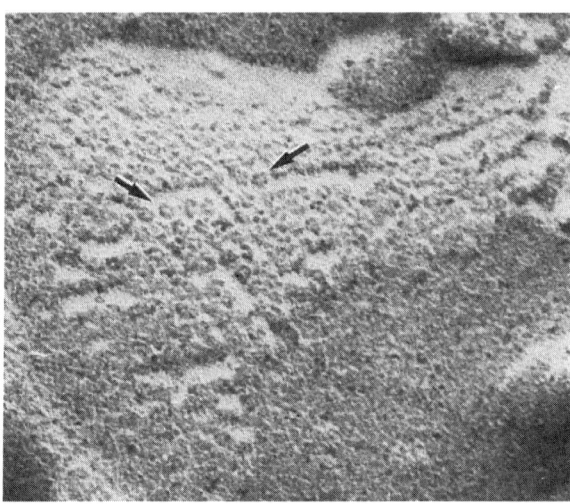

FIGURE 9.2 Clustering of nicotinic receptors in the subjunctional membrane. (A) Surface density of sites estimated to be 30,000–50,000 sites per μm^2 using snake ^3H-α-toxin on *Electrophorus electricus* membranes. (From Bourgeois et al., 1972.) (B) In subsynaptic membranes from *Torpedo marmorata* some 10,000–20,000 molecules per μm^2 are estimated, a figure that is consistent with the number of sites obtained by assuming two α-toxin sites per receptor molecule. (From Cartaud et al., 1973.)

naptic membrane. They accumulate at the tip of the fold of Couteaux's subneural apparatus, facing the active zones of the nerve terminals (Fertuck and Salpeter, 1976; Porter and Barnard, 1976). Acetylcholinesterase, on the other hand, is located mostly in the basal lamina (Barnard et al., 1971).

At the time these experiments were performed, in the 1970s, biological membranes were commonly viewed as fluid lipid mosaics, and their densely packed postsynaptic receptors as being somehow "glued together" by the nerve ending. Receptor molecules were therefore expected to disperse within hours after denervation, since denervation was assumed to remove the "glue" (McConnell et al., 1972). To test this hypothesis, an *E. electricus* electric organ was denervated and its individual electroplaques were examined for a period of 52 days. No sign of lateral diffusion of the toxin sites was noted (Bourgeois et al., 1973). The remarkable stability of receptor localization in the subjunctional membrane was subsequently confirmed for the neuromuscular junction as well (Frank et al., 1975; see Akaabounc et al., 2002). As we shall see later in this chapter, the accumulation of receptors cannot be viewed as a simple presynaptic gluing mechanism. It requires instead a quite elaborate combination of targeted biosynthesis and dynamic supra-macromolecular interactions throughout the postsynaptic domain.

Acetylcholine Receptor Biosynthesis during Endplate Formation

The intricate architecture of the postsynaptic membrane results from a long and complex series of molecular processes lasting several weeks that includes a cascade of regulatory gene transcriptions followed by multiple posttranscriptional phenomena. In this chapter we will review the initial stages of acetylcholine receptor biosynthesis and the transcriptional regulation of its muscle subunit genes, and in Chapter 10 examine the posttranscriptional assembly of the various components of the postsynaptic membrane.

The biosynthesis of the nicotinic receptor begins very early in muscle development: high levels of α-bungarotoxin-binding sites are already detected in myotomal cells from four-day-old chicken embryos (Meiniel and Bourgeois, 1982). Subsequently, but prior to the arrival of the exploratory motor axons, the fusion of myoblasts into myotubes is accompanied by the burst appearance of receptor molecules, scattered diffusely along the myotubes. These molecules exhibit significant lateral motion (Axelrod et al., 1976), undergo rapid turnover (having a metabolic half-life of 17–22 hours: see Berg and Hall, 1975; Chang and Huang, 1975; Devreotes and Fambrough, 1976), and possess a mean channel-open time of 3–10 msec (Katz and Miledi, 1972; Neher and Sakmann, 1976). By the tenth day in the embryonic chick, acetylcholine receptors are present along the entire length of the muscle fibers, though already their highest concentration is at the neuromuscular junction (Burden, 1977a). In the adult muscle, acetylcholine receptors are densely clustered at the endplate. Yet they appear to be in a stationary steady-state distribution marked by considerable receptor diffusion (Choquet and Triller, 2003), display slow turnover (having a half-life of 10 days or more), exhibit a mean channel-open time (τ) three to five times shorter than the embryonic receptor, and possess an intrinsic conductance (γ) significantly larger than the value obtained for embryonic receptors (Neher and Sakmann, 1976; Fischbach and Schuetze, 1980). The exact values and time intervals over which these changes take place vary in different muscles and in different species (in the chick, for instance, τ and γ do not change significantly throughout the course of development). Additionally, the number of receptor molecules incorporated in the subsynaptic membrane steadily increases during development (Figure 9.3), particularly after birth (Salpeter and Loring, 1985).

The initial increase in the number of acetylcholine receptors, which coincides with the fusion of myoblasts into myotubes, can be reproduced *in vitro* with cultured myoblasts. The laboratory result corresponds to *de novo* synthesis of receptor molecules, as demonstrated by the incorporation of radioactive-labeled (Merlie et al., 1975, 1978) or heavy isotope-labeled (Devreotes and Fambrough, 1975; Devreotes et al., 1977) amino acids into the receptor protein, without any significant observed change in its degradation rate (Merlie et al., 1976). Nor did chronic injection of neuromuscular blocking agents *in ovo* significantly affect this initial onset of acetylcholine receptor biosynthesis (Burden, 1977a, 1977b; Betz et al., 1980), which is thus independent of the state of activity of the muscle (Fambrough, 1979; Changeux, 1981).

A second phase in the evolution of acetylcholine receptor metabolism corresponds to the decrease of total receptor content that is observed, for example, after day 15 in chick embryo breast muscle (Betz

Genesis of the Postsynaptic Membrane

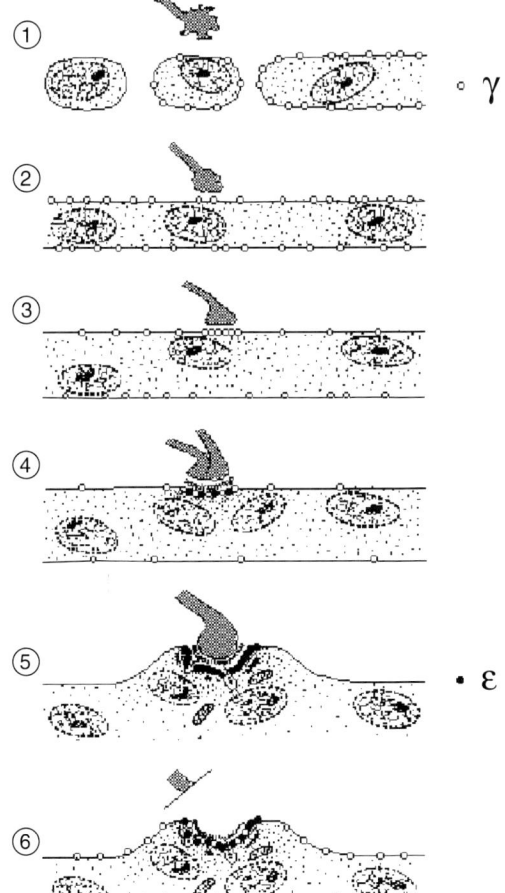

FIGURE 9.3 Schematic representation of the development of the motor endplate. This drawing shows the distribution of the acetylcholine receptor protein at successive stages of embryonic development. The two different isoforms of acetylcholine receptor found in mammals are represented by open circles (for the embryonic isoform $\alpha_2\beta\gamma\delta$) and by solid circles (for the adult isoform $\alpha_2\beta\epsilon\delta$). The distribution of the receptor protein during development in birds is qualitatively similar, except that only one isoform is present. Six stages may be distinguished: (1) the protein is expressed on committed myoblasts before fusion into myotubes; (2) the receptors increase in number and become distributed over the whole length of the myotube; (3) the motor nerve terminal comes into contact with the myotube and receptors then become clustered beneath it (after about 13–14 embryonic days in the mouse); in extrajunctional regions it is eliminated as a consequence of the onset of the neurally evoked electrical activity of the muscle; (4) around the time of birth (in the rat) several axons converge upon a unique innervation site; the metabolic lifetime of the protein increases and the adult isoform subsequently appears; (5) the adult junction is established with synaptic folds and basal lamina; the receptors persist exclusively at the junction; (6) after denervation the embryonic isoform of the receptor reappears in extrajunctional regions. (Adapted from Laufer and Changeux, 1989, and Duclert and Changeux, 1995.)

et al., 1977; Burden, 1977a, 1977b). This decline was thought to result either from enhanced degradation of the receptor protein or from repression of its biosynthesis. The metabolic degradation rate of the receptor protein throughout this period did not change, however (Betz et al., 1977, 1980; Burden, 1977a). It is therefore a *repression* of receptor biosynthesis that takes place. Parallel measurements of receptor surface density in nonjunctional areas of the muscle fiber show that the decrease of total receptor content coincides with the elimination of the acetylcholine receptor from these areas and the persistence of clustered receptors at the endplate (Betz et al., 1980). By the end of the 1970s it had become clear that the genesis of the endplate postsynaptic membrane engages patterns of gene expression that differ in the junctional and nonjunctional territories of the muscle fiber (Changeux, 1981; Salpeter and Loring, 1985; Merlie and Smith, 1986; Changeux et al., 1987a).

Spontaneous movements appear in the early stages of embryonic development (after 3.5 days in the chick embryo) that are neurogenic in character (Hamburger, 1970; Harris, 1981). To assess the effect of such neurally evoked electrical activity of the muscle fiber in repressing receptor biosynthesis, chick embryos were chronically paralyzed by injections of botulinum toxin (Giacobini-Robecchi et al., 1975), *d*-tubocurarine (Burden, 1977a, 1977b), or flaxedil (Bourgeois et al., 1978a; Betz et al., 1980). Chronic paralysis was found to maintain a high level of receptor content without altering its degradation rate. Neurally evoked activity in the embryonic muscle thus accounts for the repression of nonjunctional receptor biosynthesis. A similar activity-dependent form of regulation can be demonstrated with myotubes in culture, which exhibit spontaneous (nonneurogenic) electrical activity (Shainberg and Burstein, 1976; Betz and Changeux, 1979).

Chronic paralysis of chick embryos also caused the disappearance of acetylcholinesterase at the endplate (Giacobini et al., 1973; Betz et al., 1980), without reducing the incidence of subsynaptic receptor clusters (Betz et al., 1980; Lomo and Slater, 1980; review in Massoulie and Bon, 1982). All these studies revealed that distinct regulatory mechanisms control the expression of nicotinic receptor genes in junctional as opposed to nonjunctional areas. A *compartmentalization* of gene expression takes place during the genesis of the endplate, to which different sets of neural factors and activity-dependent processes contribute (Changeux et al., 1987a).

Compartmentalized Transcription of Genes to the Endplate

In the early 1980s, the cloning and sequencing of DNA and genes coding for the several subunits of the acetylcholine receptor from *Torpedo*, as well as from human and other vertebrate species (see Chapter 3 for details and references), supplied new tools to investigate the regulation of acetylcholine receptor gene expression. In some species, such as calves, rodents, and humans (though not chicks), a fifth subunit referred to as ε was found to replace the γ-subunit in the adult endplate receptor protein (Takai et al., 1985; Mishina et al., 1986; Witzemann et al., 1987; Brenner et al., 1990). More generally, increased understanding of the molecular biology of gene expression, and in particular of the initial conversion of DNA to messenger RNA (transcription), led to the proposal of a simple model of compartmentalized transcription (Figure 9.4). First presented at the 1985 Dahlem conference on the cellular and molecular basis of learning, this model inaugurated what may be called the "promoter approach" (Changeux et al., 1987b; Kerszberg and Changeux, 1993).

This approach has three main parts. First, in the course of embryonic development, two main classes of regulatory mechanisms at the gene level are distinguished: the *determination*, or commitment, of genes for a given cell category (e.g., skeletal muscle), which involves a local transition of *chromatin* conformation, switching a particular gene from a buried "silent" state to a "ready-to-be transcribed" state; and the actual transcription of a defined set of such "open" genes. In other words, the subjunctional "fundamental" nuclei are supposed to transcribe distinct patterns of genes by comparison with their nonjunctional counterparts.

Second, the differential transcription of acetylcholine receptor genes is asserted to be subject to the control of *at least* four distinct signaling mechanisms: "extracellular" first messengers (i.e., neural factors, as opposed to electrical activity); "intracellular" second messengers; *trans*-acting regulatory proteins that bind to specific DNA

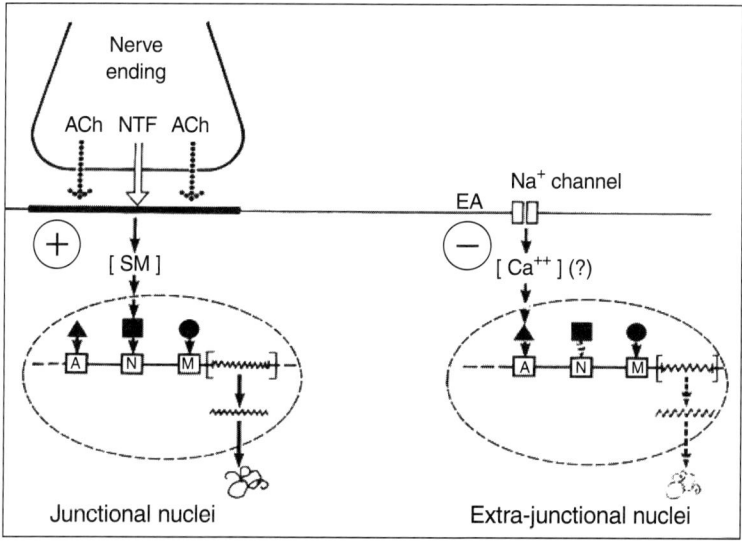

FIGURE 9.4 The promoter model, initially proposed to account for the compartmentalized transcription of genes at the endplate. Neurotrophic factors (NTF) enhance transcription in junctional nuclei, whereas electrical activity (EA) represses transcription in extrajunctional nuclei. Second messengers (SM) help to link cell membrane receptors and ion channels with specific sets of transcription factors, which bind to gene promoters. Different DNA elements are postulated to underlie transcriptional mechanisms in junctional and extrajunctional nuclei. (From Changeux et al., 1987b.)

regulatory sequences; and *cis*-acting DNA regulatory sequences. If the operation of these mechanisms does in fact differ in subneural and in nonjunctional areas, then in principle the mechanisms can be identified experimentally (Figures 9.1 and 9.3).

Third, the sharp boundary of gene transcription observed between junctional and extrajunctional areas may plausibly be supposed to be a consequence of autocatalytic switching, which would imply, for instance, that transcription factors exert a positive effect on their own formation.

Evidence in favor of this schema came first from the pioneering observations of Merlie and Sanes (1985), who cut samples of adult rat diaphragm into endplate and nonendplate sections and found, by using Northern blot hybridization, that the steady-state levels of α- and β-subunit mRNA molecules were 4 and 14 times higher, respectively, in synaptic regions than nonsynaptic regions. They suggested that this distribution could arise as a result either of directed transport of mRNA near the endplates or of increased transcription of acetylcholine receptor subunit genes by subsynaptic nuclei.

The development of the novel technique of *in situ* hybridization to localize mature spliced mRNA and its precursor unspliced transcripts (with intronic probes) offered an exceptional opportunity to test these alternative explanations in developing chick muscle (Fontaine et al., 1988; Fontaine and Changeux, 1989), as shown in Figure 9.5. First, in 15-day-old chicks, α-subunit mRNA was detected *in situ* in discrete domains, marked off by *sharp boundaries,* 80% of which colocalize to motor endplates identified by histochemical staining for acetylcholinesterase. Moreover, at this level, autoradiographic grains accumulate on and around the fundamental nuclei. Transcription levels, initially high in young embryos, decrease during the course of embryonic development. By day 11 both mature and precursor mRNAs can be detected over the entire developing muscle fiber. At day 16 the total number of grains distributed along the muscle fiber decreases, but several clusters persist. At day 19 only one grain cluster is observed per muscle fiber. The distribution of mature nicotinic receptor α-subunit mRNA becomes restricted to the newly formed endplate (Fontaine and Changeux, 1989).

A considerable body of work on adult innervated muscle since the early studies by Langley (1905) has demonstrated that denervation increases sensitivity to iontophoretically applied acetylcholine in extrajunctional areas (Axelsson and Thesleff, 1959; Miledi, 1960; Bennett et al., 1973; Salpeter and Loring, 1985; see Figure 9.3). It was therefore important to use *in situ* hybridization to test whether or not the nuclei that appear silent in adult extrajunctional areas are reactivated by denervation (Fontaine et al., 1988). Indeed, four days after denerva-

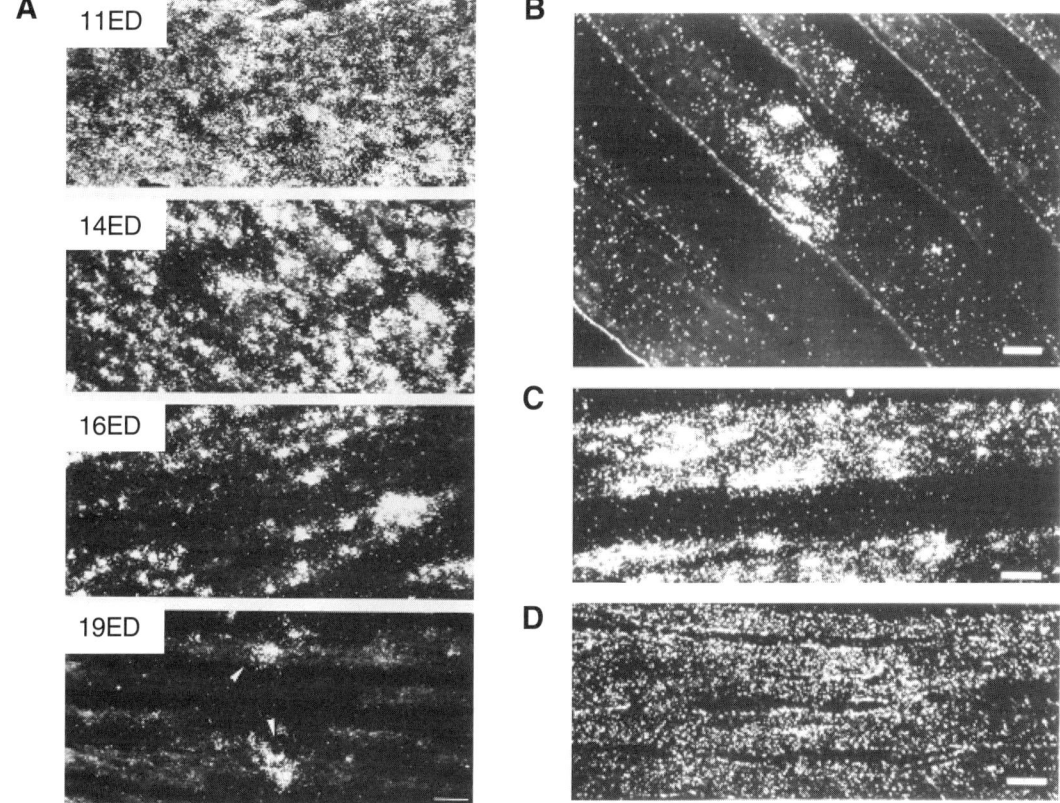

tion of 15-day-old chicks, grain clusters were seen to appear—about 10% of the number that would have been expected if the nuclei were distributed at random all along the denervated muscle fiber. A similar pattern was observed using a strictly intronic probe (Fontaine and Changeux, 1989), lending support to the notion that the transcriptional activity (see Shieh et al., 1987; Tsay and Schmidt, 1989) of individual nuclei varies under these conditions. In contrast, actin mRNA was found to be diffusely distributed all over muscle fibers during development, with no significant change noticed after denervation (Fontaine et al., 1988; Fontaine and Changeux, 1989).

These *in situ* results have been extended to observations on changes in the mRNA levels of the other nicotinic receptor subunits in the rat (Goldman and Staple, 1989; Brenner et al., 1990) that include timing differences among the various subunit mRNAs. Nuclei in different states of nicotinic receptor gene expression were found to be present within the cytoplasm of the muscle fiber, with a higher transcript accumulation observed under the nerve endings than outside the endplate. Furthermore, a "virtual" boundary of gene expression was found to separate junctional from nonjunctional nuclei. Interestingly, such discrete labeling is also observed in cultured chick myotubes, albeit in

FIGURE 9.5 Receptor synthesis during the course of endplate formation examined by *in situ* hybridization of α-subunit mRNA in the chick. (A) *In situ* hybridization during muscle development at days 11, 14, 16, and 19 of embryonic development. (B–D) Effects of denervation in a 15-day-old "young adult" chick on nicotinic receptor α-subunit mRNA. (B) Before denervation: mRNA is localized exclusively at the endplate. (C) After denervation: increase of mRNA and enhanced transcription visible in extrajunctional areas. (D) No change in actin mRNA observed following denervation. (All panels from Fontaine et al., 1988.)

the absence of innervation (Bursztajn et al., 1989; Fontaine and Changeux, 1989; Harris et al., 1989).

The spontaneous mosaic-like patterns detected in nuclei that express α-subunit mRNA in an all-or-none manner suggests that an internal switching mechanism of gene transcription is already at work in individual muscle nuclei. By virtue of being "preset" in the muscle fiber nuclei, these intrinsic all-or-none mechanisms of gene expression are ready to switch on in the actively transcribing phenotype under the nerve ending and off outside the motor endplate (Fontaine et al., 1988).

Identification of the N Box

The experimental methodology associated with the promoter approach, a critical aspect of the proposed model of compartmentalized transcription (Changeux et al., 1987a), does not proceed by first elucidating the signal transduction pathway from the receptor to the gene. It begins instead by identifying the genetic regulatory elements that control transcription in *cis,* and then goes on to characterize the transcription factors that bind to these elements in order, finally, to establish their position in the signal transduction pathway. The first step was to isolate genomic clones encoding the chicken α-subunit (Klarsfeld and Changeux, 1985; Klarsfeld et al., 1987) and to identify the 5'-upstream regulatory region (Klarsfeld et al., 1987). An 850-base pair (bp) sequence from this region was found to supervise transcription of an adjacent reporter gene, chloramphenicol acetyltransferase (CAT), in myotube cultures (Figure 9.6). This sequence contains TATA and CAAT boxes, and an Spl recognition site that shares homology with the transcriptional regulatory origin of simian virus 40 (SV40), together with other characteristic elements discussed below. Other research teams subsequently identified additional promoters from different nicotinic receptor genes exhibiting variability in the preserved distribution of their constituent elements (review in Duclert and Changeux, 1995).

In the first *in vivo* studies of the chicken α-subunit promoter, Merlie and Kornhauser (1989) introduced the CAT fusion gene constructed by Klarsfeld et al. (1987) into the mouse genome. In four independently derived lines of mice, the expression of the CAT gene was found to be specific for skeletal muscle. Moreover, an approximately 100-fold decrease in CAT steady-state levels took place over the first two to three weeks after birth—an effect reversed by denervation. Thus, in the mouse, the 842 base pairs of the chick α-subunit promoter (Klarsfeld et al., 1987) confer upon the CAT gene a capacity for activity-

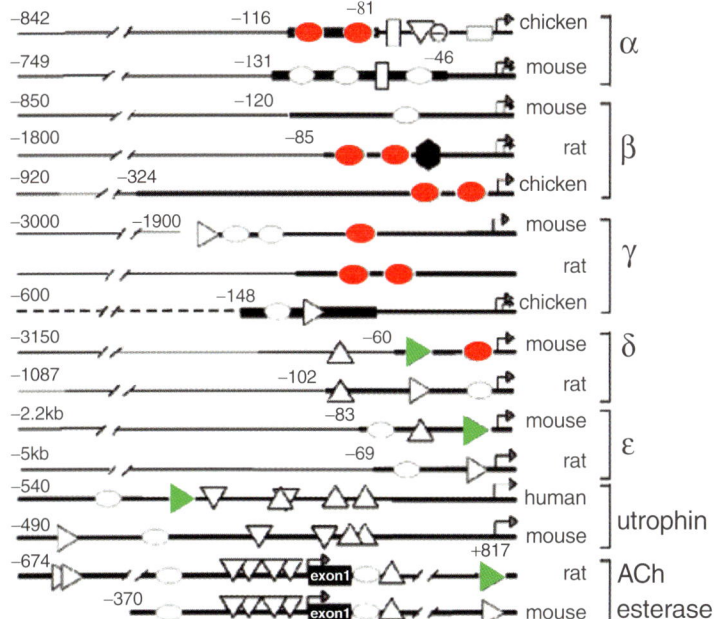
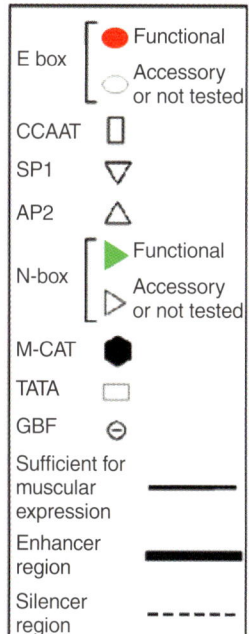

FIGURE 9.6 Promoter sequences of nicotinic receptor genes encoding α-, β-, γ-, δ-, ε-subunits and endplate-specific genes for utrophin and acetylcholinesterase. Note the presence of both the E box and the N box, which play a critical role in the regulation of transcription by electrical activity and neurotrophic factors. (From Klarsfeld et al., 1987, courtesy of American Society for Microbiology; and Duclert and Changeux, 1995.)

dependent repression of extrajunctional nicotinic receptors. But since CAT is a soluble cytoplasmic enzyme, its compartmentalized expression at the endplate could not be followed.

Transgenic mice were therefore generated using DNA constructs in which the 842-bp promoter was fused to the β-galactosidase (β-Gal) gene, which in turn was linked to a nuclear localization signal (nls) that labeled the nuclei that express the gene. Expression of β-galactosidase was first observed at day 9.5 of gestation in the somites of the embryo and subsequently in skeletal muscle. At birth, a line of stained nuclei can be discerned that coincides with the endplate alignment revealed by acetylcholinesterase detection. The staining intensity of the endplate nuclei and the contrast between junctional and extrajunctional nuclei become more striking over the first three days after birth, but the endplate staining subsequently fades (Klarsfeld et al., 1991). Moreover, denervation of adult mouse muscles produces a transient reexpression of the β-Gal gene (Salmon and Changeux, 1992). These experiments demonstrate that the chick 842-bp promoter contains DNA elements that in the mouse suffice to trigger the compartmentalized expression of the β-Gal reporter gene during the early stages of endplate morphogenesis. Similar results were obtained with mouse ε- or δ-subunit gene promoters (Simon et al., 1992; Tang et al., 1994).

The next step involved the identification of the specific DNA elements in the subunit gene promoter that target transcription to the motor endplate domain. The transgenic mice approach pioneered by Merlie and Kornhauser (1989) and Klarsfeld et al. (1991) might have been an attractive option if it did not place such heavy demands on animal facilities.

A rapid, inexpensive, and reliable method of *in vivo* promoter mapping therefore had to be devised (Figure 9.7). The assay for quantitative mapping of promoter elements (Duclert et al., 1993) consisted of directly injecting into muscle naked DNA constructs carrying the nls lacZ reporter gene, which was fused to the promoter fragments being investigated. Expression of such a construct yields a "blue spot" that either coincides or does not coincide with an endplate stained for acetylcholinesterase, corresponding respectively to synaptic or nonsynaptic events. A construct containing only 83 bp upstream of the transcription start site of the ε-subunit gene promoter, for example, was found to confer preferential synaptic expression by comparison with other promoters such as that of the muscle creatine kinase gene. Systematic 3′ and 5′ deletions from the δ-subunit promoter resulted

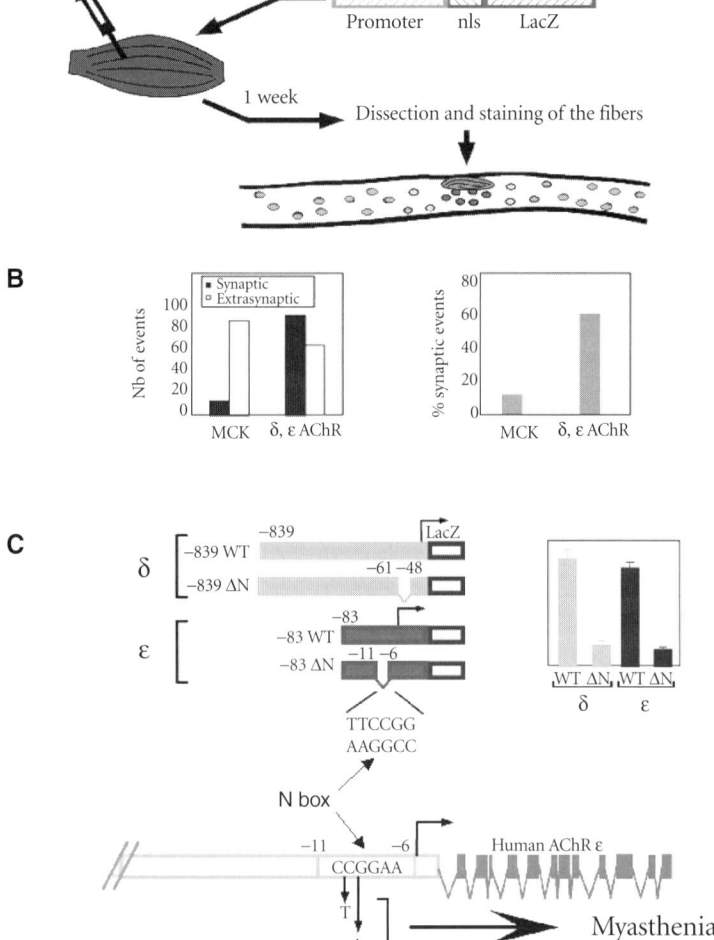

FIGURE 9.7 Promoter mapping by DNA injection into adult muscle. The plasmid DNA fuses a given promoter segment with a nuclear localization signal that targets the expression of the reporter gene lacZ to the muscle nuclei. (A) Injection of the naked plasmid into the muscle is followed by expression after one week *in vivo,* dissection, and differential staining of the nuclei (β-galactosidase) and the endplate (acetylcholinesterase). In this particular case the staining for each of the two enzymes coincides: such an "event" is said to be "synaptic." (B) The δ- and ε-promoters increase the number of synaptic (black) versus nonsynaptic (white) events as compared with the muscle creatine kinase (MCK) promoter. (C) Deletion of the N box TTCCGG abolishes synaptic targeting. *Bottom line:* mutations within the N box of the ε-promoter gene cause congenital myasthenia. *Control promoter:* muscle creatine kinase; *test promoter:* δ- and ε-subunits of nicotinic receptor. (From Duclert et al., 1993; and Schaeffer et al., 1998, 2001.)

in the identification of a single six-base pair element TTCCGG named the N box (Koike et al., 1995), which targets transcription to the endplate (Figure 9.7). This element is also present in the ε-subunit promoter (Duclert et al., 1996). When fused to a minimal promoter, the N box specifically increased the synaptic expression of a reporter gene. The extent of the expression specificity was nonetheless lower in this case than in the context of δ- or ε-subunit promoters of the acetylcholine receptor, pointing to the likelihood of cooperativity with other promoter sites or transcription factors (Koike et al., 1995).

The importance of the N box has been confirmed for transgenic mice, in which mutations in the N box of the acetylcholine receptor δ-subunit gene promoter completely abolished the expression of a reporter gene in adult innervated muscle. In these mice, expression of the transgene in denervated muscle was not affected, suggesting that although the N box is necessary for synaptic expression it is not involved in the repression of synaptic-gene expression by neurally evoked electrical activity (Fromm and Burden, 1998). Moreover, among families with congenital myasthenia, naturally occurring mutations in the N box from the ε-subunit gene ε154G → A, ε155G → A, and ε156C → T (review in Engel et al., 2003) have been shown to cause decreased levels of ε-subunit and mild muscle paralysis (Figure 9.7).

The control of transcription by the N box is not limited to the δ- and ε-subunit genes. The utrophin gene promoter contains an N box that is essential for the synaptic expression of utrophin (Gramolini et al., 1999; Khurana et al., 1999). An N box also occurs in the MuSK promoter (Lacazette et al., 2003). Furthermore, the acetylcholine esterase gene promoter contains four N boxes. One of these, located in the first intron, is crucial for the synaptic expression of acetylcholinesterase (Chan et al., 1999). The N box is therefore a critical element in targeting the transcription of several major synaptic genes to the motor endplate.

Identification of GABP Transcription Factors

The first attempts to identify the transcription factor that specifically interacts with the N-box sequence were made with extracts of adult mouse muscle or cultured myotubes (Koike et al., 1995). A factor purified on a DNA affinity column was shown to be a heterodimer composed of 58- and 43-kilodalton (kDa) polypeptides corresponding to the Ets-related transcription factor referred to as GABP (Schaeffer et al., 1998; Khurana et al., 1999). The 58-kDa GABP α-subunit binds DNA through its Ets domain. The 43-kDa GABP β-subunit, though it does not itself bind DNA, contains ankyrin repeat sequences that mediate its interaction with GABP α, thus strengthening the interaction of the α-subunit with DNA (Batchelor et al., 1998).

The binding of the GABP α/β1 complex to the N box results in transcriptional activation by the acetylcholine receptor δ- and ε-subunit promoters. Indeed, antisense oligonucleotides directed toward GABP β1 as well as dominant-negative mutants of GABP α or β block N-box-mediated transcriptional activation by these same subunit promoters in cultured myotubes (Schaeffer et al., 1998). The involvement of the GABP transcription factors in endplate targeting was further demonstrated by the reduction of ectopic synapses *in vivo* caused by a GABP β dominant-negative mutant (Ruegg and Briguet, 2000).

Finally, overexpression of an Ets transcription factor in transgenic mice, which *in vitro* produces a potent competitor of GABPα/β, drastically alters endplate morphology and the number of receptors in the postsynaptic membrane (De Kerchove D'Exaerde et al., 2002). N-box elements thus control the differentiation of the postsynaptic membrane through the recruitment of the Ets class transcription-factor GABP.

First-Messenger Signaling from the Motor Nerve

The first explanation that came to mind for the action of the motor nerve on acetylcholine receptor gene expression and cluster formation was that the neurotransmitter itself, acetylcholine, functions as a transsynaptic signal. While this mechanism may be operative for glutamate and glutamate receptors at the neuromuscular junction in the case of *Drosophila* (Schwarz et al., 2002), convergent evidence demonstrates that it is not implicated in interactions at the vertebrate motor endplate.

The first sign of endplate formation—the high-density clustering of acetylcholine receptor under the nerve ending—occurs within hours of contact being made by the growth cone with the embryonic muscle fiber (Anderson et al., 1977; Frank and Fischbach, 1979). Acetylcholine receptor clusters may, however, develop spontaneously in muscle cells in culture (review in Schuetze and Role, 1987). Moreover, acetylcholine receptor clustering in nerve-muscle cocultures takes place in the presence of curare (Cohen, 1972), and, as we noted earlier, chronic paralysis of the chick embryo *in ovo* (Giacobini et al., 1973) does not prevent the formation of acetylcholine receptor endplate clusters (Giacobini-Robecchi et al., 1975; Burden, 1977a; Bourgeois et al., 1978b). Nonphysiological stimuli such as silk threads (Jones and Vrbova, 1974) or positively charged latex beads (Peng et al., 1981) have also been found to elicit acetylcholine receptor clustering on *Xenopus* mononucleated myotomal cells. Additionally, clustering of acetylcholine receptors in the middle region of the muscle fiber may be observed at least partially *in vivo*, independently of motor innervation (Burden, 2002). Muscle cells therefore possess intrinsic mechanisms capable of aggregating acetylcholine receptors into clusters.

These mechanisms can be triggered by a variety of signals. At least three kinds of endogenous signals are thought to contribute to the differentiation of the postsynaptic membrane. One is the *calcitonin-gene-related peptide* (CGRP), a neuropeptide that is present in the motor nerve ending (Hökfelt et al., 1986) and enhances acetylcholine receptor gene expression in cultured myotubes (Fontaine et al., 1986; New and Mudge, 1986; Fontaine et al., 1987; Osterlund et al., 1989). But "knockout" mice in which the αCGRP has been deleted show neither major changes in the morphology of the neuromuscular junction nor paralysis (Lu et al., 1999; Salmon et al., 1999).

Another molecular species identified by McMahan and colleagues (Fallon et al., 1985) is *agrin*, a 210-kDa heparin sulfate glycoprotein identified as a component of the *Torpedo* electric organ extracellular matrix that stimulates acetylcholine receptor clustering in muscle cell cultures (Nitkin et al., 1987). Several forms of agrin can be generated through alternative splicing of a single agrin gene. The most potent isoform of agrin in acetylcholine receptor clustering is produced by the nerve, muscle isoforms being less efficient. Agrin immunoreactivity is high in the basal lamina. Its receptor on the muscle membrane is a complex entity: gene inactivation in mice and further experiments in cultured myotubes show that the functional agrin receptor most probably includes a tyrosine kinase known as muscle-specific kinase (MuSK). Until recently agrin was believed to act in a dedicated manner to promote the clustering of acetylcholine receptors at the neuromuscular junction, without affecting transcriptional regulation. In knockout mice deficient for either the neural form of agrin (Figure 9.8A) or MuSK, however, subsynaptic compartmentalized transcription is respectively severely affected or abolished (review in Sanes and Lichtman, 1999). Normal subsynaptic transcription thus requires the presence of both neural agrin and MuSK.

Finally, *acetylcholine receptor-inducing activity* (ARIA) was initially identified and able to be purified from chicken brain by Fischbach and colleagues (Falls et al., 1993) by virtue of its ability to stimulate acetylcholine receptor biosynthesis (rather than clustering) in cultured myotubes. ARIA is also known as *heregulin β* (neu-differentiation factor, or neuregulin β) and encodes a 42-kDa glycoprotein belonging to the neuregulin 1 family of epidermal growth factor (EGF)-like polypeptides, a group of 14 proteins that are generated from the alternative splicing of a single gene (Holmes et al., 1992). Genetic inactivation of

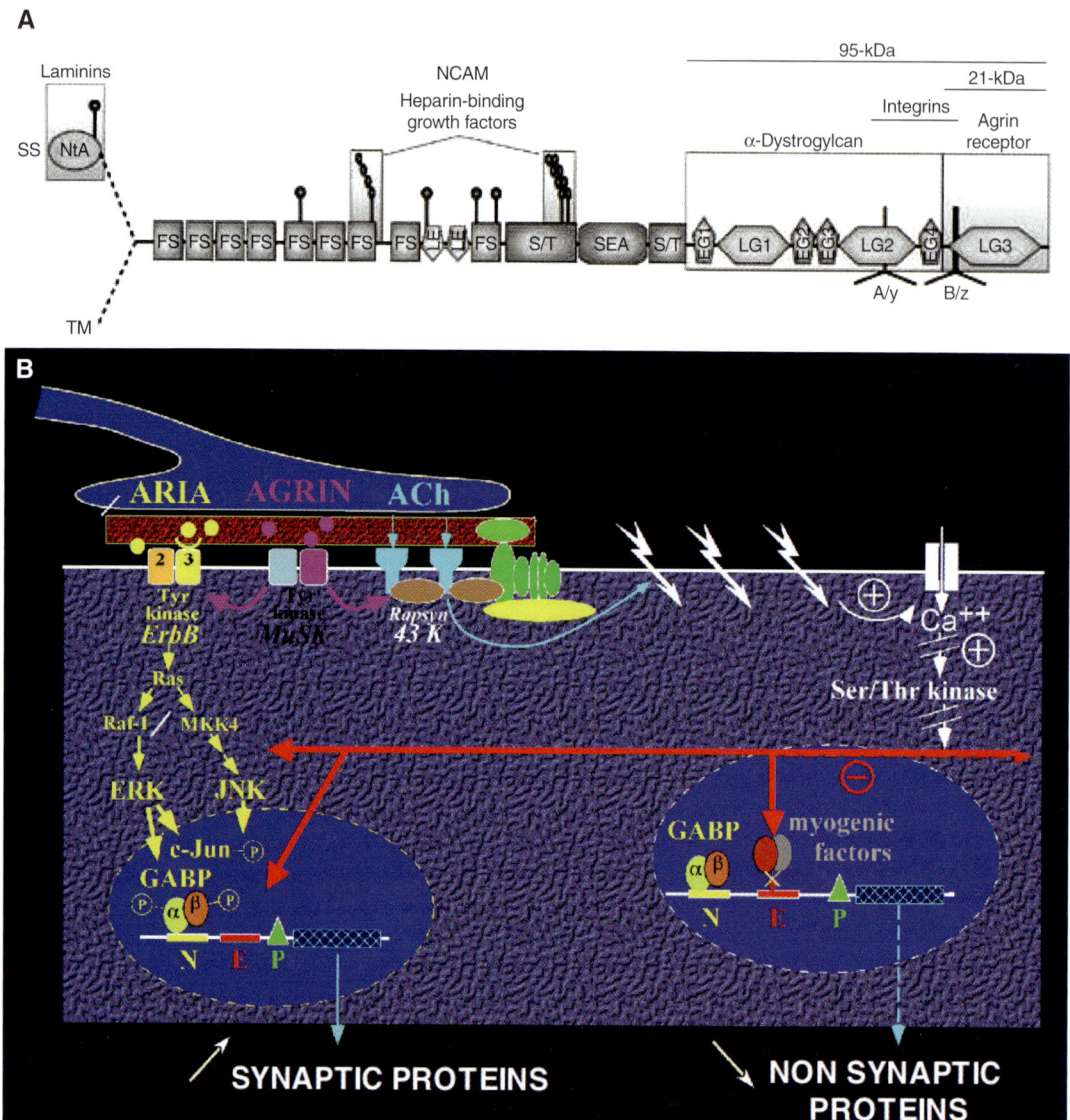

FIGURE 9.8 (A) Agrin structure and interactions. This schematic representation shows the structural domains of agrin, the sites of alternative messenger RNA splicing, and the binding regions of some of its most important partners. The signal sequence (SS) and amino (N)-terminal agrin domain (NtA) are present in an agrin isoform that is localized to the neuromuscular junction (NMJ). These two regions are responsible for the release of agrin (SS) and its binding to the laminins in basal laminae (NtA). The alternatively spliced type II transmembrane segment (TM) of agrin anchors this isoform in places that are devoid of basal laminae, such as the brain. The amino-terminal half of agrin is highly glycosylated at serine/threonine (S/T) glycosylation- and glycosaminoglycan-attachment sites. These regions are also involved in binding to neural-cell adhesion molecules (NCAM) and heparin-binding growth factors. The carboxyl terminal 95-kDa portion of agrin that starts with the first epidermal growth factor (EGF)-like domain (EG), is fully active in promoting acetylcholine receptor aggregation and contains binding sites for α-dystroglycan, heparin, some integrins, and an agrin receptor that remains to be identified. (From Bezakova and Ruegg, 2003.) (B) Model of gene transcription regulation in synaptic and extrasynaptic nuclei. In the adult innervated muscle, depolarization of the muscle fiber leads to a calcium influx and activation of a serine/threonine protein kinase. Under these conditions the kinase is assumed to phosphorylate and inactivate the "myogenic factor," a critical factor(s) of the myogenic family that is required for the transcription of genes through an E box DNA element. This inactivation both decreases gene transcription (shown here) and breaks a positive autoregulatory loop consisting of the activation of the myogenic gene by its own product (not shown). Gene transcription in endplate nuclei proceeds via independent pathways involving Erb B tyrosine kinase and the Ras pathways ERK and/or JNK, causing phosphorylation of the GABP transcription factor. (From Schaeffer et al., 2001.)

a subset of neuregulin isoforms—the form containing an immunoglobulin (Ig)-like domain—is lethal prior to the formation of neuromuscular junctions. Analysis of heterozygous mutant mice revealed that these animals exhibit signs of skeletal muscle paralysis, however, and show a 50% decrease in the RNA coding for acetylcholine receptors (Sandrock et al., 1997). The receptors for neuregulins belong to the same family as the EGF receptors, the erbB tyrosine kinase family. ErbB 1, 2, 3, and 4 all form active chimeras and cause their autophosphorylation. Several of these forms are present in innervated muscle and clustered at the motor endplate (Altiok et al., 1995; Altiok et al., 1997; review in Trinidad et al., 2000). Confocal microscopy reveals that, of the four kinases, only erb B2 and erb B4 are clustered in rat postsynaptic membrane, but these clusters lie deep in the secondary junctional folds. Their spatial distribution differs from that of MuSK, which codistributes with acetylcholine receptor at the tip of the postsynaptic folds (Trinidad et al., 2000).

Recently it was reported that acetylcholine receptor clustering and compartmentalized transcription in some mutant mice was preserved following denervation (Lin et al., 2001; Yang et al., 2001; Pun et al., 2002). It appears that MuSK, but not agrin, is required to initiate this prepatterning of acetylcholine receptors. On the other hand, agrin may be necessary for their maintenance. Alternatively, it is possible that different forms of agrin exist and differentially contribute to both "aneural" and "neural" distributions of the receptor protein.

Taken together, these results suggest that under conditions of normal development agrin causes MuSK activation, which together with neuregulins, originating in either the nerve or the muscle, causes the clustering of erbB receptors; and that subsequent activation of downstream intracellular signaling pathways ultimately leads to the transcriptional activation of synaptic genes.

Intracellular Signaling Pathways in the Junctional Domain

The second messenger signaling pathways that mediate the effect of neural signals on endplate formation at this stage have not been completely elucidated. Neuregulins binding to erb B tyrosine kinase receptors are known to initiate a signal transduction cascade that, in skeletal muscle, includes an adaptor molecule (SHC), the Ras signaling pathway, and the phosphatidyl inositol 3′ kinase (PI3K) (Tansey et al., 1996; Altiok et al., 1997; see Figure 9.8B). ErbB receptors also interact with another kinase, the cyclin-dependent kinase cdk5, and its activator p35. These two proteins are concentrated at the neuromuscular junction, and cdk5 activity is both stimulated by, and necessary for, erbB activation by neuregulins (Fu et al., 2001). Downstream Ras activation results in the activation of the MAP kinase (or only ERK2) or JNK pathways. Blocking either the MAP kinase pathway (or only ERK2) or the JNK pathway is sufficient to arrest neuregulin-dependent activation (Altiok et al., 1997; Burden, 1998; Si et al., 1999). It is possible that the rapid activation of the ERK and JNK kinases is followed by activation of the immediate early gene c-Jun expression and the phosphorylation of both c-Jun and GABP (Si et al., 1999). Phosphorylated GABP and c-Jun, either sequentially or in concert (through a still-unknown mechanism), subsequently activate transcription via GABP binding to the N box.

Convergent evidence further demonstrates that the N box and GABP play a role in the responses of muscle cells to neuregulin. Indeed, in myotube cultures, neuregulin-mediated activation of acetylcholine receptor ε-, δ-, or utrophin promoters requires an intact N box (Fromm and Burden, 1998; Sapru et al., 1998; Schaeffer et al., 1998; Khurana et al., 1999). Similar results have been obtained *in vivo* with the utrophin gene promoter (Gramolini et al., 1999).

The role of GABP as an intermediary in neuregulin action has been investigated using dominant-negative GABP β and Ets-2 mutants. Such mutations of either GABP subunit prevent activation of the acetylcholine receptor δ- and ε-subunit promoters by neuregulin in cultured myotubes (Schaeffer et al., 1998; Sapru et al., 1998). The GABP β mutant used by Schaeffer et al. (1998) has also been expressed *in vivo* and prevents agrin-induced activation of the acetylcholine receptor ε-promoter (Ruegg and Briguet, 2000). Finally, overexpression in muscle of wild-type GABP *in vivo* mimics the action of neuregulins on reporter gene expression through the utrophin gene promoter (Gramolini et al., 1999).

The exact mechanism by which neuregulins potentially regulate the transcriptional activity of GABP is not fully understood. Neuregulin causes a twofold increase in GABP α levels and enhances the phosphorylation of both GABP subunits by MAP kinases (Schaeffer et al., 1998; Khurana et al., 1999). The neuregulin-elicited increase in GABP α levels may account for the stronger expression of GABP α transcripts in synaptic nuclei (Schaeffer et al., 1998). Neuregulin-elicited phosphorylation of GABP regulates its ability to mediate transcriptional activation without, however, affecting its DNA-binding activity (Sunesen et al., 2003).

In summary, differentiation of the postsynaptic compartment is most likely set in motion by the switching on of a preset mechanism that controls the activity of intrinsic regulatory loops present in muscle nuclei. Their transition to the "on" state is probably due to the activation of MuSK, itself the result either of autoactivation, activation by unknown factors, or the binding of neural agrin. Then, as we shall see in the following chapter, a clustering of several components of the postsynaptic domain occurs, beginning with acetylcholine receptors and 43K-rapsyn together with the neuregulins in the synaptic basal lamina and their erbB receptors in the plasma membrane. Neuregulins are now able to trigger the phosphorylation cascade, which results in the activation of GABP and the enhanced transcription of acetylcholine receptor and other synaptic protein genes.

Extrajunctional Regulation of Genes by Electrical Activity

The elimination of extrajunctional acetylcholine receptors in the course of embryonic development corresponds, as we have already mentioned, to a repression of acetylcholine receptor gene transcription (which persists until the adult state). Conversely, denervation of adult muscle causes an enhanced sensitivity to acetylcholine (Axelsson and Thesleff, 1959; Miledi, 1960), which is reversed by direct electrical stimulation of the muscle (Lomo and Rosenthal, 1972; Lomo and Westgaard, 1976). This "hypersensitivity" following denervation results from an increase in the number of receptors that arises from a derepression of acetylcholine receptor gene transcription consecutive to the blocking of electrical activity (Fontaine and Changeux, 1989; Klarsfeld et al., 1989; Osterlund et al., 1989; Tsay and Schmidt, 1989).

The approach suggested by the model of compartmentalized gene expression that involves identification of the *cis*-acting DNA elements and associated transcription factors (Changeux et al., 1987b) has also been followed in trying to account for the regulation of extrajunctional receptors by electrical activity (see review in Duclert and Changeux, 1995). It was observed that transgenic mice bearing an α 850-bp nls/β-gal construct of the α-subunit promoter display a decrease both in transgene expression in extrajunctional areas during development and in transgene reexpression in the adult following denervation (Klarsfeld et al., 1991; Salmon and Changeux, 1992). The component sources of electrical regulatory activity are thus to be found within the 850-bp length of the promoter. Examination of this sequence revealed the presence of two CANNTG consensus boxes. These CANNTG sites, also known as E boxes, are targets of DNA-binding proteins of the helix-loop-helix (HLH) family known as "myogenic proteins." These transcription factors were initially discovered by virtue of their ability to convert nonmuscle cell types such as fibroblasts to myoblasts (Lassar et al., 1989; Olson, 1990; Sassoon, 1993). Four members of the HLH family exhibiting myogenic properties have been identified: MyoD, myogenin, myf5, and MRF4. These myogenic factors share a basic homology region (DNA-binding region) and an HLH homology region (dimerization region), where they associate with other ubiquitous factors of the HLH family and jointly bind to a target CANNTG site. Factors belonging to this family that are able to dimerize with myogenic factors include E12/E47, ITF2 (E2–2) and HEB (see references in Duclert and Changeux, 1995). The expression of myogenic factors is almost exclusively restricted to skeletal muscle in vertebrates, in which they are involved in the activation of many muscle-specific genes, among them the muscle-specific creatine kinase gene and the myosin light chain l/3 gene.

Purified MyoD fusion proteins produced in *Escherichia coli* have been found to bind to both CANNTG consensus sites in the chicken α-subunit gene enhancer (Piette et al., 1990). Furthermore, mutations of the CANNTG sites that greatly diminish *in vitro* affinity for the MyoD protein also decrease the level of expression in primary myotube cultures (Piette et al., 1990). E boxes present in the α-subunit enhancer are therefore critical for muscle-specific expression of this gene. Similar results regarding the importance of myogenic factor binding sites for muscle-specific expression have been reported for other acetylcholine receptor subunit genes (see references in Duclert and Changeux, 1995), and recently *in vivo* for the promoter of the 43K-rapsyn gene (see Chapter 10) as a result of analysis of congenital myasthenic patients (Ohno et al., 2003).

The importance of the two E boxes found in the chicken α-subunit gene enhancer for the regulation of the acetylcholine receptor α-subunit promoter by electrical activity has been tested using an adenovirus-based strategy both *in vivo* (upon denervation) and *in vitro* (upon TTX treatment of muscle cells in primary cultures). Mutation of the two E boxes was observed to differentially affect transcription (Bessereau et al., 1994): *in vivo*, the right proximal E box enhanced the activity of the α-subunit promoter after denervation, while the left distal E box appeared to have somewhat less effect; *in vitro*, the right E box was required for α-subunit promoter activity to be enhanced following TTX treatment, whereas the left E box was not absolutely necessary. These data conclusively show that myogenic binding sites (E boxes) play a critical role in the electrical regulation not only of the α-subunit promoter but also of other acetylcholine receptor subunit genes (see Tang et al., 1994).

The possibility was then considered that electrical activity may also regulate the transcription of the acetylcholine receptor genes indirectly, via the myogenic factor itself. It has been shown, for example, that the continuous synthesis of one or more protein factors is necessary to obtain a high level of acetylcholine receptor α-subunit mRNA after TTX treatment *in vitro* (Duclert et al., 1990), as well as *in vivo* after denervation (Tsay et al., 1990). Positive regulatory factors with a short metabolic half-life are therefore involved in the early expression of the acetylcholine receptor α-subunit gene and in its up-regulation following the blocking of electrical activity. Interestingly, the myoblast determination protein (MyoD) has a short metabolic half-life (~30 min to 1 hr) in mouse myogenic cells (Thayer et al., 1989). It is also known that levels of myogenic protein mRNA decrease with that of α-subunit mRNA during postnatal development and increase following denervation, with the strongest up-regulation being seen in the case of myogenin mRNA (Duclert et al., 1991; Eftimie et al., 1991; Witzemann and Sakmann, 1991). Moreover, the presence of E boxes in the promoter of myogenic protein genes suggests that autocatalytic regulation may be involved (see references in Duclert and Changeux, 1995). It is therefore possible that myogenic promoters play a crucial role—still largely undeciphered—in the regulatory cascade linking membrane electrical activity and acetylcholine receptor gene transcription, as well as in autocatalytic gene switching responsible for setting boundaries and gene-expression patterns (Kerszberg and Changeux, 1993, 1994).

Studies of the signaling pathways that mediate the effect of electrical activity on acetylcholine receptor gene expression have made wide use of *in vitro* muscle cultures, which spontaneously fire action potentials despite the absence of innervation. Both neurally evoked and spontaneous depolarization of the muscle fibers cause a Ca^{2+} influx, presumably through voltage-sensitive L-type Ca^{2+} channels and possibly Na^+ channels as well (see reviews in Duclert and Changeux, 1995). In chick cultures blocked by TTX, the observed higher level of α-subunit mRNA was decreased by the calcium ionophore A-23187. Conversely, exposure of spontaneously active myotubes to the Ca^{2+} channel blocker verapamil was observed to produce increased levels of both acetylcholine receptor protein and spliced or unspliced α-subunit mRNA, but little effect was noticed with dantrolene, which blocks Ca^{2+} efflux from the sarcoplasmic reticulum (Klarsfeld et al., 1989). This result supports the view that Ca^{2+} influx contributes to repression of acetylcholine receptor gene expression by electrical activity. Early studies, using protein kinase C activators such as phorbol esters (Fontaine et al., 1987) and the rather nonspecific kinase inhibitor staurosporine (Klarsfeld et al., 1989), had pointed to a contribution by conventional protein kinase C (PKC), activation of which by Ca^{2+} ions could cause a repression of acetylcholine receptor biosynthesis. Recent experiments show that the addition of phospholipase D and C inhibits acetylcholine receptor expression in a manner that is mimicked by okadaic acid, a specific inhibitor of protein phophatases 1 and 2A that also mimics the effects of electrical activity. Okadaic acid was found to cause an increase in threonine phosphorylation of the atypical PKC isoform PKCζ and the activator of transcription factors ATF2 together with a *decrease* in phosphorylation of SP1 (Altiok and Changeux, 2001). Furthermore, the phosphorylation level of myogenin is known to unbind it from the E box (Mendelzon et al., 1994). It is therefore plausible to conclude that the regulation of the transcription of acetylcholine receptor genes and other synaptic proteins involves the concerted action of several transcription factors, including Sp1 and myogenic proteins. If so, the phosphorylation of myogenic proteins and/or the dephosphorylation of Sp1 would furnish evidence of the repression of acetylcholine receptor gene transcription by electrical activity. This tentative hypothesis remains to be confirmed, however.

Conclusion

Further investigation is needed if we are fully to understand the complex networks of interactions that regulate the compartmentalized transcription of acetylcholine receptor and other synaptic proteins. Nevertheless, the data so far obtained appear to be consistent with the model first proposed to account for these interactions, which has recently been updated (Schaeffer et al., 2001; Figure 9.8). The latest version contains several novel features and emphasizes a number of important conclusions:

1. DNA elements in the promoter of acetylcholine receptor genes and other synaptic genes (N box and E boxes) differentially regulate the transcription of these genes in junctional and nonjunctional compartments.
2. Distinct sets of transcription factors (GABP and myogenic protein) interact with these DNA elements.
3. Components (e.g., agrin, MuSK, heregulin) of the cell surface signaling system that directs the morphogenesis of the subsynaptic compartment clearly differ from those involved in the process of extrajunctional repression by electrical activity (e.g., L-type Ca^{2+} channels).
4. The network of signaling pathways linking the cell membrane to the nuclear transcription factors, while still incompletely understood, differs significantly in the junctional (e.g., Ras, ERK, JNK) and extrajunctional (e.g., Ca^{2+}, Ser/Thr kinase) compartments.
5. In agreement with the formal model proposed by Kerszberg and Changeux (1993, 1994), an autocatalytic loop may form at the

level of transcriptional regulation (e.g., myogenic proteins), thus helping to sharpen the "initial" gene-expression boundary between the junctional and nonjunctional compartments.

The proposed mechanism for the differential transcription of synaptic and nonsynaptic protein genes stresses the selective role of electrical activity, as opposed to trophic factors, in the morphogenesis of the motor endplate. As a reasonable model of synaptic plasticity in its own right, it warrants comparison with alternative molecular mechanisms of synaptic plasticity currently being investigated in *Aplysia* (Pittenger and Kandel, 2003) and the vertebrate hippocampus (Bliss et al., 2003).

CHAPTER 10

Supramolecular Assembly of the Postsynaptic Membrane

The neuromuscular junction is a highly differentiated organelle exhibiting a sophisticated molecular architecture that is specifically adapted to rapid cellular communication. In the last chapter we examined its genesis at the transcriptional level (Figure 9.1). The targeted transcription of nicotinic receptor genes to the subjunctional domain, together with their activity-dependent repression in extrajunctional territories, give rise to enhanced biosynthesis and accumulation of receptor proteins in muscle fiber at the synapse. Nonetheless, these properties do not suffice to account for the highly restricted and stable distribution of receptor proteins in the mature synapse. Additional protein components are necessary to make up a subneural apparatus that is notable for its precise differential distribution of synaptic proteins, particularly the receptor protein's restricted localization at the tip of the synaptic folds.

A rich repertoire of molecular species has been identified in the subjunctional membrane (see reviews in Duclert and Changeux, 1995; Sanes and Lichtman, 1999, 2001). The entire length of the muscle fiber is covered by a basal lamina. At the endplate, however, this extracellular matrix possesses a characteristic biochemical composition marked by the presence of several distinct forms of the enzyme acetylcholinesterase, which help to bring about the termination of signal transmission by hydrolyzing acetylcholine (reviews in Massoulié et al., 1993; Perrier et al., 2002). The synaptic basal lamina is similarly enriched by a variety of proteins, among them laminin $\alpha 2$, $\alpha 4$, $\alpha 5$, $\beta 2$, $\gamma 1$, and collagen IV $\alpha 3$, $\alpha 4$, $\alpha 5$, as well as agrin and neuregulin. Within the postsynaptic membrane, at the tip of the folds, nicotinic receptors are found alongside the integrins α7A, B, β1, the dystroglycans α, β, and the sarcoglycans α, β, γ, δ, together with a tyrosine kinase (MuSK) and ErB 2,3, and four receptors for trophic factors; in addition, Na^+ channels and neural cell adhesion molecule (NCAM) factors accumulate at the bottom of the subsynaptic folds (see review in Sanes and Lichtman, 1999).

On the cytoplasmic side, filamentous material is associated with the postsynaptic membrane of *Torpedo* electroplaque (Rosenbluth, 1975; Heuser and Salpeter, 1979; Cartaud et al., 1981; Kordeli et al., 1986) and of the vertebrate neuromuscular junction (Couteaux, 1981; Hirokawa and Heuser, 1982). A constellation of cytoskeletal proteins (actin, α-actinin, talin, vinculin, and filamin) has been identified at this junction (Bloch and Hall, 1983), together with 43K-rapsyn, utrophin α, dystrobrevin 1, and β2 syntrophin, as well as ankyrin, dystrophin, α-dystrobrevin 2, and α1- and β1-syntrophin (reviews in Duclert and Changeux, 1995; Sanes and Lichtman, 1999). The postsynaptic membrane is a complex structure built out of many different proteins. All of these protein species are plausible candidates for the supramolecular assembly of postsynaptic architecture.

The formation of the subneural domain does not take place all at once, but develops progressively from about day 10 of embryonic development until about two weeks or so after birth in chicks and rodents. In particular, the number of postsynaptic folds and receptor molecules steadily increases postnatally (Salpeter and Loring, 1985). During this period a precise coincidence between the presynaptic active zones of acetylcholine release and the receptor-rich mouths of the postsynaptic folds also is established, indicating that transsynaptic mechanisms contribute to the developing neuromuscular junction. What not so long ago was called a "problem of the future" (Changeux, 1981) has now clearly become an experimentally tractable research program. Prospects for molecular explanation are promising in the view of some researchers (see, for example, Sanes and Lichtman, 2001), but considerably more work remains to be done if the underlying mechanisms of supramolecular assembly are to be fully elucidated (Duclert and Changeux, 1995; Sanes and Lichtman, 1999, 2001; Schaeffer et al., 2001; Huh and Fuhrer, 2002; Marchand and Cartaud, 2002).

43K-Rapsyn and the Postsynaptic Scaffold

The 43K-rapsyn protein, a fundamental component of the postsynaptic membrane, plays a crucial role in "assembling" its various elements (reviews in Cartaud and Changeux, 1993; Sanes and Lichtman, 2001; Marchand and Cartaud, 2002). This protein was first identified using high-resolution SDS-polyacrylamide gel electrophoresis (Sobel et al., 1977, 1978) as an element of nicotinic receptor-rich *Torpedo* membranes that, after purification with detergent extracts, no longer associates with the receptor protein (Figure 10.1). Its amino acid composition was determined by Sobel et al. (1978). It was initially called "43K protein," after its apparent molecular weight, but following its cloning and sequencing it was renamed "rapsyn," a shorthand for receptor-associated protein at the synapse (Carr et al., 1987; Frail et al., 1987). As a reminder of its origins we refer to it here as the 43K-rapsyn protein, or simply 43K-rapsyn.

Selective proteolysis (Wennogle and Changeux, 1980), labeling with isotopic iodine (St. John et al., 1982), and gold labeling with specific monoclonal antibodies (Nghiem et al., 1983; Sealock et al., 1984) have shown that the 43K-rapsyn protein faces only the cytoplasmic side of the membrane and is distributed along with the receptor protein in approximate equimolar ratios (Sobel et al., 1978; LaRochelle and Froehner, 1986). Brief exposure of the receptor-rich membrane to pH 11 (Neubig et al., 1979) or lithium di-iodosalicylate (Elliott et al., 1980) releases 43K-rapsyn and other peripheral proteins without significantly changing the receptor's functional properties, as analyzed by ion-flux measurements or rapid binding of agonists and noncompetitive blockers (Neubig et al., 1979; Heidmann et al., 1980; Elliott et al., 1980).

Contrary to earlier suggestions (Sobel et al., 1978), 43K-rapsyn seems not to directly contribute to the ionic response to acetylcholine,

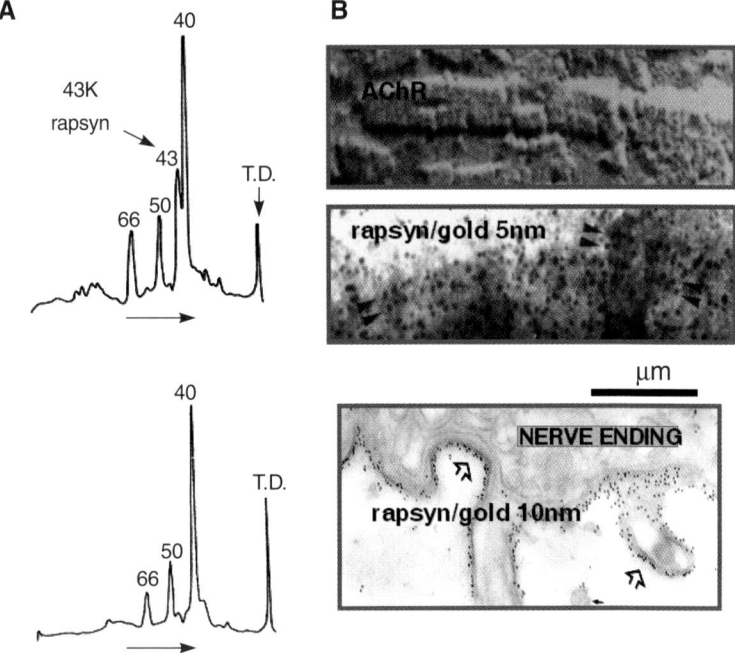

FIGURE 10.1 Characterization of 43K-rapsyn, a cytoskeletal protein anchoring acetylcholine receptor molecules. (A) Profiles from purified membranes (top) and purified receptors (bottom) showing the separation of the protein from the receptor subunits (indicated here by 40, 50, 66). (From Sobel et al., 1977b.) (B) Distribution of 43K-rapsyn in the postsynaptic membrane by immunocytochemistry as revealed by electron microscopy. (From Nghiem et al., 1983.)

although recent findings with a zebra fish mutant suggest that it does exert an indirect effect on synaptic strength regulation (Ono et al., 2002). Electron microscopy at high resolution confirms the direct interaction of 43K-rapsyn with the receptor molecule on its cytoplasmic face (Toyoshima and Unwin, 1988), but, as we noted in Chapter 7, its precise relationship to nicotinic receptor density has subsequently been reevaluated (Unwin, 2000). 43K-rapsyn can in any case be chemically cross-linked to the β-subunit of the nicotinic receptor in *Torpedo* receptor-rich membranes (Burden et al., 1983).

The 43K-rapsyn protein performs a structural role as well (Figure 10.2). Its elimination destabilizes the receptor under heat treatment (Saitoh et al., 1979) or proteolytic attack (Klymkowsky et al., 1980), and enhances its rotational and translational mobilities, monitored with spin-labeled (Rousselet et al., 1982) or phosphorescent (Lo et al., 1980) derivatives or α-bungarotoxin, and also by electron microscopy (Barrantes, 1980; Cartaud et al., 1981). On the basis of these data it was hypothesized that 43K-rapsyn "anchors" and stabilizes the nicotinic receptor in the postsynaptic membrane (Cartaud et al., 1981; Rousselet et al., 1982). This function was subsequently confirmed and amplified (see review in Huh and Fuhrer, 2002).

A

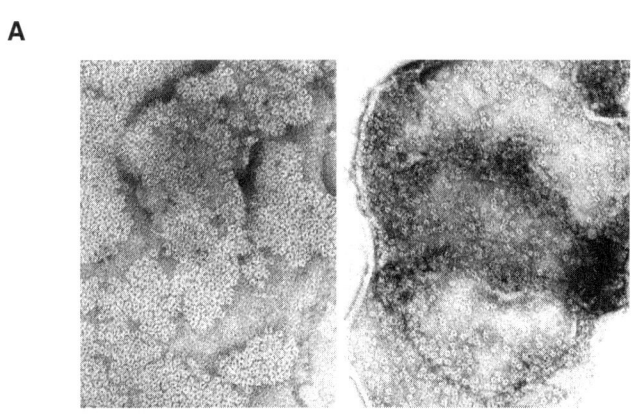

B

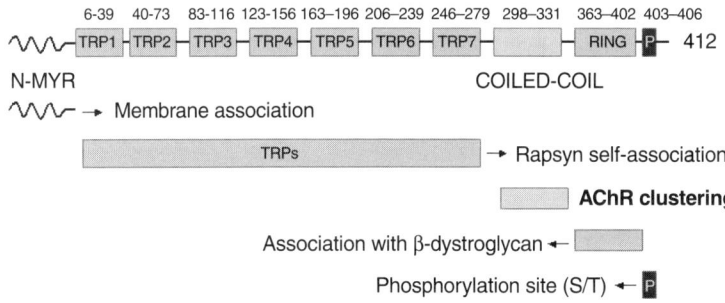

FIGURE 10.2 Characterization of 43K-rapsyn and nicotinic receptor clustering. (A) Dispersion of nicotinic receptors after the elimination of 43K-rapsyn. (From Cartaud et al., 1981, courtesy of The Rockefeller University Press.) (B) Functional domains of 43K-rapsyn involved in membrane association, self-association, and clustering. (From Bartoli et al., 2001.)

Immunochemical studies had previously revealed that a protein immunochemically related to 43K-rapsyn is present at mammalian endplates (Froehner, 1981, 1984) and that it codistributes exactly with the receptor protein (Burden, 1985; Peng and Froehner, 1985; Bloch and Froehner, 1987; Flucher and Daniels, 1989). The 43K-rapsyn protein has been cloned and sequenced in *Torpedo* (Carr et al., 1987; Frail et al., 1987), *Xenopus* (Baldwin et al., 1988), and mouse (Frail et al., 1988). Its primary sequence reveals distinct structural domains (Figure 10.2): the amino terminus is myristylated (Musil et al., 1988), and followed by as many as eight tetratricopeptide repeats (codons 6–319), a coiled-coil domain (codons 298–331), a cysteine-rich domain (codons 369–402) that is predicted to be a RING-H_2 domain, and a consensus sequence for phosphorylation by protein kinases A and C (codons 403–406).

These various domains differentially contribute to the clustering of receptor proteins. Mutation of the myristoylation site prevents membrane targeting, with the result that 43K-rapsyn accumulates in the nucleus (Phillips et al., 1991; Ramarao and Cohen, 1998). The tetratricopeptide repeats and the coiled-coil domain have been found to contribute to 43K-rapsyn self-association (Ramarao et al., 2001), while a mutant lacking the coiled-coil domain (codons 288–348) failed to cluster receptors (Ramarao et al., 2001). The coiled-coil domain interacts with the large M3-M4 cytoplasmic loop of each individual receptor subunit to promote its clustering (Bartoli et al., 2001). On the other hand, the RING-H_2 domain, which is not necessary for self-association or clustering, interacts only with β-dystroglycan. Finally, 43K-rapsyn contains a serine phosphorylation site (codon 406) at which it is phosphorylated *in vivo* and can be phosphorylated *in vitro* by protein-kinase A (Hill et al., 1991). Specific phosphorylation of *Torpedo* 43K-rapsyn by a new type of kinase using thiamine triphosphate as the phosphate donor was recently discovered in *Torpedo* receptor-rich membranes (Nghiem et al., 2000). Phosphoamino acid and chemical stability suggest that the phosphorylated residues are histidines. This rare type of phosphorylation (which also occurs in the brain) may perhaps play a particular role in the regulation of synaptic function, but this remains to be demonstrated.

The 43K-rapsyn protein plays a central role in the clustering of nicotinic receptors *in vivo*. Spontaneously formed receptor clusters on *Xenopus* muscle cells in culture are known to coincide with accumulations of 43K-rapsyn (Burden, 1985; Peng and Froehner, 1985). Acetylcholine receptor clusters elicited on these cells either by nerve terminals (Burden, 1985) or by artificial signals such as polypeptide-coated latex beads (Peng and Froehner, 1985) also contain high levels of 43K-rapsyn. Accumulations of 43K-rapsyn are detected in junctional nicotinic receptor clusters *in vivo* as soon as they form during development in the mouse (Noakes et al., 1993). In some situations, however, a nicotinic receptor aggregation can occur without the presence of 43K-rapsyn. Indeed, this protein is undetected in certain spontaneously occurring nicotinic receptor clusters in chicken primary muscle cultures (Tsui et al., 1990). Moreover, during *Torpedo* electrocyte development, the receptors accumulate at the ventral pole while 43K-rapsyn is still largely cytoplasmic (Kordeli et al., 1989; Nghiem et al., 1991).

Reconstitution experiments reveal a uniform distribution of nicotinic receptors introduced in *Xenopus* oocytes or fibroblasts, but coexpression of the nicotine receptor and 43K-rapsyn results in the formation of small clusters of each protein (Froehner et al., 1990; Phillips et al., 1991). Conversely, 43K-rapsyn-null ($^{-/-}$) mice lack aggregates of nicotinic receptor, ErbB receptors, dystroglycan, and utrophin; nor do cultured myotubes from 43K-rapsyn$^{-/-}$ mice form clusters spontaneously, even in response to agrin (Gautam et al., 1999). In a subset of patients presenting congenital myasthenia syndromes but no mutation in the nicotinic receptor, several mutations in 43K-rapsyn have been identified that show diminished coclustering of acetylcholine receptor and simplified endplate regions (Ohno et al., 2002). Two mutations have been identified in E-box elements (see Chapter 9) of the 43K-rapsyn promoter in patients with congenital myasthenic syndrome as well (Ohno et al., 2003). The 43K-rapsyn is therefore a necessary component of all forms of nicotinic receptor clustering.

Yet things turned out not to be as simple as they first appeared. Quite surprisingly, MuSK, which as we noted in the last chapter is a functional component of agrin membrane receptors, continues to be localized synaptically at the endplate of 43K-rapsyn knockout mice, suggesting that 43K-rapsyn acts downstream of MuSK in the assembly of nicotinic receptors. Constitutively activated MuSK was also found to cluster in the absence of agrin and to stimulate the formation of ectopic postsynaptic-like membranes (Jones et al., 1999). In the absence of agrin, then, it may be that a MuSK complex is preassembled as a kind of "primary synaptic scaffold" (Huh and Fuhrer, 2002). In that case it may further be supposed that agrin subsequently comes into play, triggering a higher-order assembly of the MuSK complex in which 43K-rapsyn is associated with the nicotinic receptors and the other components of the postsynaptic apparatus (Apel et al., 1997; Fuhrer et al., 1999; Sander et al., 2001). A cascade of phosphorylations (Teichberg and Changeux, 1977; Teichberg et al., 1977; Saito and Changeux, 1981), including the tyrosine phosphorylation of the β- and δ-subunits of the nicotinic receptor, would then regulate its integration to the postsynaptic cluster and its association with the subneural cytoskeleton (Huh and Fuhrer, 2002). This complex macromolecular network (Figure 10.3A) includes utrophin, which on the

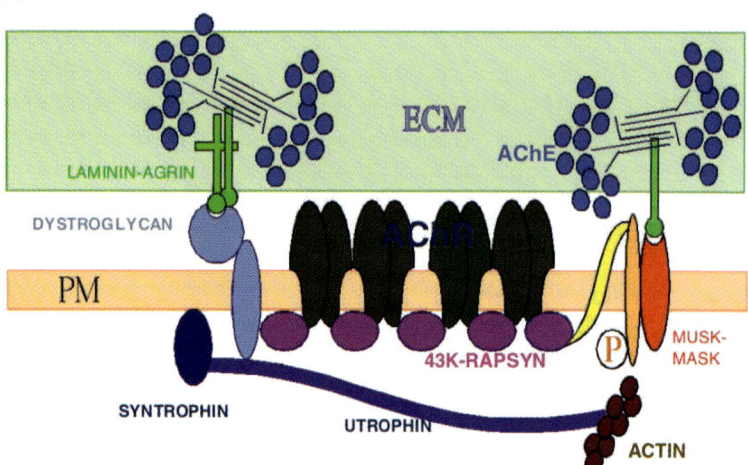

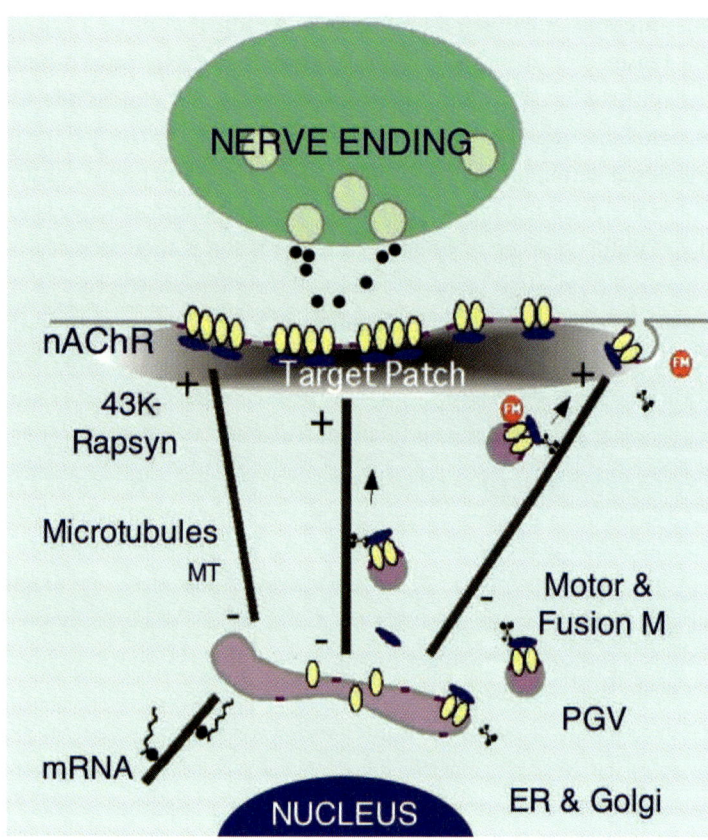

FIGURE 10.3 Supramolecular architecture of the postsynaptic membrane (PM) and targeting of the acetylcholine receptor and the 43k-rapsyn protein to the postsynaptic membrane. (A) Molecular organization of the postsynaptic apparatus. The functional significance of the different components is described in the text. ECM: Extracellular matrix. (B) Targeted delivery of nicotinic receptors to the subsynaptic membrane via the microtubule-based secretory pathway. MT: Microtubules; PGV: Post-Golgi vesicles; ER: Endoplasmic reticulum. (Courtesy of J. Cartaud.)

cytoplasmic side establishes links with the actin cytoskeleton of the muscle fiber, and β-dystroglycan, which links these cytoskeletal elements with the basal lamina (Huh and Fuhrer, 2002).

In short, the nicotinic receptor molecules become trapped within a complex supramolecular scaffolding that sandwiches the postsynaptic membrane between the basal lamina and the cytoskeleton, creating a robust and stable structure that assures failure-free transmission of nerve signals to the muscle fibers.

Passage of Receptor Oligomers from the Nucleus to the Postsynaptic Membrane

As described in Chapter 9, the targeted transcription of nicotinic receptor genes to the subjunctional "fundamental" nuclei results in a discrete localization of nicotinic receptor subunit mRNAs (Fontaine et al., 1988; Fontaine and Changeux, 1989), whose sharp boundaries may result from autocatalytic loops involving myogenic transcription factors (Kerszberg and Changeux, 1993). RNA levels of the four α-, β-, γ-, and δ-subunits are not strictly coregulated, but rather are compartmentalized at an early stage of development (embryonic day 13.5 in the mouse diaphragm; see Piette et al., 1993). On the other hand, ε-subunit mRNA is first detected only at about the time of birth (Mishina et al., 1986; Witzemann et al., 1989) and is restricted to synaptic sites (Brenner et al., 1990). Despite these discordant transcriptional regulations, only heterooligomers containing the four types of subunits—α, β, γ/ε, δ—are assembled, whereas the incorrectly folded, processed, or assembled intermediates are differentially degraded (Merlie et al., 1983; Blount and Merlie, 1991).

The assembly of nicotinic receptor pentamers takes place in the perinuclear endoplasmic reticulum through a multistep process (Figure 10.3B). In the adult synapse, the receptor and its companion proteins are then channeled by a series of specialized processes to the postsynaptic membrane. Properly assembled oligomers are first addressed to the Golgi compartment, where they undergo further processing by Golgi enzymes and, after sorting, are delivered to the cell surface (Merlie and Smith, 1986). In embryonic myotubes, nicotinic receptors are distributed over the entire cytoplasmic membrane and the Golgi apparatus is detected with a specific antibody over all sarcoplasmic nuclei in a perinuclear manner. In adult muscle fiber (from 15-day-old chicks), on the other hand, the Golgi apparatus is visibly restricted to the neuromuscular junction (Jasmin et al., 1989).

Denervation results in a redistribution and fragmentation of the Golgi apparatus over the entire muscle fiber (Jasmin et al., 1995). Its spatial restriction at mature endplates therefore requires an intact nerve. Moreover, Ralston et al. (1999) have shown that the structure of the Golgi apparatus varies in extrajunctional areas with the type of fiber, and confirmed that it is uniformly organized at the junction in a perinuclear manner as a stack of characteristic compartments. Specialized forms of the Golgi apparatus are thus concentrated in the subneural domain and devoted to the processing of synaptic proteins (Jasmin et al., 1989).

Microtubules are known to establish close relationships with the Golgi apparatus (review in Kreis, 1990; Cole and Lippincott-Schwartz, 1995). In highly polarized cells, such as myotubes, microtubules make a sparse and labile network parallel to the axis of the fiber (Jasmin et al., 1990). On the other hand, the subneural domain contains a dense and specialized network of cold-stable, acetylated microtubules (Jasmin et al., 1990) that are radially organized around the subjunctional nuclei and extend from the nuclear periphery toward the postsynaptic membrane. Trains of noncoated vesicles are frequently observed in their vicinity, for example in *Torpedo* electrocytes (Jasmin et al., 1991). It was therefore proposed that a specialized "secretory pathway" serves to process and carry synaptic proteins to the postsynaptic domain (Cartaud and Changeux, 1993; see Figure 10.3B).

This notion has been confirmed and extended by the observation of trains of vesicles containing nicotinic receptors (Camus et al., 1998), 43K-rapsyn (Marchand et al., 2002), and other components of the postsynaptic membrane traveling along the axis of the microtubules. Interestingly, some of these components, such as β-dystroglycan, follow the same path as nicotinic receptor and 43K-rapsyn proteins, while others, such as dystrophin and dystrobrevin, proceed directly to the postsynaptic membrane following cytoplasmic synthesis (Marchand et al., 2001). Time-lapse imaging of vesicles containing fluorescent 43K-rapsyn-GFP in Cos7 cells (Marchand et al., 2002) further revealed linear trajectories of labeled material that advanced at a rate of 0.2 mm/sec, a rate concordant with that of motor-guided movement. The same study also demonstrated an association of nicotinic receptors and 43K-rapsyn-GFP with cholesterol-sphingolipid-enriched microdomains referred to as *lipid rafts*.

Lipid rafts are domains that selectively incorporate or exclude proteins, and so may function as membrane platforms for the assembly and sorting of proteins, in particular within the *trans*-Golgi network (Ikonen, 2001). Lipid rafts have been shown to be necessary for the maintenance of α7 neuronal nicotinic receptor clusters in ciliary neuron spines (Bruses and Rutishauser, 2001). Interestingly, analysis of the lipid composition of purified receptor-rich membranes from *Torpedo* highlighted the prominence of cholesterol and long-chain fatty acids (Popot et al., 1978), which are abundant in lipid rafts. Flotation gradient centrifugation further revealed that, in cultured cells, nicotinic receptors and 43K-rapsyn are recovered in low-density fractions that are enriched in two raft markers, caveolin-1 and flotilin-1. These observations raise the possibility that the raft machinery participates in the targeting of nicotinic receptors and companion molecules to the postsynaptic membrane (Marchand et al., 2002).

Once incorporated in the subjunctional membrane, the nicotinic receptor molecules form stable and dense clusters that persist after denervation with a nearly constant density (Bourgeois et al., 1973, 1978b; Frank et al., 1975). The metabolic degradation half-life of these junctional receptors was estimated by following the decay of ^{125}I-α-bungarotoxin complex to be as much as 10 days, while the half-life of nonjunctional receptors was put at 17–22 hours (Berg and Hall, 1975; Chang and Huang, 1975; Devreotes and Fambrough, 1975; Merlie et al., 1975, 1976, 1978; Burden, 1977; Bourgeois et al., 1978). Moreover, muscle denervation was shown to destabilize junctional nicotinic receptors and to produce an increase in their turnover (Loring and Salpeter, 1980). Blocking neuromuscular transmission by botulinum toxin A or α-bungarotoxin had a similar effect (Avila et al., 1989). Conversely, electrical stimulation of the muscle (Brenner and Rudin, 1989; Fumagalli et al., 1990) and Ca^{2+} influx associated with activity reversed the effect (Rotzler et al., 1991; Caroni et al., 1993). Drastic changes in the molecular properties of nicotinic receptors thus occur within subsynaptic clusters without any apparent accompanying change in the number of receptors (see review in Duclert and Changeux, 1995). Moreover, a series of classic nerve-muscle degeneration and regeneration experiments showed that nicotinic receptor clusters reappear at their former location on the muscle fiber, even in the absence of innervation, so long as the basal lamina is present (Marshall et al., 1977; McMahan et al., 1978; Burden et al., 1979; Loring and Salpeter, 1980). These experiments revealed the existence of an "organizing matrix" involving the basal lamina, which possesses much higher metabolic stability than the receptor with which it coaggregates.

Subsequent research has confirmed these early findings, in particular the presence of a rapidly constructed, permanent subjunctional scaffold and the "organizing" role of agrin as one of its components, as we saw above. Furthermore, it has revealed that a group of proteins referred to as the dystrophin-glycoprotein complex, linking the extracellular matrix of muscle fibers to the cytoskeleton, includes 43K-rapsyn and dystroglycan, which interact with each other at the endplate and trap the nicotinic receptor (Apel et al., 1995; Cartaud et al., 1998; Bartoli et al., 2001).

Consistent with these results was the important discovery that central neurotransmitter receptors (Vannier and Triller, 1997; Choquet and Triller, 2003), as well as junctional nicotinic receptors (Akaaboune et al., 1999, 2002), have greater mobility than previously suspected. Time-lapse imaging that relies upon sequential unbinding and relabeling with differently colored fluorescent derivatives of α-bungarotoxin showed that synaptic nicotinic receptors become completely intermingled over the course of about four days and exchange continuously with perisynaptic receptors. Individual nicotinic receptors have been found to remain at a single site within an adult neuromuscular junction for as little as eight hours (Akaaboune et al., 1999, 2002). The effect of activity on receptor stability noted above appears to be associated with this behavior: in the active synapse, nicotinic receptors seldom escape to the perisynaptic region, despite significant mobility within this domain; within one hour after neurotransmission has been blocked, nicotinic receptors disperse from the crest of the postsynaptic folds into the perisynaptic domains, where they intermingle and degrade. Nicotinic receptor molecules are lost at a rate of ~4% per hour, with no detectable addition of new receptors. In the active endplate, by contrast, about 9% of receptor proteins are lost per day (i.e., 0.4% *per hour*) and replaced by ongoing biosynthesis of the receptor protein without net loss (Akaaboune et al., 1999). In mice that lack α-dystrobrevin, the intermingling of receptors increased four to five times (Akaaboune et al., 2002), suggesting that this component of the subsynaptic scaffold regulates the tethering of nicotinic receptors to the cytoskeleton.

These observations show that the organization of the nicotinic receptors in the postsynaptic membrane depends not only on their degradation but also, via the secretory pathway, on the machinery of biosynthesis. Moreover, the postulated regulation of nicotinic receptor stability and distribution as a function of synaptic activity provides a particularly useful way to investigate *synaptic plasticity* at the molecular level, with implications we shall examine later for central synapses (Chapter 12).

Conclusions and Pathologies of the Neuromuscular Junction

As we have seen in this chapter and the preceding one, the formation of the neuromuscular junction is a highly complex process, combining subtle regulation of transcription with interactions among a number of specialized proteins, that serves to stabilize the supramolecular architectures of individual synapses. Recent progress in this area can be attributed largely to the availability of powerful new observational techniques.

Nicotinic receptor molecules have been found to be closely packed at the tip of the postsynaptic fold, where they face the nerve ending. They are trapped within a constellation of molecules that constitutes an intricate, stable, and rigid supramolecular scaffold. During embryonic development, the transcription of receptor proteins, initially uniform in the myotubes, becomes restricted to a few fundamental nuclei (Chapter 9). Neural (and nonneural factors such as agrin and heregulin) contribute to this targeted transcription. Outside the junctional region, electrical activity reverses the inhibition of transcription of the receptor genes. In agreement with an earlier model (Changeux et al., 1987a), distinct elements from nicotinic receptor gene promoters that make up the N box and E box have been shown to play a role in the control of junctional and nonjunctional transcription, respectively. Different sets of transcription factors—GABP factors and myogenic proteins—also interact with these elements. They are regulated in turn by nonoverlapping second messenger signal transduction pathways, which are themselves controlled by agrin and heregulin at the junctional level and by Ca^{2+} influx at the extrajunctional level (Chapter 9).

An intracellular secretory pathway (Cartaud and Changeux, 1993; Marchand and Cartaud, 2002) is hypothesized to channel the assembled receptor oligomers to the postsynaptic membrane via a specifically adapted Golgi apparatus and a network of stable microtubules. In the subjunctional membrane, receptor molecules are cross-linked by a specialized cytoskeletal molecule, the 43K-rapsyn protein (Sobel et al., 1977b), that interacts with the large M3-M4 cytoplasmic domains of the receptor protein (Bartoli et al., 2001). A sophisticated supramolecular scaffold, the dystrophin-glycoprotein complex, binds together the intracellular matrix and the cytoskeleton, trapping the nicotinic receptor molecules. These molecules nevertheless preserve a remarkable freedom of movement (Choquet and Triller, 2003). The putative regulation of the local properties of receptor proteins by junctional activity may provide the basis for a fruitful new model of synaptic plasticity.

Last but not least, genetic alterations of these components of the junctional molecular machinery are associated with a variety of neuromuscular pathologies. In the last chapter we reported that mutations occurring in the N box of the ε-subunit promoter have been shown to cause congenital myasthenia (Nichols et al., 1999; Ohno et al., 1999). These patients carry mutations at the same position occupied by mutations that abolish synaptic targeting of ε-subunit gene transcription in the mouse (Schaeffer et al., 1998). Patients suffering from congenital myasthenia also present mutations in the 43K-rapsyn protein (Ohno et al., 2002) or in the E box of its promoter (Ohno et al., 2003; review in Engel et al., 2003). Furthermore, a zebra fish mutant containing an amino acid substitution in the tetratricopeptide domain displays lowered synaptic strength together with frequency-dependent depression (Ono et al., 2002)—evidence that 43K-rapsyn plays a role in regulating synaptic function in this species. Finally, mutations of several of the components of the dystrophin-glycoprotein complex (for example, dystrophin, α-sarcoglycans, α-dystrobrevin, and α2-laminin) lead to muscular dystrophies in both humans and experimental animals (Straub and Campbell, 1997; Engel et al., 2003a). Such "experiments of nature" unambiguously demonstrate the functional role of these molecular mechanisms in generating serious human pathologies. Understanding the mechanisms responsible for the clustering of acetylcholine receptors in the postsynaptic membrane is a first step to finding a cure for these serious neuromuscular diseases.

The identification of the nicotinic receptor brought into existence a new field of research, the chemistry of neurotransmitter receptors. Analysis of the biochemical and molecular mechanisms of neuromuscular junction development is now laying the groundwork for an understanding of the formation of neuronal networks and of synaptic plasticity at the molecular level (see Changeux, 1981). In this respect it is worth mentioning that, since the identification of 43K-rapsyn (Sobel et al., 1977b; Frail et al., 1987), "scaffolding proteins" have been identified in interneuronal synapses with a variety of central neurotransmitter receptors: the 93-kDa protein gephyrin for glycine receptors and $GABA_A$ receptors; MAP_{1B} for $GABA_A$ and $GABA_C$ receptors; the PSD_{95} family of proteins for the NMDA receptor; GRIP and $PICK_1$ for AMPA receptors; and Homer for group I metabotropic receptors (Lee and Sheng, 2000; Marchand and Cartaud, 2002). Many of these proteins—notably gephyrin and PSD 95—have been shown to be involved in a number of synaptic plasticity processes in the central nervous system at the molecular level that are relevant to learning and memory (Collingridge, 2003). We now go on, then, to examine the contribution made by neuronal nicotinic receptors to brain functions.

CHAPTER 11

Molecular Biology of Brain Nicotinic Receptors

In vertebrates, acetylcholine acts as a neurotransmitter operating upon the neuromuscular junction, where it triggers skeletal muscle contraction under the command of spinal cord motor neurons (Katz, 1966). Acetylcholine also acts as a major neurotransmitter in the autonomous sympathetic and parasympathetic nervous systems. In all these instances intercellular communication follows the standard scheme of fast chemical transmission (see Chapter 1). In the brain, however, the situation appears to be rather different. While similar mechanisms may operate there, brain cholinergic neurons also contribute to a distinct process of signal transmission referred to as *neuromodulatory signaling*. Indeed, unlike the pattern displayed by focal innervation of voluntary muscles, the projections of cholinergic neurons spread over extensive encephalic territories where they modulate the state of activity of large populations of neurons (Descarries et al., 1997; Perry et al., 1999) through a diffuse release of acetylcholine known as paracrine (or "volume") transmission (Zoli et al., 1999).

The cell bodies of brain cholinergic neurons are distributed in two principal ensembles over the ventral brain stem from the spinal cord to the telencephalon (Figure 11.1). In the posterior portion they form the cranial nerve nuclei, the medullary tegmentum, and the pontomesencephalic tegmentum, which includes the important pedunculopontine and dorsolateral tegmental nuclei. In the anterior portion, within the telencephalon, two major cholinergic subsystems may be observed: the basal forebrain cholinergic nuclei, which send out projections through the cortex and hippocampus, and distributed cholinergic interneurons, whose prolongations provide very rich local innervation

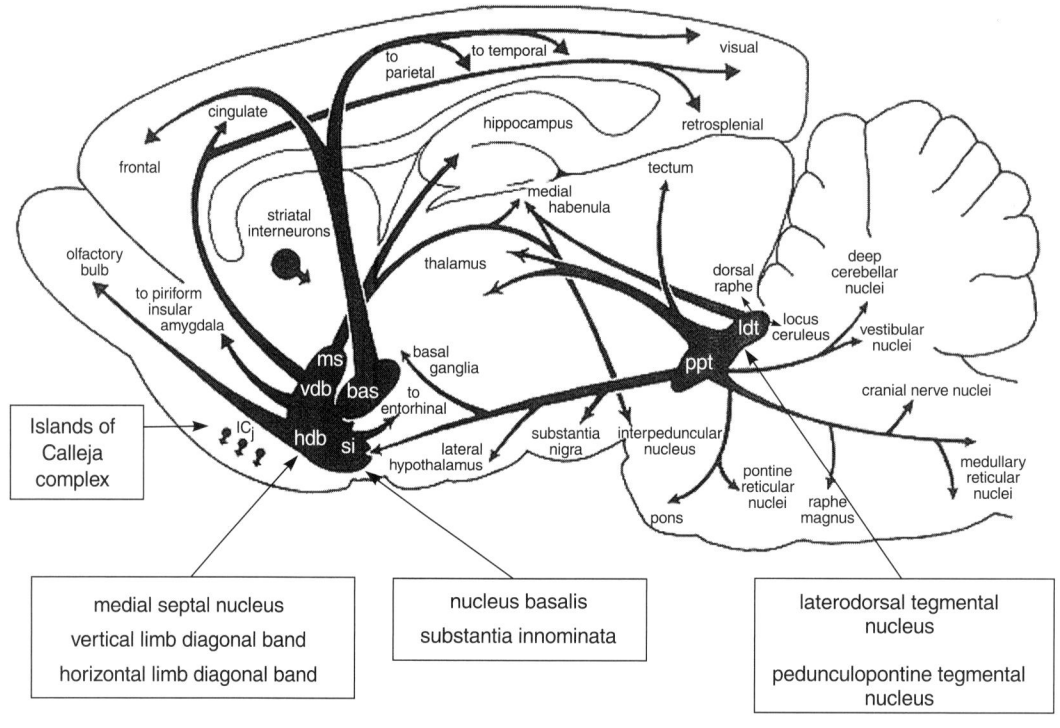

FIGURE 11.1 Cholinergic nuclei and projections in the rat brain. The central projections of cholinergic cells are schematically represented in a parasaggital section. The entire cortical mantle is innervated by the basal forebrain subsystem and the subcortical mass innervated by the pontomesencephalic subsystem, which includes the laterodorsal and the pedunculopontine tegmental nuclei. (From N. J. Woolf, 1991.)

(see Mesulam, 2000). Through these wide-ranging pathways, cholinergic neurons release acetylcholine over distributed populations of nicotinic receptors throughout the brain, sometimes with an apparent "mismatch" between nerve endings and receptors (references in Descarries et al., 2004).

In this chapter we will emphasize the importance of molecular biology in trying to understand the functional significance of the various nicotinic receptor subunits present in the brain (Boulter et al., 1986a; Couturier et al., 1990a; review in Role and Berg, 1996; Le Novère et al., 2002a). We will first consider the distribution of neuronal nicotinic receptor subunit mRNAs and proteins throughout the brain. In contrast with the restricted focal localization of muscle nicotinic receptors at the endplate (Chapters 9 and 10), brain nicotinic receptor subunit mRNAs and protein oligomers exhibit a variety of distinct patterns at the regional, cellular, and subcellular levels (Clarke, 1993; Sargent, 1993; Zoli et al., 1995; Figure 11.2A.) We will then describe what is known about the molecular mechanisms that regulate nicotinic receptor gene expression in the course of brain development and, finally, take up the issue of the functional significance of the diverse patterns of nicotinic receptor subunit distributions at the cellular and subcellular levels.

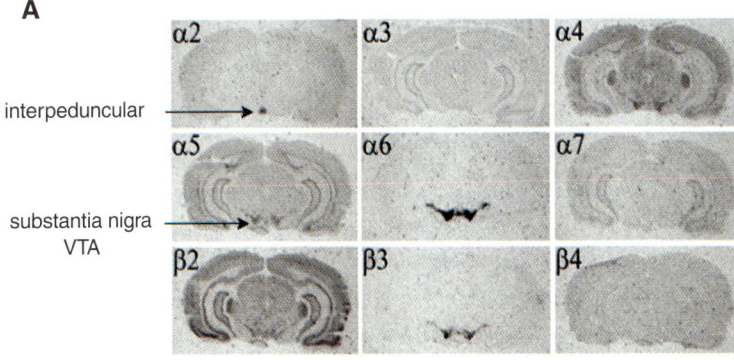

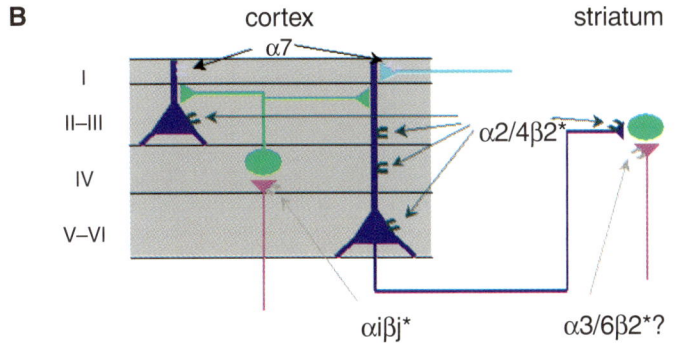

FIGURE 11.2 Localization of nicotinic receptor subunit mRNA by *in situ* hybridization in the brain. (A) Sections of mouse brain showing autoradiograms with subunit-specific labeling. The figure illustrates the widespread distribution of α4- and β2-subunit mRNA and the restricted distribution of the α6- and β3-subunits among dopaminergic nuclei (the substantia nigra and the ventral tegmental area). (B) Layer-specific distributions of selected receptor protein subunits in mouse and monkey. (From Han et al., 2003.)

Distribution of Nicotinic Receptor Oligomers

The subunit composition of skeletal muscle nicotinic receptor subunits throughout the vertebrate phylum is well established. The corresponding composition of neuronal and epithelial nicotinic receptors remains a subject of debate, however, in particular the precise stoichiometry of subunits within a given heteropentamer. The combinatorial diversity of the pentameric assemblies that in principle may be formed from the seventeen known subunit gene products among amniotes is indeed very large (C_n ~270). While only a few of these combinations have so far been identified in the brain, definite rules of subunit assembly are known to limit their number.

As we noted in Chapter 4, two major categories of subunits can be defined: the archaic α7-, α8-, and α9-subunits, which form functional homopentamers; and the more recently evolved α2–6, α10, and β2–4, which—though they do not form functional oligomers when expressed individually—may give rise to nicotinic responses when co-expressed in pairs or as part of larger assemblies. The simplest stoichiometry would be 2[α]:3[β] for such heteropentameric receptors, with two binding sites at the α-β interfaces (Anand et al., 1991). More complex forms of subunit organization are possible, however, with binding sites at α-α interfaces (Grailhe, unpublished results). Moreover, the α5- and β3-subunits do not carry the critical residues necessary to form either the principal or the complementary components of the acetylcholine-binding site; they may serve instead as "structural" subunits, participating in α3β4α5- or α3β4β3-oligomers, for example, with one subunit per pentamer (as in the case of β1 in muscle receptor heteropentamers). The α6-subunit carries all the residues necessary to form the principal component of the binding site, but it appears not to be readily incorporated into functional oligomers. The α6β2- and α6β4-subunits coexpressed in transfected BOSC 23 human cells can form functional oligomers, however (Fucile et al., 1998; Kuryatov et al., 2000). The α6-subunit (with other subunits) has also been identified as a necessary component of functional nicotinic receptor oligomers present in dopaminergic neurons (Champtiaux et al., 2002, 2003), a point we will discuss further in Chapter 12.

Tables 11.1 and 11.2 summarize the principal rules of neuronal nicotinic receptor assembly into functional oligomers, together with the various nicotinic ligand-binding properties of the known combinations of subunits. Some of these combinations exhibit a rather low affinity for acetylcholine (e.g., α7, α3β4), whereas others bind acetylcholine more tightly (e.g., α4β2, α4β4, and α3α5β2). As a consequence, *in situ* receptor autoradiography (currently used with tritiated nicotine: see Clarke, 1993; Sargent, 1993) may reveal only a fraction of nicotinic receptor binding sites. High-affinity nicotine-binding sites therefore

Table 11.1

Assembly Rules for Functional Nicotinic Receptors in Oocytes

	Homooligomeric[a]	Heterooligomeric		
		Principal subunits α	Complementary subunits β	Structural subunits β*
Subunit	α7, α8, α9	α2, α3, α4, α6	β2, β4	β3, α5, α10[b]
Stoichiometry[c]	5α	2(α)3(β) or 2(α)2(β)β*		
Binding sites	5	2 (at interface between subunits α and β)		

[a]Certain homooligomeric subunits can form functional nicotinic receptors in association with heterooligomeric subunits (for example, α7β2 or α7α5).

[b]The α10-subunit possesses the characteristics of a structural subunit in terms of the sequence of its N-terminal domain. It is found in functional nicotinic receptors only in association with α9.

[c]The stoichiometry of heterooligomeric receptors may not be as strict as suggested. Moreover, within a given receptor, the two α-subunits and the two (or three) β-subunits are not necessarily identical, e.g., α3β2β4 (Colquhoun and Patrick, 1997).

Table 11.2
Pharmacology of Human Nicotinic Receptors Reconstituted in Frog Oocytes[a]

	Acetylcholine (EC_{50}, µM)	Nicotine (EC_{50}, µM)	Cytisine (EC_{50}, µM)
α7	179.6	113.3	71.4
α9	~30	Antagonist	ND
α9α10	~30	Antagonist	ND
α2β2	68.7	19.2	25.4
α3β2	442.9	132.4	67.1
α4β2	68.1	5.5	2.6
α2β4	82.6	20.7	38.9
α3β4	203.1	80.3	72.2
α4β4	19.7	5	ND
Effect of structural subunits (Wang et al., 1996)			
α3β2	26	6.8	ND
α3α5β2	0.5	1.9	ND
α3β4	163	106	76
α3α5β4	122	105	20
α6-containing nicotinic receptors (Kuryatov et al., 2000)			
α3β2	76	91	ND
α6α3β2	373	30	ND
α4β2	2	0.34	ND
α6α4β2	2.6	1.3	ND
α6β4	18	7.1	9.2
α6β3β4	33	10	ND
α6β2α5	2.9	ND	ND

[a] The EC_{50} values in the first part of the table are from Chavez-Noriega et al., 1997; and Elgoyhen et al., 2001.

consistently coincide throughout the brain with the distribution of α4 and β2 mRNAs and of α4β2 protein oligomers (Picciotto et al., 1995; Marubio et al., 1999; see also Figure 12.4).

In contrast, α-bungarotoxin toxin selectively tags the α7 low-affinity homopentamers in the brain, and the overall labeling pattern of this toxin resembles that of α7 mRNA (Seguela et al., 1993). In the case of both α4β2- and α7-oligomers, however, detailed studies have revealed that binding sites near axon terminals may be located quite far from the cell bodies containing their mRNA (see following paragraphs). Among the many toxins synthesized by the marine snails of the genus *Conus,* α-conotoxin MII (Cartier et al., 1996), which initially was thought to label α3-containing receptors (Cartier et al., 1996), was discovered to selectively tag α6-containing receptors (Champtiaux et al., 2002, 2003), and possibly the α6-β3 interface as well. Epibatidine, a highly toxic alkaloid recently isolated from the skin of a South American frog, has also been widely used to label high-affinity nicotinic receptors in the brain (Picciotto et al., 1995; Zoli et al., 1998; Marubio et al., 1999). Most of these receptors correspond to α4β2 oligomers, but in certain discrete brain regions (inferior colliculus, tractus solitarius nucleus, area postrema, and within the habenulo-interpeduncular system) epibatidine may also label α4β4-, α3β4-, and α2β4- receptor oligomers (Champtiaux and Changeux, 2002). In summary, we may say that a great variety of nicotinic receptor oligomers exists in the brain, with rather distinct physiological and pharmacological properties. Their distribution varies to a considerable extent according to the type and combination of subunits of which they are composed. A critical question, then, is how the regulatory mechanisms of gene expression operate to produce this distribution.

Expression Patterns of Nicotinic Receptor Subunits in the Brain

One of the first major consequences of the cloning and sequencing of brain nicotinic receptor genes (Le Novère and Changeux, 1999) was the availability of highly specific oligonucleotide probes for *in situ* hybridization of specific subunit mRNAs. Several expression profiles have been identified in rat and mouse brains (Figure 11.2) using this method, which we will now briefly review. The most common subunit mRNAs, α4 and β2, which are known to contribute to high-affinity nicotinic binding, are widely expressed in the rodent central nervous system, including the retina (Wada et al., 1989; Hoover and Goldman, 1992; Zoli et al., 1995). Yet in the cerebellum and hippocampus, by contrast with the striatum, α4 mRNA is not detected, whereas β2 mRNA is frequently observed. The mRNA for α7, which codes for the low-affinity oligomer, displays a different pattern. It is abundantly expressed in some telencephalic and diencephalic nuclei (notably the amygdala and olfactory bulb), but also in the cortex of hippocampus. In the mouse it is found in the superior colliculus, the interpeduncular nucleus, *and* in dopaminergic neurons. The structural α5-subunit mRNA also displays a broad expression profile: it is coexpressed with α4 and β2 in cortex, hippocampus, and dopaminergic nuclei, but with α3 and β4 in the pineal gland and autonomous ganglia. These last two subunit mRNAs are heavily expressed in the autonomous ganglia, where the subunits they code play a critical role in autonomous transmission (Rust et al., 1994). In the brain, however, α3 and β4 are discretely expressed in only a few structures. In catecholaminergic nuclei such as the dopaminergic neurons and the locus ceruleus, α6 and β3 mRNAs are coexpressed (Denerìs et al., 1989; Le Novère et al., 1996), and α6 mRNA is also abundantly found in retinal ganglion cells (Champtiaux and Changeux, 2002).

Interestingly, the pattern of expression of α2, a still rather enigmatic subunit, varies considerably from one species to the other. In the rat brain α2 mRNA is expressed exclusively in the olfactory bulb and in the interpeduncular nucleus (Wada et al., 1989). In the chick brain it is restricted to only one small nucleus: the lateral spiriform nucleus (Daubas et al., 1990). In the rhesus monkey, on the other hand, α2 mRNA is expressed throughout the brain (Han et al., 2000, 2003). In humans α2 expression has been observed in various parts of the brain and spinal cord (Martin-Ruiz et al., 2002; Keiger et al., 2003b), as well as the nasal mucosa (Keiger et al., 2003a).

Finally, nicotinic receptor subunit gene expression is not restricted to the brain. Subunit genes α9 and α10 are expressed exclusively in epithelial cells, such as the hair cells of the cochlea (Elgoyhen et al., 1994, 2001). Furthermore, several brain nicotinic receptor subunit mRNAs have been reported in lymphocytes, thymocytes, and keratinocytes, and even in muscle cells (Corriveau et al., 1995; Cormier et al., 2004).

With a few notable exceptions (e.g., α2), this rich variety of expression patterns is conserved across species. It is encoded, as we shall see, in the promoters and transcription factors that selectively regulate the expression of individual subunit genes.

Transcriptional Regulation of Brain Nicotinic Receptor Genes

The molecular mechanisms that account for pattern formation in the brain are far more complex than those involved in the compartmentalized expression of muscle nicotinic receptor genes at the motor endplate, and still much less well understood. The functional organization of the brain is orders of magnitude more complex than that of the skeletal muscle. The models and methods used in connection with the neuromuscular junction, particularly the ones associated with the promoter approach (Changeux et al., 1987b; Kerszberg and Changeux, 1993; Chapter 9) have nevertheless been usefully extended to the differential transcription of neuronal nicotinic receptor genes in the brain. Let us begin by considering two extreme cases: the α2-subunit, whose distribution in the chick is restricted to a single nucleus; and the β2-subunit, which is very broadly distributed in the brains of all species examined.

The chick α2-gene promoter (Bessis et al., 1993) contains several consensus sequences for transcription factors. Quite surprisingly, they include three Ets sites that are analogous to the ones found in the muscle ε- and δ-promoters (Chapter 9). These sites are found together with other sites (Ap2, ATF, Ap4, Myc, CIIS, SCG10, NF-M) and weakly degenerate OCT-like sites; no TATA or CAAT boxes have been observed (for the structure and functional significance of these DNA elements, see the Molecular Biology Web Book at http://www.web-books.com/MoBio/). Promoter sequences of up to 7 kb were unable, however, to stimulate expression of a reporter gene in any cell line (either neuronal or nonneuronal) tested. Dissection of this 5′ sequence revealed that an 11-bp motif (CCCCATGCAAT) capable of binding POU transcription factors is repeated six times in the $-130/-30$ region of the gene. High-resolution mapping further revealed that these six tandem repeats exert a strong *silencing effect* on an SV40 promoter construct. Paradoxically, this *silencer element* was able to act as a transcription activator if only four copies, or two copies, or even one copy of the 11-nucleotide sequence—rather than six copies—were placed upstream of the basal promoter fragment (Bessis et al., 1993). One of several repressing transcription factors from the POU family, Brn 3b, behaved as an activator, however, unlike its homologues Brn 3a and 3c (Milton et al., 1995). Remarkably enough, then, it turns out that complex interactions between DNA regulatory elements and transcription factors, including strong repressors as well as efficient activators, contribute to the highly restricted process of expression supervised by the α2-subunit promoter.

By contrast, β2-subunit mRNA is widely expressed throughout the brain in rats and other species examined (Wada et al., 1989; Hill et al., 1993). What is more, its appearance occurs very early in development, and in a way that closely parallels the timing of neuronal differentiation (Zoli et al., 1995). Within the 1.2-kb sequence constituting its promoter region upstream of the transcription start site, several consensus DNA elements can be recognized: an Sp1 site, a cAMP responsive element, a nuclear response element, an OCT-like motif (-522), and—quite unexpectedly—an E box (-118) analogous to the one found in the proximity of muscle genes (Chapter 9). The transcribed domain contains a neural restrictive silencer element (NRSE) sequence ($+18$ to $+38$) (Bessis et al., 1995). The β2-promoter sequence gives rise to efficient transcriptional activation in pheochromocytoma and neuroblastoma cell lines, but not in nonneuronal cell lines. The Sp1 and E-box elements, which have been shown to contribute to α1 subunit gene expression in muscle (Chapter 9), also behave as powerful sources of activation in the context of the β2-promoter sequence. Moreover, a point mutation of the NRSE element within the 1163-bp promoter results in a more than 100-fold increase in transcriptional activity in fibroblasts, but only a 3.2-fold increase in neuroblastomas. This result strongly suggests that the NRSE element contributes to the *neuron-specific expression* of the β2-subunit gene, primarily by repressing its transcription in nonneuronal cells. *In vivo*, in transgenic mice, point mutation of NRSE within the same promoter context strikingly affects the pattern of expression of the lacZ reporter gene (Figure 11.3). At approximately day 13 of embryonic development, signs of lacZ staining disappear in the posterior and peripheral nervous system while at the same time becoming intensely visible in the anterior part of the brain. In other words, a switch in reporter gene expression from posterior to anterior takes place along the anteroposterior axis of the nervous system (Bessis et al., 1997). This constitutes preliminary, but nonetheless striking, evidence in favor of the hypothesis that a combination of defined promoter elements and transcription factors plays a significant role in the formation of topologically defined *patterns* of brain nicotinic receptor gene expression in the brain (Bessis et al., 1995, 1997).

As we noted in Chapter 3, examination of the chromosomic localization of nicotinic receptor genes revealed that some of the identified neuronal genes form definite clusters: [α3, α5, β4] on mouse chromosome 9 and human chromosome 15, and [α6, β3] on mouse and human

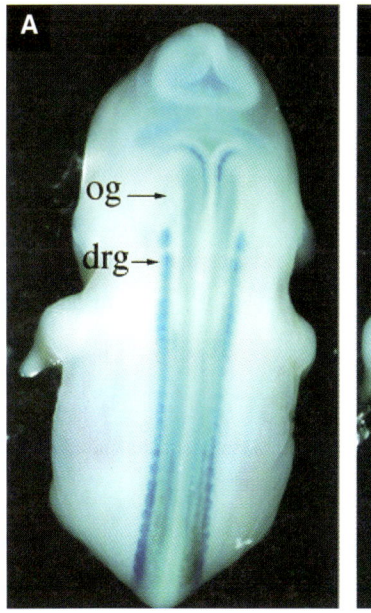
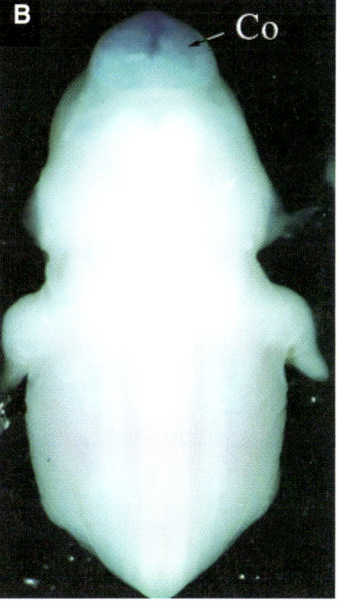

FIGURE 11.3 Dissection of the β2-promoter. (A, B) Whole-mount coloration of transgenic mouse embryos at day 13.5 of development expressing (A) a wild-type promoter and (B) a promoter with a point mutation in NRSE. og: orthosympathetic ganglia; drg: dorsal root ganglia; Co: colliculus. (C) The overall structure of the mutated promoter. As a consequence of the mutation, the mostly posterior pattern of transgene expression under the control of the β2-promoter is converted into an anterior one. (From Bessis et al., 1997.)

chromosome 6 (Bessis et al., 1990; Boulter et al., 1990; Tseng et al., 2001). The genes in these clusters are typically coexpressed, an indication of common regulatory elements. Indeed, an enhancer sequence in the [α3, α5, β4] cluster named β4-3′ has been located in the 3′ noncoding sequence of the β4-gene, upstream of the α3-gene (Boyd, 1994; McDonough and Deneris, 1997). This enhancer sequence also contains Ets-binding sites, within which mutations are known to abolish a loss of β4-3′ activity (Yang et al., 1997; McDonough et al., 2000). Here again, combinational *interactions* of a variety of transcription factors in addition to Ets, such as Sox 10 (Liu et al., 1999) and POU (Yang et al., 1994), contribute to neuron-specific expression. Other promoter sequences for α7- (Criado et al., 1997), α4- (Watanabe et al., 1998), and β3- (Hernandez et al., 1995) subunit genes have also been investigated. Some of these contain known transcription factor elements, among them a functional E box (β3) and a CAAT box (β3) (Roztocil et al., 1998), as in the case of other nicotinic receptor genes.

Although still in their infancy, all these studies reveal a sophisticated interplay among transcription factors and DNA elements in the patterned expression of neuronal nicotinic receptors in the brain. They have further shown that, for a given promoter, only a minor sequence homology exists between species (in the case of β4, for example, between bovines and humans; see Valor et al., 2002). This suggests that promoters—rather than the actual number and sequence of protein-coding genes, which are very similar in mouse and human genomes (see Changeux, 2002, 2004)—have been the primary targets of factors responsible for the differential expression of genes within the brains of different species during the course of biological evolution, to which the human brain owes its origin (see Changeux, 2002, 2004; Le Novère et al., 2002a). These studies also illustrate, although they do not yet completely explain, the complex assembly of transcription factors that control gene transcription in conjunction with the core RNA polymerase II complex (Mannervik et al., 1999; Pennisi, 2003). These allosteric oligomers may constitute critical "regulatory nodes" in the intracellular networks of gene regulation that control pattern formation in the embryo and, more specifically, in the brain (Kerszberg and Changeux, 1994, 1998; Davidson et al., 2002).

Subcellular Compartmentalization of Brain Nicotinic Receptor Oligomers

In the adult skeletal muscle, one molecular species of nicotinic receptor protein is expressed at a single locus: the endplate postsynaptic membrane. As described in Chapter 10, a unique secretory pathway targets the expression of the receptor to the postsynaptic domain. In the brain the situation is far more complex, since the rather elaborate transcriptional mechanisms we have just discussed operate upon the single nucleus of the nerve cell. This transcriptional activity is succeeded by a phase of sophisticated, multidimensional *posttranscriptional* processing that assures that the right combination of nicotinic receptor subunits is established at the right place in a given nerve cell. Investigation of the subcellular compartmentalization of nicotinic receptors at the nerve cell level—carried out mainly by electrophysiological recording techniques, either on brain slices or *in vivo*—is a first step toward understanding these quite complicated phenomena.

Nicotinic Receptors in Axon Terminals

Early *in vitro* work on synaptosomes (review in Wonnacott et al., 2000) revealed the presence in the rat striatum of functional nicotinic receptors that are *not* blocked by the channel blocker tetrodotoxin (TTX), and thus are authentic "terminal" nicotinic receptors. Electrophysiological experiments were conducted using slices of the prefrontal cortex, an area of the brain known to play a key role in higher brain functions (see Chapter 12 and Granon et al., 2003). Initial extracellular recordings of field potentials and unit discharges in the prelimbic area of the prefrontal cortex revealed excitatory effects from nicotine (Vidal and Changeux, 1989; Figure 11.4). Intracellular recordings were subsequently made (Vidal and Changeux, 1993) of pyramidal cells located predominantly in layer II/III, where nicotine-binding sites are concentrated (review in Clarke, 1993). Iontophoretic applications of nicotine near the recording site *increased* the amplitude of the postsynaptic potentials evoked by stimulation of the superficial cortical layers (comprising 14% of the cells). Muscarinic agonists, on the other hand, *decreased* the amplitude of the postsynaptic response. In every case, the early postsynaptic potentials were blocked by a specific inhibitor of non-NMDA glutamate receptors (CNQX), but not by a selective blocker of NMDA glutamate receptors, APV (2-amino-5-phosphonopentanoic acid). It follows that functional *presynaptic* nicotinic receptors are found in the neocortex, in particular in the prefrontal cortex, which, as we shall see in the following chapter, plays a critical role in higher brain functions (Dehaene et al., 2003). In this area they were found to potentiate excitatory *glutamatergic* synapses, a result subsequently confirmed and extended in different systems (McMahon et al., 1994b; McGehee et al., 1995). Nicotinic receptors are also present, however, in inhibitory nerve endings, which are known to be crucially involved in the central processing of information.

CHAPTER 11

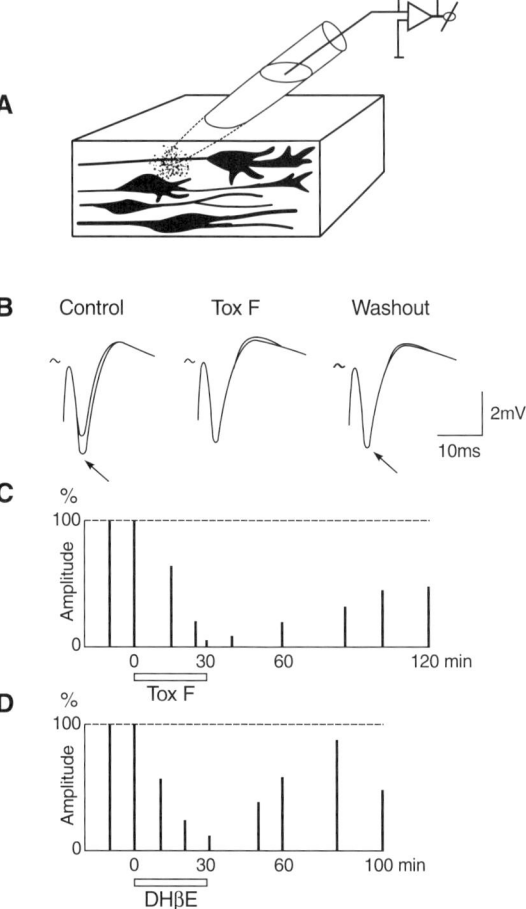

FIGURE 11.4 Effects of iontophoresis-applied nicotine on field potentials evoked by white matter stimulation in slices of rat prefrontal cortex. (A) Glass microelectrode recording of extracellular field potentials. (B) Effects of nicotine and toxin F, a brain nicotinic receptor antagonist, on field potentials: the arrow indicates the application of nicotine. (C, D) Time course of the effect of two antagonists, toxin F and dihydro-β-erythroidine DHβE. (From Vidal and Changeux, 1989.)

Alpha-7 nicotinic receptors have positive effects on the efficacy of immature connections, as has been recently discovered in slices of the newborn rat hippocampus (Maggi et al., 2003). A significant number of Schaffer collaterals-CA1 glutamatergic synapses are in a latent ("silent") state at birth; but brief application of nicotine is sufficient to cause them to switch from a silent state to a functionally active one. Interestingly, the action of nicotine on α7-receptors facilitates this transition. The current balance of evidence suggests that the α7-receptors involved are presynaptic (Maggi et al., 2003).

Nicotinic agonists acting upon thin slices of the ventrobasal complex and the dorsolateral geniculate nucleus from the sensory thalamus of the mouse (Léna and Changeux, 1997) enhance the frequency of miniature *GABA currents* and decrease the failure rate of evoked inhibitory synaptic currents. This nicotine-enhanced release results either directly from the entry of Ca^{2+} ions through nicotinic receptor channels (in the dorsolateral geniculate nucleus) or indirectly from the activation of voltage-dependent Ca^{2+} channels (in the ventrobasal nucleus) by the depolarization of the nerve-ending membrane.

The "Preterminal" Axonal Compartment

The idea that nicotinic receptors are found along axons, proposed more than four decades ago by David Nachmansohn (1959), was resisted by many distinguished electrophysiologists (Eccles, 1964; Katz, 1964). In the interval, the presence of these receptors was empirically confirmed, though in a rather different conceptual context. As anticipated, whole-cell recordings (Léna et al., 1993) made on thin slices of the interpeduncular nucleus or on acutely isolated neurons whose synaptic contacts remain attached to their cell bodies (as indicated both by synaptophysin immunoreactivity and by spontaneous GABA and glutamate release) showed that nicotine dramatically increases the frequency of large gabaergic currents in these cells (Figure 11.5).

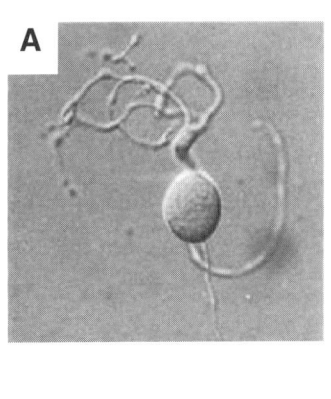
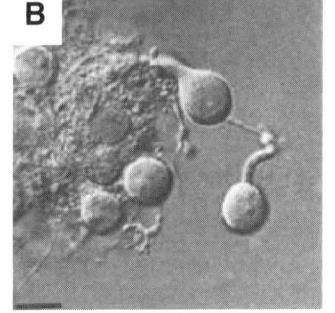
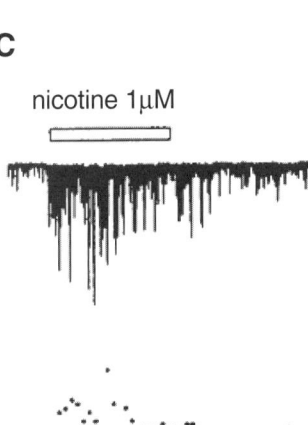
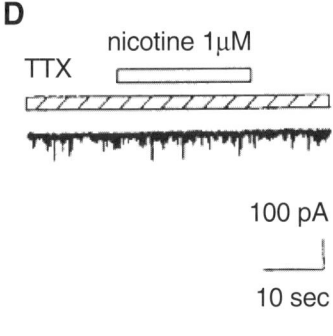

FIGURE 11.5 Electrophysiological properties of nicotinic brain receptors. (A, B) Small cells freshly dissected from the habenular nucleus of a 17-day-old rat are a useful model system for measuring calcium influx though the neuronal nicotinic receptor channel. (From Mulle and Changeux, 1990.) (C, D) Evidence for "preterminal" nicotinic receptors. While abundant evidence exists for the occurrence of nicotinic receptors at nerve endings, where they are resistant to the Na-channel blocker tetrodotoxin (TTX), measurements made with slices of interpeduncular neurons demonstrate that some presynaptic receptors are sensitive to TTX, suggesting that they are located on axons some distance away from the nerve endings. (From Léna et al., 1993.)

Rather unexpectedly, however, the Na$^+$ channel blocker tetrodotoxin (TTX) was found to *inhibit* this effect. Since the action of TTX is strictly limited to Na$^+$ channels, these channels must assist signal transmission between the nicotinic receptor and the GABA release site. The location of these nicotinic receptors therefore could not be the same as that of the "terminal" receptors mentioned earlier in connection with synaptosomal preparations, which are insensitive to TTX. The term "preterminal" was coined by Léna et al. (1993) to specify the distribution of these nicotinic receptors along the axon. These observations were subsequently confirmed, and its term adopted, by other groups working with different systems (McMahon et al., 1994a, 1994b; Role and McGehee, 1996). The exact function of these preterminal nicotinic receptors on axon branches still remains elusive. One possibility is that they are the targets of cholinergic terminals that have yet to be identified. Another possibility, having important theoretical implications for the understanding of brain communication, is that they exert "global" control of neurotransmitter release by a general depolarization of afferent axons due to volume transmission of acetylcholine (Beaudet and Descarries, 1978; Descarries et al., 1997; Agnati et al., 1995; Zoli et al., 1999).

The Somatodendritic Compartment

Medial habenula neurons constitute a particularly convenient system for studying nicotinic currents since, as we have noted earlier, single channels can be recorded using patch-clamp techniques, either from freshly dissociated neurons of juvenile rats (Mulle and Changeux, 1990; Mulle et al., 1991, 1992a, 1992b) or from thin slices of the adult brain (Connolly et al., 1995; Picciotto et al., 1995). In the ventral region of the medial habenula, neurons express mRNAs encoding the α3-, α4-, α7-, β2-, β3-, and β4-subunits. Moreover, neurons from the medial habenula project to the interpeduncular nucleus via the fasciculus retroflexus. Since patch-clamp recording can be performed on interpeduncular neurons both after acute isolation and in thin slices of brain tissue, the physiology and pharmacology of nicotinic receptors located at both pre- and postsynaptic sites can be compared within the same system. The data show that the rank order of efficacy for agonists of *presynaptic* nicotinic receptors (nicotine > cytisine > acetylcholine > DMPP) differs from the order observed for the soma of *postsynaptic* interpeduncular neurons (cytisine > acetylcholine > nicotine > DMPP) (Mulle et al., 1991). In other words, the physiological properties, and probably the subunit composition of pre- and postsynaptic nicotinic receptors as well, are apt to differ within a given synaptic connection. Although the recording techniques were not the same, evidence also exists for pharmacological differences (IC_{50} values for antagonists) between somatic and axonal nicotinic receptors within the same habenular neuron.

Different nicotinic receptor species thus not only coexist within a given nerve cell, but they may also segregate into different compartments (somatodendritic and axonal) on the neuronal surface. Similar conclusions have been reached with regard to peripheral neurons in, for example, the chick ciliary ganglion (Bey and Couroy, 2002) and to autonomic neurons (Rosenberg et al., 2002).

Conclusion

Acetylcholine was initially recognized as a neurotransmitter that faithfully transmits signals from the motor nerve to skeletal muscles by means of "quantal" transmitter release (see Katz, 1996; Kuffler and Yoshikami, 1975; reviews in Changeux, 1981, and this book). Acetylcholine is also abundantly present in the brain along with many other excitatory neurotransmitters, such as glutamate and aspartate, and inhibitory neurotransmitting substances, such as GABA. Yet a wealth of data suggests that acetylcholine may also serve as a neuromodulator that is released in the wake of paracrine (or volume) transmission (Beaudet and Descarries, 1978; Agnati et al., 1995; Zoli et al., 1998). The molecular biology of nicotinic receptors, inaugurated by biochemical studies first carried out with the electric organ of fish (review in Changeux, 1980), led directly to the identification of the various genes (seventeen in all) that code for central and peripheral nicotinic receptor proteins (Heinemann et al., 1990). The overall distribution and function of brain nicotinic receptors is consistent with the posited neuromodulatory role of acetylcholine, which is supposed to coexist with its familiar "wiring" mode.

Even though the evidence in favor of this hypothesis is still fragmentary, it seems probable that a compartmentalized targeting of nicotinic receptor oligomers takes place in central neurons within the somatodendritic, preterminal, and terminal domains of the cell membrane. The posttranscriptional mechanisms involved in this process remain to be identified. Nevertheless the widespread distribution of different species of nicotinic receptor oligomers among these loci is consistent with the view that, in addition to classical synaptic transmission, the diffuse paracrine release of acetylcholine plays a critical role in helping to process information and to regulate plastic phenomena in the brain, particularly those associated with cognition and learning. These matters form the subject of the final chapter.

CHAPTER 12

Nicotinic Receptors and Brain Functions

Cholinergic neurons in the brain have long been known to modulate cognitive behavior (Bartus et al., 1982; Levin, 1992), but research focused initially on muscarinic receptors as the targets of acetylcholine (Levin, 1992). Only in recent decades has the capacity of nicotinic receptors been recognized to exert a role in neuroprotection or in brain functions such as reward processes, addiction, the central processing of pain signals, working memory, selective attention, anxiety, sleep and wakefulness, and what might be called "access to consciousness" (Dehaene et al., 1998, 2003; Perry et al., 1999). Moreover, nicotinic receptor alterations or deficits have been implicated in cognitive impairments associated with Alzheimer's disease, schizophrenia, attention deficits, hyperactivity disorders, and other types of neurological dysfunction, notably Parkinson's disease (Levin and Rezvani, 2002; Picciotto and Zoli, 2002) and the autosomal dominant nocturnal frontal lobe epilepsy discussed earlier in Chapter 8. Finally, nicotine is largely responsible for the addictive behavior of tobacco smoking, along with the dramatic consequences it entails for the development of cancer.

The overall impact of current research on nicotinic receptors and brain functions is far too broad to be dealt with in a single chapter. We confine ourselves here to examining the role of brain nicotinic receptors on the basis of ongoing experiments being carried out with a simple animal model. Despite the rather limited character of its cognitive abilities, the mouse has been selected because it can be engineered for specific nicotinic receptor subunit genes—deleted, in the case of knockout mice (Picciotto et al., 1995, 1998); mutated, in the case of knockin mice (Labarca et al., 2001; Orr-Urtreger et al., 2000); or selectively reexpressed in defined brain regions (Maskos et al 2005). In recent years, mice lacking the following specific nicotinic receptor subunits have been generated by means of homologous recombination technologies: α3 (Xu et al., 1999a), α4 (Marubio et al., 1999; Ross et al., 2000), α5 (Wang et al., 2002), α6 (Champtiaux et al., 2002), α7 (Orr-Urtreger et al., 1997), α9 (Vetter et al., 1999), β2 (Picciotto et al., 1995, 1998), β3 (Cui et al., 2004), or β4 (Xu et al., 1999b) (review in Champtiaux and Changeux, 2002). Striking differences can be observed between mutant and wild-type mice, as in the case of the β2 knockout, which results in a complete loss of high-affinity ^{3}H-nicotine binding throughout the brain, including the prefrontal cortex (Figure 12.1). As we shall see, analysis of the physiological, pharmacological, and behavioral phenotypes of these mutant mice has succeeded in clarifying the contribution made by individual subunits, or combinations of subunits, to specific aspects of nicotinic transmission and modulation of behavior at different levels of organization. We shall conclude by discussing the implications of this research on the contribution of neuronal nicotinic receptors to human higher brain functions and to their pathologies.

Nicotinic Receptors in Brain Development and Aging

A possible role for nicotinic receptors in the early development of the nervous system is a regulation of the *proliferation and/or survival of neuroblasts* (see, for instance, Buss and Oppenheim, 2004). Quite interestingly, infusion of the competitive antagonists *d*-tubocurarine or α-Bgt in chick embryos was found to rescue spinal cord motoneurons from naturally occurring cell death (Renshaw et al., 1993). Similarly, the number of newborn neurons in the adult olfactory bulb has been found to be significantly larger in β2-deficient ($^{-/-}$) mice compared to their controls (Mechawar et al., 2004), possibly because of a decreased rate of cell death (Buss and Oppenheim, 2004).

Conversely, excessive stimulation of nicotinic receptors enhances neuronal cell death as illustrated by studies of knockin animals expressing mutant forms of α4- or α7-subunits. In these mice, both the gain-of-function mutation α4Leu-9′Ser and the mutation at the homologous position, α7L250T, have been shown to cause a selective degeneration of the particular neurons that express the α4- or α7-subunits at high levels (Labarca et al., 2001; Orr-Urtreger et al., 2000; Broide et al., 2001). Also, in tests of an experimental model of cortical lesions elicited in newborns by the glutamate receptor agonist ibotenate, the size of the lesion was found to be significantly smaller in α7-deficient

CHAPTER 12

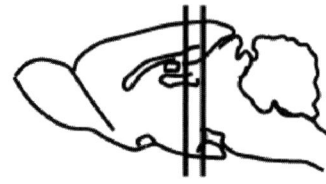

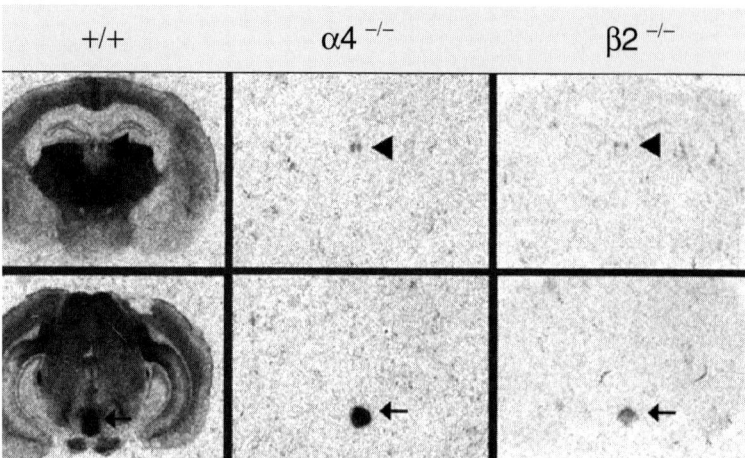

FIGURE 12.1 Binding of ^{3}H-cytisine in $\alpha 4^{-/-}$ and $\beta 2^{-/-}$ mutant mice. Deletion of $\alpha 4$- and $\beta 2$-subunit genes causes an extensive loss of high-affinity nicotine and cytisine (another nicotinic agonist) binding throughout the brain, confirming the major role $\alpha 4 \beta 2$ nicotinic receptors are presumed to play in brain functions. (From Picciotto et al., 1995; Marubio et al., 1999; and Zoli et al., 1998.)

mice by comparison with their wild-type controls (Laudenbach et al., 2002). These findings suggest that endogenous acetylcholine, acting at α7-receptors, potentiates the excitotoxic effects of ibotenate in wild-type animals, possibly as a consequence of massive calcium entry.

Subsequently in the development of the nervous system, nicotinic receptors may affect the *formation, stabilization, and elimination of synaptic connections*. In vertebrates, particularly mammals, interneuronal synaptic connections do not become established as the result of a programmed one-to-one matching. They seem instead to follow a kind of gradual trial-and-error process exhibiting phases of exuberant growth, characterized by maximal variability, that are followed by phases of selective stabilization and a concomitant elimination of developing connections (Ramón y Cajal, 1909–1911; Changeux et al., 1973; Changeux and Danchin, 1976; Changeux, 1983b, 2002, 2004; Lichtman and Sanes, 2003). Consistent with the theory that these selection processes in developing networks play a critical role in shaping adult connectivity (for discussion see Changeux, 1983a, 1985), the issue today is to what extent the activation of nicotinic receptors regulates the selective stabilization of synapses in the course of brain development. Here, again, genetically engineered mice have served as useful experimental models.

The classic experiments of Hubel and Wiesel over forty years ago demonstrated the crucial influence of early experience on the development of the visual system. The role played by nicotinic receptor activation in this system has been examined in detail in the case of β2-deficient mice. Refinement of retinothalamic projections in the mouse visual system (P4–P9) occurs during early postnatal development (So et al., 1990). The retinal ganglion cells from each eye initially create intermingled projections to their thalamic target, the dorsolateral geniculate nucleus, that subsequently become segregated into eye-specific layers. The segregation of retinothalamic projections is a competitive process which is endogoneously controlled by what may be called an "internal" experience triggered by the spontaneous wavelike activity arising from retinal ganglion cells (Penn et al., 1998). Drug-elicited blocking of these spontaneous patterns is known to inhibit the segregation process (Feller et al., 1996; Penn et al., 1998). These pharmacological observations have been confirmed and extended by the further discovery that spontaneous retinal waves are absent in mice lacking the β2-subunit (Bansal et al., 2000), which implies that nicotinic receptors in the retinal network form part of the genetic basis for such oscillations. In β2-knockout animals at P4 (before the normal onset of binocular segregation), the pattern of inputs from the two eyes in the dorsolateral geniculate nucleus is normal. Adult mutant mice, on the other hand, exhibit disrupted lamination of their retinogeniculate projections, comparable to the overlap observed in the immature geniculate body of wild-type neonates (Rossi et al., 2001; Grubb et al., 2003; Muir-Robinson et al., 2003; Figure 12.2). Nonetheless, fine-structure segregation of ipsilateral and contralateral axons does in fact develop (Muir-Robinson et al., 2003), with normal gross retinotopy in the geniculate nucleus but a lack of fine-scale mapping along the nasotemporal visual axis accompanied by a gain in the spatial segregation of on/off cells (Grubb et al., 2003). In addition, mutant mice were found to exhibit reduced visual acuity as well as a functional expansion of the binocular visual cortex (Rossi et al., 2001; Grubb et al., 2003). These experiments, which are still in progress, unambiguously demonstrate that nicotinic receptors contribute in a positive and necessary way to the epigenetic maturation of the visual system.

In Alzheimer's disease, a selective decrease in the number of telencephalic nicotinic receptors has been observed (Aubert et al., 1992; Nordberg, 2001), and the severity of associated cognitive deficits correlated to the loss of cholinergic function (Mega, 2000). The impact of nicotinic dysfunction in $β2^{-/-}$-mutant mice has also been found to intensify the neurodegeneration that normally accompanies aging (Zoli et al., 1999). The performance of aged $β2^{-/-}$ mice turned out to be significantly impaired by comparison with age-related wild-type animals in the Morris water-maze learning test, for example, and also in a fear-conditioning task (Zoli et al., 1999; Caldarone et al., 2000). While a decrease in neocortical thickness and a loss of pyramidal hip-

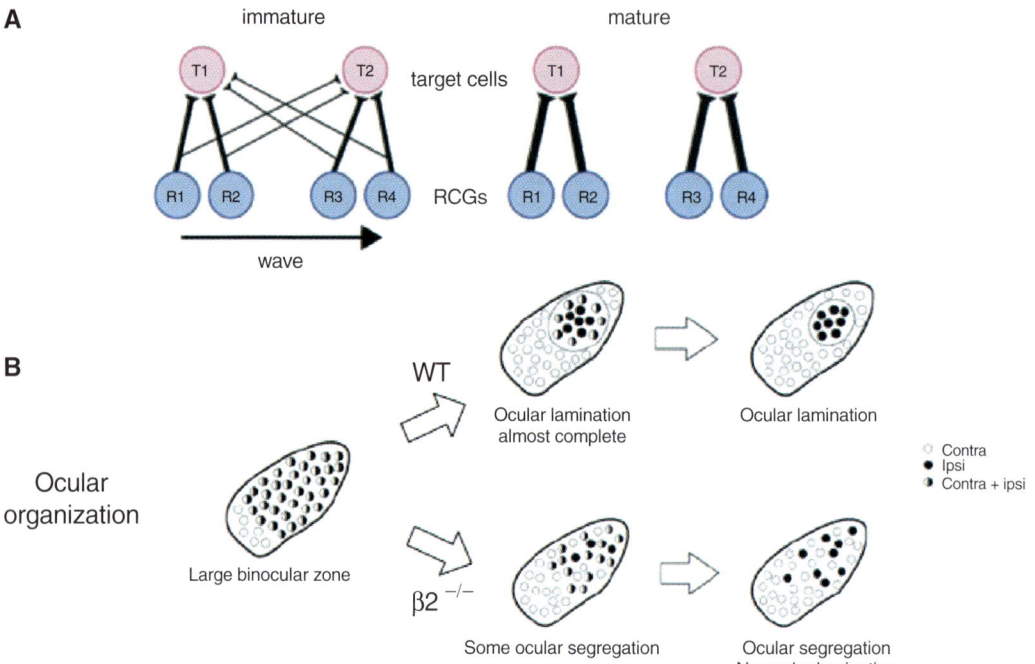

FIGURE 12.2 Consequences of β2-subunit deletion for the development of retinogeniculate projections in the visual system. (A) The selective-stabilization hypothesis as proposed by Changeux, Courrège, and Danchin (1973) and redrawn by Eglen et al. (2003). The waveform spontaneous activity of the retina contributes to synaptic selection but is superseded in the newborn by externally evoked activity. (B) Ocular organization of the geniculate body in wild-type and β2$^{-/-}$ mice. Spontaneous activity of the retina is absent in the latter type, and the lamination of the retinal projections is disorganized. (From Grubb et al., 2003.)

pocampal neurons were also observed in aged wild-type animals, neuronal atrophy was accelerated in β2-mutant mice.

Even if the analogy between Alzheimer's disease and the β2-mutant mice phenotype is still debatable, the role of β2 nicotinic receptors in promoting the neuroprotective effects of nicotine during aging is now well established. These studies give rise to hope that therapies focusing on α4β2-containing nicotinic receptors may one day prevent, or at least offset, the impact of cholinergic loss upon cognitive functions in Alzheimer's patients. Chronic treatment by nicotine or nicotinic agents may simply be an efficient form of such therapy.

Epidemiological studies have suggested that smoking may protect against the evolution of neurodegenerative diseases such as Alzheimer's disease and Parkinson's disease; that nicotine may defend against several neurotoxic agents in animal models, both *in vivo* and *in vitro*; and that the number of nicotinic binding sites is differentially decreased by comparison with muscarinic binding sites in patients suffering from dementia or other aging-related conditions (reviews in Zoli et al., 1999; Picciotto and Zoli, 2002). On the other hand, at early stages of development, activation of nicotinic receptors may have the opposite effect of enhancing cell death and/or synaptic elimination.

Nicotinic Receptors in Reward and Addiction

Ever since the pioneering studies of Edward L. Thorndike (1911) and Clark Hull (1943) on animal learning it has been realized that the storage of learning and memory traces does not take place in a passive, non-selective, instructive manner like a "blind" imprinting process (Kandel, 1976). On the contrary, *evaluation* mechanisms operating by means of externally and/or internally represented *reward* processes are thought to determine the selection of the stored (or rejected) representations (review in Dehaene and Changeux, 2000). Processes of selection by reward have been initially modeled by artificial networks (Barto et al., 1983), together with neurally plausible models incorporating neuronal mechanisms of reinforcement (Dehaene and Changeux, 1989, 1991) and reinforcement learning (Schultz et al., 1997). In some of these models, molecular mechanisms for the selection of "brain representations" are hypothetically implemented as allosteric receptors—in particular, nicotinic receptors—in order to detect the temporal coincidence between internal states of activity and externally evoked reward signals (Heidmann and Changeux, 1982; Dehaene and Changeux, 1991; see also Chapter 6).

In parallel with these theoretical studies, an important body of neuropharmacological experiments has emphasized the role of the mesolimbic dopaminergic system—and particularly its projection to the nucleus accumbens, which belongs to the striatum—as a critical shared component of reward systems, together with other neuromodulatory neurons employing neurotransmitters such as norepinephrine (Drouin et al., 2002) and/or serotonin (Olausson et al., 2001). Dopaminergic neurons have also been identified as the target of the reinforcing effect of drugs of abuse such as cocaine, amphetamines, morphine, and, of course, nicotine, whose addictive action is thought to result from their "diversion" to the "benefit" of endogenous reward systems. In other words, these drugs cause addiction because they produce *plastic changes* (both within and outside the mesolimbic-accumbens circuits) that subsequently alter the state of activity of neurons regulating motivated behavior and hedonic feelings (Hyman, 1996; Koob, 1996). The *shell* of the nucleus accumbens (specifically its ventromedial part) is believed to contribute to the integration and expression of reward and emotions through its projection to the amygdala, lateral hypothalamus, and central gray matter. The *core*, on the other hand, is supposed to be concerned more directly with somatomotor functions (review in Di Chiara, 2000, 2002).

Cholinergic nicotinic transmission has been proposed to play a critical role in reinforcement by modulating the activity of the main dopaminergic reward pathways. Three sets of findings support this hypothesis. First, nicotine acts as a positive reinforcer, as shown by place-preference and self-administration studies (references in Di Chiara, 2000). Second, lesions of the pedunculopontine nucleus, the major cholinergic input to mesencephalic dopaminergic neurons, have been reported to impair learning and the expression of behaviors reinforced by nonnicotinic addictive drugs (Olmstead and Franklin, 1994), brain stimulation (Lepore and Franklin, 1996), or natural rewards (Stefurak and van der Kooy, 1994; but see Olmstead et al., 1999).

Finally, like many drugs of abuse, nicotine stimulates dopamine release in the nucleus accumbens (Pontieri et al., 1996; Di Chiara, 2000). This effect has been attributed, at least in part, to the direct activation of nicotinic receptors located on dopamine neurons of the ventral tegmental area; indeed, nicotine is known to increase the firing rate of these neurons, both *in vitro* (Pidoplichko et al., 1997; Picciotto et al., 1998) and *in vivo* (Grenhoff et al., 1986). Moreover, nicotinic antagonists infused into the ventral tegmental area prevent nicotine-elicited dopamine release in the nucleus accumbens and disrupt systemic nicotine self-administration in rats (Corrigall et al., 1994; Nisell et al., 1994).

Knockout mice are a valuable tool in trying to identify the subunit composition of brain nicotinic receptors involved in these processes, and hence to describe the receptor target(s) for nicotine-addictive behavior (Picciotto et al., 1998; review in Champtiaux and Changeux, 2002). *In vivo*, β2-containing nicotinic receptors have been shown to be necessary for the stimulation of striatal dopamine release by systemic nicotine, and for the maintenance of nicotine self-administration (Picciotto et al., 1998; Figure 12.3). Moreover, electrophysiological recordings of $β2^{-/-}$ mice demonstrated that somatic β2-containing nicotinic receptors are required for nicotine to affect the firing rate of dopamine neurons *in vitro* (Picciotto et al., 1998). And in synaptosomes prepared from the striatum of $β2^{-/-}$ animals, nicotine failed to stimulate dopamine release by direct action upon nicotinic receptors in dopamine terminals (Grady et al., 2001).

While these studies have shown that nicotine reinforcement is dependent on the activation of nicotinic receptors containing β2, other studies have sought to determine the nature of the subunits associated with β2. Some authors, relying on pharmacological evidence, have proposed that the α4- and α3-subunits are likely partners of β2 in dopamine neurons (Sharples et al., 2000). This hypothesis, however, has only been partially confirmed by studies on $β2^{-/-}$ mice. Patch-clamp recordings of mesencephalic neurons from α4-deficient mice show that the majority of slowly desensitizing nicotine-elicited currents disappear in these mice (Klink et al., 2001), demonstrating the existence of α4-containing nicotinic receptors in these neurons. Moreover, the release of dopamine due to systemic injections of nicotine is abolished, *in vivo*, in $α4^{-/-}$ mice (Marubio et al., 2003). And studies of α3- and α6-deficient mice using the snail toxin α-conotoxin MII have shown that, in dopamine neurons, α6 is more likely than α3 to associate with the β2-subunit to form functional nicotinic receptors (Champ-

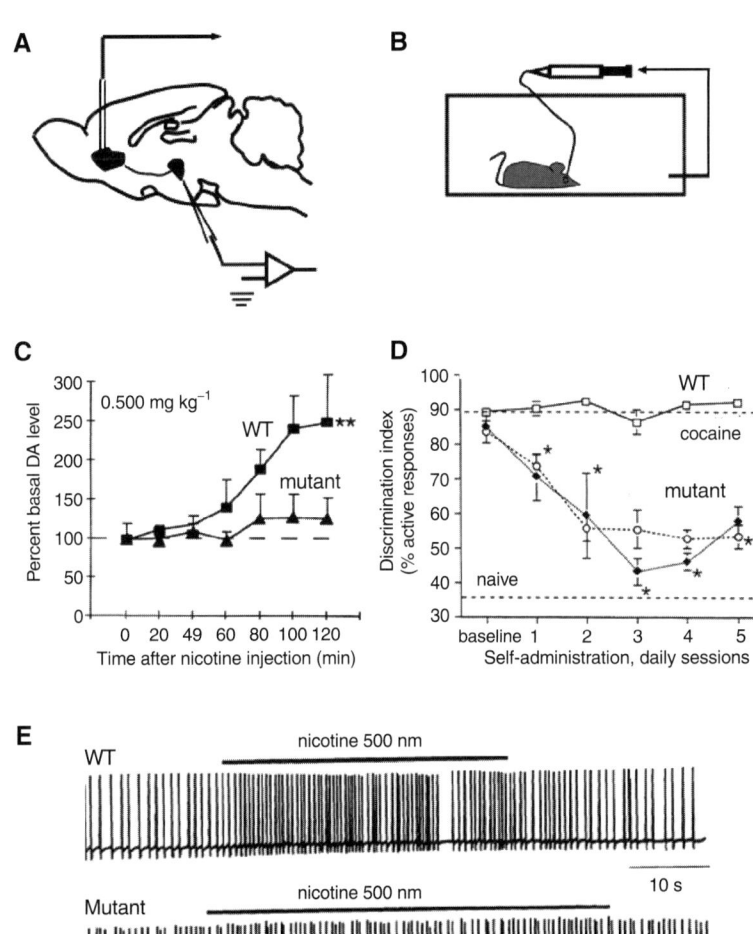

FIGURE 12.3 Absence of response of dopaminergic neurons from the ventral tegmental area and substantia nigra to nicotine and lowered nicotine self-administration in β2-deficient mice. (A, B) Electrophysiological recordings and method for measuring self-administration. (C, D) Level of nicotine-elicited dopamine release measured by microdialysis (C) and of nicotine self-administration (D) as a result of the deletion of the β2-subunit. (E) The electrophysiological response of dopaminergic neurons to nicotine is no longer present in β2$^{-/-}$ mice. (From Picciotto et al., 1998.)

tiaux et al., 2003). Taken together, these results demonstrate that the main nicotinic receptor oligomers expressed in dopaminergic neurons include the α4β2- and α6β2-subtypes (Champtiaux et al., 2002, 2003; Figure 12.4). A still unresolved issue, however, is whether or not the reinforcing properties of nicotine are caused by direct or indirect stimulation of nicotinic receptors present on dopaminergic neurons (Nomikos et al., 2000; Corrigall et al., 2001; Laviolette and van der Kooy, 2004). Recent stereoselective reexpression experiments involving the β2-subunit in the VTA of β2-deficient mice tend to show that it is indeed the case (Maskos et al., 2005).

Taken together, these results demonstrate that cholinergic control of the dopaminergic system contributes to the mechanisms of reinforcement involved in learning behavior and nicotine addiction, and that β2-containing receptor oligomers play a pivotal role in this control.

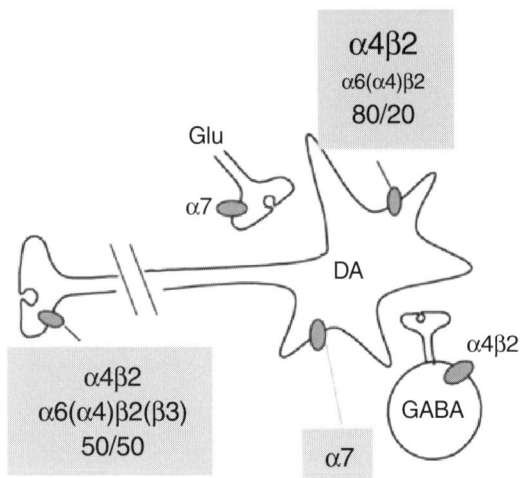

FIGURE 12.4 Summary of nicotinic receptor and dopamine systems. The α4β2-receptor is abundant in dopamine neurons, but the α6-subunit is predominant in the axonal compartment. (From Champtiaux et al., 2003.)

This conclusion is of major practical importance because nicotine, the principal compound present in tobacco smoke, is known to cause tobacco addiction (possibly together with other substances), and because tobacco addiction in Western countries is responsible for one-third of the annually preventable deaths due to cancer (Sasco et al., 2004).

Nicotinic Receptors and Pain

The earliest theories of learning by reward (see Thorndike, 1911) provided roles for both positive and negative (or punishment) reinforcement mechanisms. Negative reinforcement mechanisms mobilize pain systems, among other functional networks. Interestingly, nicotine has long been known to antagonize pain and thus to exert analgesic effects.

As part of an attempt to analyze nicotinic analgesia, the antinociceptive properties of nicotine were studied in mice lacking α4 and β2 nicotinic receptor subunits (Marubio et al., 1999), using either the tail-flick response for *peripheral* response or the hot-plate test for *supraspinal* response. The tail-flick response to nicotine is still observed in $\beta 2^{-/-}$ and $\alpha 4^{-/-}$ mice, but the dose-response curve is shifted to the right indicating a decreased sensitivity. The still unidentified nicotinic receptors involved are likely to be located in the spinal cord and/or in peripheral ganglia. Presynaptic modulation of serotonin release in the spinal cord was observed to be largely mediated by the tonic modulation of inhibitory neurons by acetylcholine, and controlled by nicotinic receptors that do not contain α4- or β2-subunits (Cordero-Erausquin and Changeux, 2001). On the other hand, in the hot-plate test, both nicotine and epibatidine failed to produce analgesia in either type of mutant mice. Patch-clamp recordings from serotoninergic neurons in the raphe magnus showed a loss of nicotine-elicited currents in α4- and β2-deficient animals as well. These findings are consistent with the notion that the supraspinal analgesic effects of nicotine are due to the activation of neurons in this nucleus, which depends in turn on the activation of α4β2 nicotinic receptors. All the evidence indicates, then, that cholinergic nicotinic systems function as antagonists of the central pain system and thus may contribute as well to positive reinforcement. These studies have also stimulated the design and development of new types of analgesic molecules whose structure departs from the familiar opiate model.

Molecular Mechanisms of Synaptic Plasticity Involved in Nicotine Addiction

A characteristic feature of tobacco addiction is the difficulty to quit smoking and the high frequency of relapse. Chronic exposure to nicotine has been proposed as the main cause for these effects. On the other hand, most learning theories proceed from the assumption of long-term changes in synaptic efficiencies as a function of the experience of the organism, and the long-term effects of nicotine have often been compared to these plastic phenomena. Evidence suggesting that nicotinic receptors contribute to synaptic plasticity is drawn directly from the phenomenon of nicotine addiction itself, which requires a strong, slowly reversible alteration of signal processing in the reinforcement circuits following chronic exposure to nicotine.

In Chapters 4–6 we noted that considerable progress has been achieved in understanding the functional organization of brain nicotinic receptors as targets of nicotine (Le Novère et al., 2002a), and that the molecular mechanisms of *activation* and *desensitization* elicited by nicotine are now beginning to be understood within the general framework of the allosteric transitions of the receptor protein (Changeux and Edelstein, 1998). These molecular processes, which develop over the millisecond-to-minute timescale and are rapidly reversible, therefore need to be consolidated or extended if we are to be able to account for the elementary mechanisms of synaptic plasticity that underlie nicotine addiction.

A related, though less well understood, molecular process known as "up-regulation" develops over longer timescales (minutes, hours, days) under conditions of long-term exposure to nicotine. It manifests itself as an increase in the total number of high-affinity nicotine-binding sites in the brains both of smokers (Benwell et al., 1988) and of rodents chronically exposed to nicotine (Blat et al., 1991; Flores et al., 1992; Marks et al., 1992; Rowell and Li, 1997). This phenomenon is not due to a regulation of nicotinic receptor gene transcription. It can be reproduced in reconstituted systems (oocytes and mammalian cell cultures, for example) with α4β2- and α3β2-receptors, but not with α3β4-receptors (Wang et al., 1998; Buisson and Bertrand, 2001). Systematic β2–β4 chimera experiments suggested by this difference in subunit role have delineated a domain of the extracellular N-terminal moiety of the receptor protein that appears to be critical for up-regulation. This microdomain includes loops of the β-subunit that are located in the vicinity of the acetylcholine/nicotine-binding site and directly concerned by the assembly of subunits (Sallette et al., 2003). Recent studies have elucidated the role of this microdomain. It was noticed first, both *in vivo* and *in vitro,* that during the course of its biosynthesis the nicotinic receptor protein accumulates within the cytoplasm, which is where important posttranscriptional processes take place (including subunit folding, assembly of subunits, and glycosylation). Second, and rather unexpectedly, it was found that nicotine—like other drugs of abuse, but unlike acetylcholine—penetrates the cell. Finally, analysis of the intracellular process of receptor glycosylation indicated that nicotine (as well as choline) binds to early receptor precursors and improves the chances that the receptor protein will mature into a high-affinity binding form exposed to the surface of the cell. Behaving in effect, then, as a "maturational enhancer" of the receptor protein, nicotine may be supposed to promote its own up-regulation (Sallette et al., 2005). This maturational action may constitute a novel elementary mechanism for neural plasticity. The actual mechanisms of synaptic plasticity and the neuronal targets directly responsible for the long-term effects of nicotine addiction *in vivo* nonetheless remain to be identified (Laviolette and van der Kooy, 2004).

Nicotinic Receptors, Sleep, and Wakefulness

Executive functions in humans (Fuster, 2001) are often said to require a "conscious effort" (Dehaene et al., 1998, 2003). Even though the use of such terms may at first glance appear rather far-fetched in connection with mice, the standard tasks used to investigate their behavior require that these laboratory animals be in a state of wakeful awareness. One often forgets that during the day mice alternate between episodes of quiet (or slow-wave) sleep and vigilance. Interestingly, β2-containing high-affinity receptors have been found to regulate these states of wakefulness (Cohen et al., 2002; Léna et al., 2004). A role for the cholinergic system in regulating sleep and wakefulness had been postulated for decades by a number of researchers—Jasper, Jouvet, and Hobson among them (review in Jones, 2003)—but, again, for the most part only muscarinic receptors and their ligands were taken into consideration. Comparison of breathing measurements of wild-type and β2$^{-/-}$ mice revealed a rather unanticipated and highly selective contribution by nicotinic receptors, however (Cohen et al., 2002; Figure 12.5). Ventilation measurements made using whole-body plethysmography readily distinguish regular large-amplitude patterns during sleep from highly irregular movements of variable amplitude during arousal. Placing mice in a low-oxygen environment creates a familiar stress response that normally provokes powerful, protective cardiorespiratory excitation and, in particular, an *arousal response*. Interestingly, arousal from sleep was found to be diminished in β2$^{-/-}$ mice, while breathing drive was accentuated. And while brief exposure to nicotine significantly reduced breathing drive in sleeping wild-type mice, it had no effect in mutants.

In vivo polygraphic electrophysiological recordings made with chronically implanted electrodes confirmed the contribution of β2-containing acetylcholine receptors to the transition from sleep to wakefulness (Lena et al., 2004). Nicotine was observed to transiently increase arousal in wild-type, though not β2$^{-/-}$, mice. In the absence of the drug, and in agreement with the ventilation data, episodes of slow sleep were prolonged and more stable in β2$^{-/-}$ mice than in wild-type mice. The duration of REM sleep was also found to be longer in the mutant mice. Furthermore, it was noticed that slow-wave sleep was interrupted by brief "micro-arousals," and that these events decreased

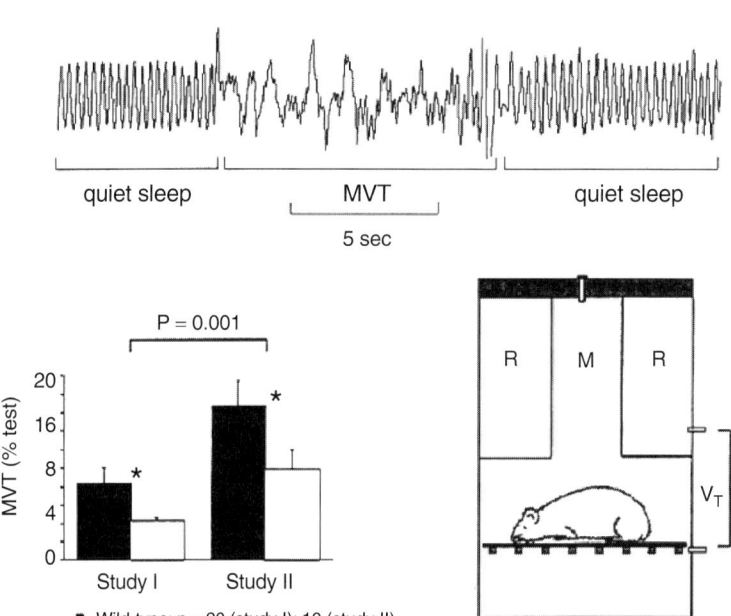

FIGURE 12.5 Attenuation of the arousal response to hypoxic stress in β2$^{-/-}$ mice. Ventilation is measured by whole-body plethysmography: quiet sleep is accompanied by regular slow movements, while arousal gives rise to movement artifacts (MVT). The arousal response is consistently lower in mutants (white box) than in wild-type (black box). (From Cohen et al., 2002.)

in frequency in $\beta2^{-/-}$ mice. In this connection it is worth noting that the nocturnal seizures of patients suffering from ADNFLE (see Chapter 8) coincide with such microarousals. In any case, the fact that arousal response from slow sleep (as well as from REM sleep) was diminished in mutant mice indicates that β2-containing nicotinic receptors are implicated in the transition to wakefulness (Léna et al., 2004).

In conclusion, from the cognitive perspective, these findings clearly illustrate the involvement of nicotinic receptors in regulating the transition between states of consciousness. Moreover, from a neuropathological perspective, it is also possible that such sleep-disordered breathing underlies the well-established association between smoking by pregnant women and an increased risk of sudden infant death syndrome.

Nicotinic Receptors and Cognition

A rich body of pharmacological and behavioral observation demonstrates the importance of cholinergic pathways for working memory and for attentional processes, both of which depend in large part on the integrity of the prefrontal cortex (Bartus et al., 1982). Moreover, nicotine acting on nicotinic receptors selectively improves working memory and attention in both experimental animals and humans (review in Levin and Rezvani, 2002).

Delayed-response tasks were initially designed to test for the acquisition of cognitive patterns (or rules) of behavior that selectively mobilize the prefrontal cortex (review Shallice et al 1982; Dehaene and Changeux, 1989, 1991). In a first approach to the problem, the nicotinic antagonists neuronal bungarotoxin and dihydro-β-erythroidine were assayed with respect to the performance by rats of a delayed matching-to-sample (MTS) task (Granon et al., 1995). A less difficult task, the non-matching-to-sample task (NMTS), was selected as a control since it brings out the rat's innate tendency to spontaneously alternate its choice of routes, going on a second run to the branch it *did not* visit on the first run. As anticipated, lesions of the prefrontal cortex selectively impaired performance of the MTS task, while preserving the capacity to successfully perform the NMTS task (Granon et al., 1994). Interestingly, injections of *neuronal* bungarotoxin into the prelimbic area of the prefrontal cortex produced a significant decrease in working-memory performance in the MTS task, but not in the NMTS task. In contrast, scopolamine, an inhibitor of the muscarinic receptor, was found to impair working memory in both tasks (Granon et al., 1995). These findings were consonant with the observation that nicotinic receptors mediate enhancement of glutamate release in the prefrontal cortex of rats, while muscarinic receptors exert the opposite effect (Vidal and Changeux, 1993). It remained, then, to investigate the precise contribution of defined nicotinic receptor subunits on these effects.

Although the behavior of mice is generally more difficult to analyze than that of rats, novel insights were obtained by comparing the behavioral responses of wild-type mice and their β2-deficient knockout partners (Picciotto et al., 1995, 1998, Granon et al., 2003). In addition to the biochemical and physiological differences observed, there is a striking loss of electrophysiological response to nicotine by neurons in the anterior thalamus. Subjecting young adult $\beta2^{-/-}$ mice to the passive avoidance task further revealed a clear-cut difference of behavioral response to nicotine (Picciotto et al., 1995; Paylor et al., 1998; Caldarone et al., 2000). The test consists in measuring the time elapsed before a mouse placed in a well-lit compartment enters a dark adjacent chamber. In the sample run, the mouse is administered an electric shock upon entering the dark chamber and immediately thereafter is injected with a nicotine solution or the vehicle alone. Twenty-four hours later, *retention* of the electric-shock *memory* was assessed by measuring the delay in entering the dark chamber. Low nicotine was consistently found to aid retention in wild-type animals, but it was completely ineffective in $\beta2^{-/-}$ mice (Picciotto et al., 1995; Figure 12.6).

In addition, slight modification of the learning behavior of the $\beta2^{-/-}$ mice was noticed *in the absence* of nicotine. For instance, it was found that in a cross-maze paradigm in which a place strategy was reinforced, the $\beta2^{-/-}$ mice learned the task *more rapidly* than wild-type mice but without exhibiting significant alteration of food reward and/or anxiety behavior (Granon et al., 2003). Automatic quantitative decomposition of mice trajectories into successions of so-called symbols representing both the velocity and position of the animals in the arena (Faure et al., 2003) offered an explanation of this paradoxical behavior. It disclosed a subtle though nonetheless striking modification in the displacement patterns of $\beta2^{-/-}$ mice. Placed in an open field, wild-type mice gradually and spontaneously shift away from rapid navigation at the periphery to slow exploration nearer the center. It was discovered that β2-deficient mice, on the other hand, do not modify their movements over time as they became acquainted with the environment (Figure 12.7). Similarly, when confronted with objects, $\beta2^{-/-}$ mice did not spontaneously adapt their behavior by progressively moving to the center, but instead continued to run along the outer wall of the area. In a different novelty paradigm, $\beta2^{-/-}$ mice confronted with "soft" conflict situations did not modify their routine behavior, while the presence of new objects or reconfiguration of the maze elicited longer exploration periods for wild-type animals. Beta-

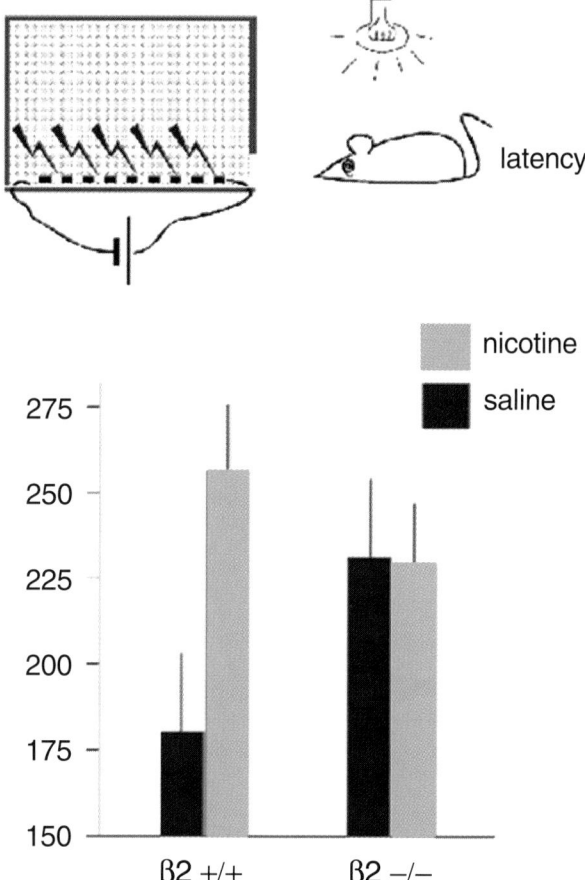

FIGURE 12.6 Alteration of the performances of β2-deficient mice in a passive avoidance test. The mouse spontaneously enters the dark compartment after a short delay (or latency period). On initially entering this compartment, it receives an electrical foot shock. On the next run, 24 hours later, the mouse remembers the shock and enters the dark compartment after a longer delay. If the mouse is injected with nicotine immediately after administration of the electric shock, the length of the latency period increases. Deletion of the β2-subunit gene abolishes the effect of nicotine on latency, but causes an unanticipated increase in latency in the absence of nicotine. (From Picciotto et al., 1995.)

$2^{-/-}$ mice did not display any sensitization to novelty, however, failing to react even to drastic changes in maze configuration (Granon et al., 2003). In short, the balance between *navigation* behaviors (rapid movements aimed at acquiring general information about environment) and more "cognitive" *exploration* behaviors (slow movements devoted to more precise investigation of the environment) was modified as a consequence of the mutation. In a somewhat metaphorical sense, one might say that the $β2^{-/-}$ mice have lost their "curiosity"!

The behaviors of $β2^{-/-}$ and wild-type mice were also compared with respect to a social interaction paradigm requiring organized sequences of actions that place a high premium on conflict resolution, mainly because it forces the animal to adapt rapidly to the unpredicted behavior of another animal. Whereas wild-type mice exhibited an alternation between approach and escape behaviors, $β2^{-/-}$ mice disproportionately initiated approach behaviors, but rarely escape behaviors (Granon et al., 2003). While once again acting in a "rigid" manner, the knockout mice were nonetheless fully able to memorize the environment and the objects in it, as we noted earlier.

Taken together, these results demonstrate that elimination of β2-containing nicotinic receptors causes a clear-cut dissociation between the "executive" organization of locomotor behavior, in which planning plays a crucial role (Roberts et al., 1998), and lower-level elementary behaviors, such as recognition, memory, and anxiety. Consistent with this conclusion, as we have already mentioned, planning and flexible behaviors in the rat are subject to nicotinic receptor regulation and depend also upon prefrontal cortex integrity (Granon et al., 1994, 1995; review in Robbins, 2000; Fuster, 2001).

In humans, a wealth of neuropsychological observations and brain-related imaging data have shown that these executive (or supervisory) functions mobilize the dorsolateral prefrontal cortex, the anterior and posterior cingulate, and the posterior parietal cortex (see review in Tanji and Hoshi, 2001). To account for these supervisory processes, which rely on the integrity of the prefrontal cortex (Shallice et al., 1982), a network model known as the "global neuronal workspace" has been proposed (Dehaene, Kerszberg, and Changeux, 1998) that mobilizes sets of neurons equipped with long-range axons, whose relative contribution to brain organization has increased dramatically in the course of the recent evolution of the mammalian brain. On this hypothesis, the neuronal workspace is the seat of intense spontaneous activity, to which sensory inputs have access in all-or-none fashion. The workspace generates distributed representations that differentially modulate specialized and modular perceptual, motor, long-term

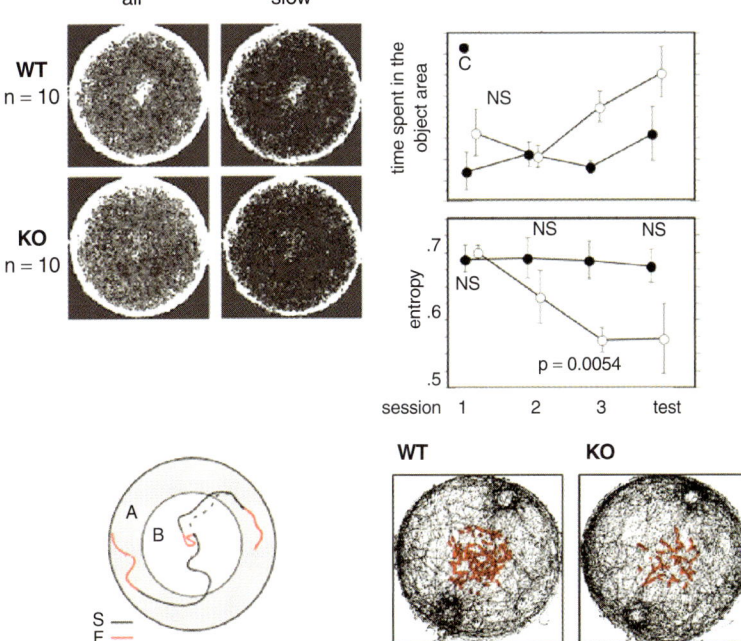

FIGURE 12.7 Selective impairment of executive functions in the mouse caused by deletion of the β2-subunit gene. *Left panels:* The exploratory (cognitive) and navigatory (automatic) behaviors of wild-type and β2$^{-/-}$ mice are measured in an enclosed circular field with an automated method by decomposing mice trajectories into successions of "symbols" representing both velocity (fast or slow) and position in the arena (periphery or center). A dense accumulation of slow-trajectory segments in the center of the arena is found for wild-type but not β2$^{-/-}$ mice. *Right panels:* Similarly, exploration for two different objects located in the field displays a pattern of habituation as a function of the number of sessions for wild-type mice (open circles) that is absent in their β2-deficient counterparts (filled circles). (From Granon et al., 2003.)

memory, evaluative, and attentional processors in a top-down manner (Dehaene et al., 1998, 2003; Figure 12.8). These processor neurons are abundant in cortical layers II and III of the prefrontal and cingulate cortices, where, as we noted above, efficient responses to nicotine are recorded (Vidal and Changeux, 1989). Furthermore, both the monitoring of shifts in behavior and the temporal organization of sequential actions are known to enhance the activation of these territories (Dias et al., 1997; Gehring and Knight, 2000; Fuster, 2001). In short, all of these data are consistent with the view that the binding of acetylcholine to high-affinity β2-containing nicotinic receptors "gates" access to the workspace and thus serves to regulate these executive processes. In addition, it offers a plausible neural explanation for the role of nicotine in cognitive functions generally (Granon et al., 2003).

These observations recall the discussion earlier in this chapter of nicotinic receptors in development, neuroprotection, and aging. There we saw that Alzheimer's patients show marked deficits in working memory and planning behavior, and display differential loss of β2-containing nicotinic receptors. Autistic children, who suffer from a dramatic impairment of the ability to use language and to interact with others, selectively exhibit a lack of high-affinity α4β2-receptors in the prefrontal cortex as well (Perry et al., 2001; Martin-Ruiz et al., 2004). In both cases, the genetic evidence suggests that either neuronal development or aging and/or stabilization of synaptic connections has been altered. One interesting possibility is that neurons having long-range axons, which are assumed to populate the neuronal workspace, are a selective target of the disease process. This would appear to be a plausible hypothesis in view of the considerable metabolic and energetic demands entailed by a system of very long and heavily branched axons. Nicotinic receptors would then exert a neuroprotective effect in helping maintain the integrity of the extensive arborizations that support the cognitive functions of the neuronal workspace. Once again, chronic treatment by nicotine or nicotinic agents may be an efficient form of therapy (Perry and Perry, 2003).

General Conclusion: The Role of Nicotinic Receptors in Brain Function and Dysfunction

In addition to the evidence offered by the mouse model, there is reason to believe that nicotinic transmission regulates a number of important brain processes in humans and mammals. Nicotine is known to improve performance in various cognitive tasks involving spatial and associative learning, working memory, and attention; and mecamylamine, a general nicotinic antagonist, has been demonstrated to impair memory performance (Levin, 1992). Furthermore, lesions of the cholinergic forebrain system result in cognitive deficits that are reversed by nicotine (Tilson et al., 1988; Decker et al., 1992; and Levin et al., 1993; but see Gallagher and Collombo, 1995, and Chappell et al., 1998). These findings, together with observations that nicotinic functions are impaired in the brains of both Alzheimer's and autistic patients, and that the incidence of Alzheimer's and Parkinson's disease is significantly reduced in smokers (reviewed in Fratiglioni and Wang, 2000), make it probable that nicotinic receptors are selectively affected in these human pathologies.

The molecular biology of brain nicotinic receptors has opened up many new avenues for research into the neurobiology of brain functions and cognitive processes. In particular, it has illuminated the precise distribution of receptor species at strategic "nodal points" in signaling networks which are anticipated to play a decisive role in brain plasticity (see Le Novère et al., 2002a; Laviolette and van der Kooy, 2004).

In this respect the attempt to devise formal but nonetheless neurally plausible networks capable of performing delayed-response and other cognitive tasks, and also of suggesting how access to consciousness is achieved, has the virtue at least of demonstrating that an understanding of the role of neurotransmitter receptors in learning and consciousness requires that their intrinsic functional properties be understood. But it also requires, as we have sought to emphasize, a description of their precise localization within a quite complex but nonetheless well-defined architecture that provides for several levels of organization, multiple sets of interconnected groups of neurons, and access to reward mechanisms (see Figure 12.8).

Studies conducted on strains of mice carrying selective deletions in various nicotinic receptor genes have confirmed and extended these preliminary models, making it possible now to associate particular subunit types with an observed phenotype (Figure 12.9). This is especially important in exploring complex behaviors, particularly when the concentrations of agonists or antagonists used *in vivo* are often much higher than those required to ensure specificity. Gene inactivation studies have demonstrated that the inactivation of a single nicotinic receptor subunit gene abolishes the pharmacological response to specific nicotinic agents in the central nervous system, and have permitted identification of the particular nicotinic receptor subtypes involved in cognitive executive functions and nicotine reward processes, as well as in mediating the analgesic and neuroprotective effects of nicotine. Studies of knockout mice have also revealed the importance of nicotinic receptor activation in normal development and aging. Further research will nonetheless be required in order to fully under-

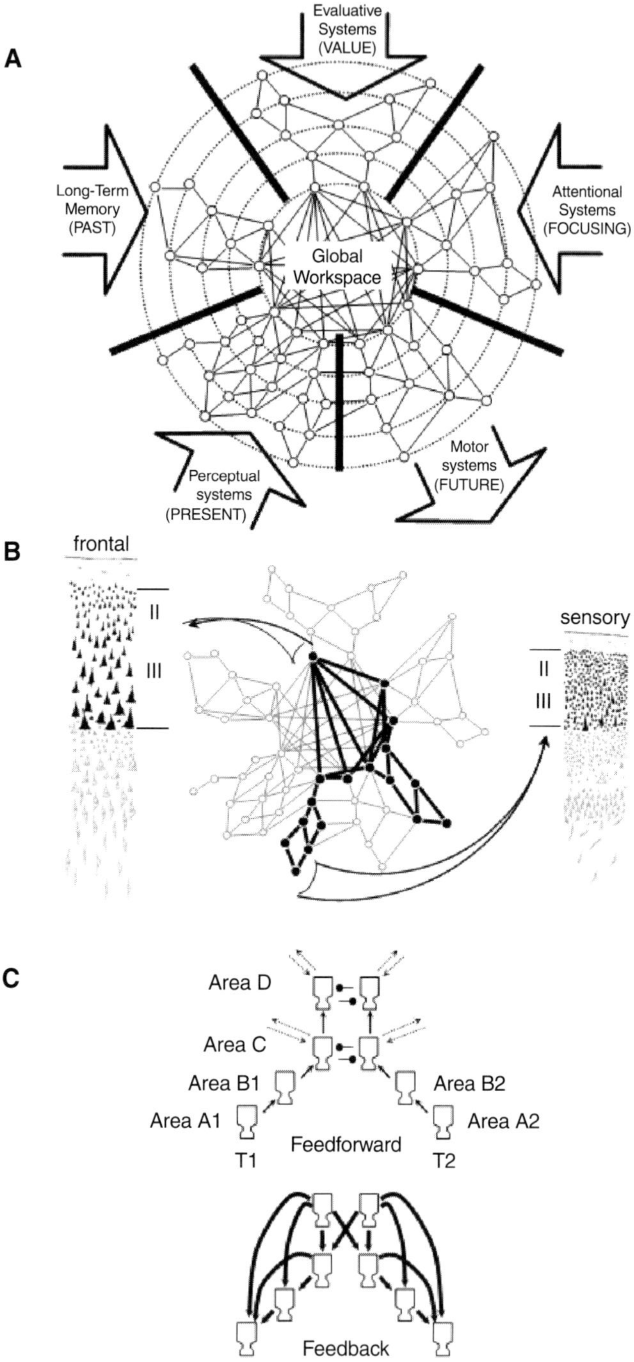

FIGURE 12.8 A neuronal model of a global workspace engaged in effortful "conscious" tasks. (A) Schematic representation of the brain's principal (automatic) processors and of the global (conscious) workspace, which is composed of layer II-III pyramidal neurons with long-range axons. (B) A coherent link between two processors is established through the activation of distributed workspace neurons. (C) Application of the workspace model to the problem of all-or-none access to consciousness: feedforward and feedback connectivity give rise to thresholds in the processing of sensory data. (From Dehaene et al., 1998; and Dehaene et al., 2003.)

stand the role of nicotinic receptors in cognitive function, making it necessary in turn to devise still more refined tools. The development of tissue-specific and inducible knockout and knockin strategies, together with the stereotaxic reexpression or inactivation of defined subunits (or sets of subunits) in well-defined brain territories using lentiviral vectors (Maskos et al., 2005), may be expected to help to

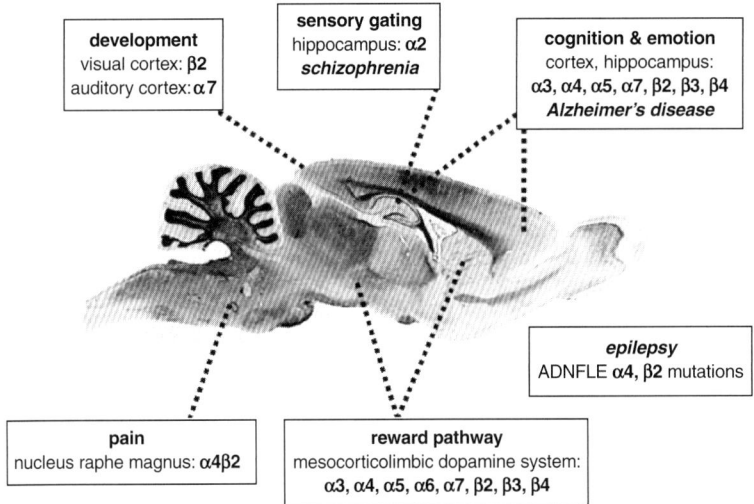

FIGURE 12.9 Nicotinic receptor subunits in the rodent brain and their tentative assignment to defined brain functions and pathologies.

overcome the obstacles currently faced by researchers investigating the function of neuronal networks in the brain.

Nicotinic receptors, along with ligand-gated ion channels and ion channels in general, are *allosteric membrane proteins* that integrate multiple kinds of signals at several types of topographically distinct sites through multiple and discrete conformational transitions. As a consequence, nicotinic receptors mediate signal transduction in intercellular communication, as well as *higher-order forms of regulation* that are thought to control synaptic efficiency and neuronal functions. They may thus serve as coincidence detectors in Hebbian chemical synapses and as targets of nicotine in the plastic processes responsible for nicotine addiction. If, in addition, nicotinic receptors are crucially involved in various aspects of access to consciousness, as a growing body of evidence suggests, they may also contribute to the gating of executive functions.

Finally, a detailed knowledge of nicotinic receptors, their functional organization, biosynthesis, and targeting in the brain from the subcellular to the regional levels holds out the prospect that novel pharmacological tools and therapeutic techniques may be developed that can help patients suffering from a range of neurological and psychiatric disorders, among them Alzheimer's and Parkinson's diseases, Tourette's syndrome, nocturnal frontal lobe epilepsies, and possibly schizophrenic syndromes and autism as well—to say nothing of addiction to nicotine itself. For forty years now the nicotinic receptor has served as an outstanding model of signal transducers and pharmacological receptors.

APPENDIX

Models of Acetylcholine Receptor Dynamics

Our current understanding of the mechanisms of the acetylcholine receptor is due not only to specific studies of the receptor itself, but also to general conceptual developments arising from the study of other proteins, notably hemoglobin, which has long been a fertile object of experimental investigation with respect to the kinetics of ligand binding and conformational transitions. Major advances in the study of hemoglobin were achieved in the 1920s by Gilbert Adair, whose work identified hemoglobin as a tetrameric molecule. Interestingly, Hill had published two papers in 1910 dealing with the binding of oxygen by hemoglobin and the effects of nicotine (and "curari") on muscle contraction, respectively, without appearing to suspect that the two topics were in any way related.

The modeling of receptors initially involved a purely pharmacological approach consisting of dose-response measurements of a given physiological phenomenon. Implicit in this approach was the assumption that dose-response effects could be directly related to an equilibrium binding reaction of agonist to receptor, along the lines developed by Hill, Langmuir, and other researchers noted in Chapter 1. A key advance was the model developed by Del Castillo and Katz (1957), which specified the fundamental binding reaction together with an activation step:

$$X + R \underset{}{\overset{K_D}{\rightleftharpoons}} XR \underset{}{\overset{E}{\rightleftharpoons}} XR^*. \qquad (A.1)$$

where X is the agonist, R is the receptor, and XR^* refers to the receptor in the active form, $K_D = [X][R]/[XR]$ and $E = [XR^*]/[XR]$. The hyperbolic dose-response curve can be characterized by the ligand's concentration at half-maximal response, $EC_{50} = K_D/(E + 1)$. This formulation has inspired attempts to distinguish between "binding" and "efficacy" (see Colquhoun, 1998; Kalbaugh et al., 2004), but where E is large (>100) the formal distinction is extremely difficult to evaluate experimentally. As a practical matter, information can be obtained only when E is <100, since then the observed response will have a maximum amplitude smaller than that observed for the most efficacious ligands of the same receptor. Indeed, it was diminished responses of this sort that gave rise to the notion of a "partial" agonist, with the maximal response given by $1/(1 + 1/E)$. The complete fractional response is given by $1/(1 + 1/E + K_D/E[A])$.

The model described above is based on a single ligand-binding site, but cooperative interactions between the sites may occur when more than one site is present. For ligand binding in the single-site case, a dose-response curve with hyperbolic shape will occur and the data can be represented in terms of the fractional response, f_R, versus [A], by the equation $f_R = [A]/([A] + K_D)$. In the case of interacting sites on the receptor, the dose-response curve may have a sigmoidal appearance and the Hill equation can be applied. For n molecules of an agonist interacting with the receptor, the reaction becomes

$$nA + R \underset{}{\overset{K_D}{\rightleftharpoons}} A_n - R^* \tag{A.2}$$

with fractional response now defined as:

$$f_R = \frac{[A]^n}{[A]^n + K_D}. \tag{A.3}$$

Applied to the binding of oxygen by hemoglobin, the Hill equation yields a value of $n = 2.8 \pm 0.2$, which could correspond to three sites on hemoglobin that bind oxygen in an all-or-none reaction. Adair established, however, that there are four sites on hemoglobin. The Hill coefficient, though it is a convenient index of cooperativity, therefore does not give a precise mechanistic description.

A first mechanistic model for hemoglobin (Hb) was proposed by Adair (1925) based on four successive binding reactions with increasing affinity for a ligand, X (see Figure A.1A). The corresponding equation has four unique binding constants, K_1–K_4. In the case of a tetramer in a single conformational state with four identical sites (characterized by **K**), the equation reduces to the simple form $Y = X/(X + K)$, as shown in Figure A.1B. Although the Adair model does invariably provide a satisfactory description of oxygen binding by hemoglobin, the individual K values vary considerably from one experiment to another and often do not follow a monotonic progression.

Mechanistic interpretations were therefore difficult. A novel solution to the problem was proposed by Monod, Wyman, and Changeux (1965), whose model (conventionally abbreviated as MWC) was based on conformational transitions between symmetric, noncooperative states. In its simplest form, cooperative behavior of hemoglobin can

A. Adair model: four successive binding steps for hemoglobin:

$$Hb_0 + 4X \underset{}{\overset{K_1}{\rightleftharpoons}} Hb_1 + 3X \underset{}{\overset{K_2}{\rightleftharpoons}} Hb_2 + 2X \underset{}{\overset{K_3}{\rightleftharpoons}} Hb_3 + X \underset{}{\overset{K_4}{\rightleftharpoons}} Hb_4$$

$$\bar{Y} = \frac{\tfrac{1}{4}P_1 + \tfrac{2}{4}P_2 + \tfrac{3}{4}P_3 + P_4}{P_0 + P_1 + P_2 + P_3 + P_4}$$

$$\bar{Y} = \frac{\frac{X}{K_1} + \frac{3X^2}{K_1 K_2} + \frac{3X^3}{K_1 K_2 K_3} + \frac{X^4}{K_1 K_2 K_3 K_4}}{1 + \frac{4X}{K_1} + \frac{6X^2}{K_1 K_2} + \frac{4X^3}{K_1 K_2 K_3} + \frac{X^4}{K_1 K_2 K_3 K_4}}.$$

B. Adair equation for four identical binding steps ($K = K_1 = K_2 = K_3 = K_4$):

$$\bar{Y} = \frac{\frac{X}{K}\left(1 + \frac{X}{K}\right)^3}{\left(1 + \frac{X}{K}\right)^4} = \frac{X}{X + K} = \frac{1}{1 + K/X}.$$

C. MWC Model: two states (T and R) each with identical sites K_R and K_T:

$$T_0 + 4X \overset{K_T}{\rightleftharpoons} T_1 + 3X \overset{K_T}{\rightleftharpoons} T_2 + 2X \overset{K_T}{\rightleftharpoons} T_3 + X \overset{K_T}{\rightleftharpoons} T_4$$

$c = K_R / K_T$

$L_0 = T_0/R_0 \quad L_1 = L_0 c \quad L_2 = L_0 c^2 \quad L_3 = L_0 c^3 \quad L_4 = L_0 c^4$

$$R_0 + 4X \overset{K_R}{\rightleftharpoons} R_1 + 3X \overset{K_R}{\rightleftharpoons} R_2 + 2X \overset{K_R}{\rightleftharpoons} R_3 + X \overset{K_R}{\rightleftharpoons} R_4$$

$$\bar{Y} = \frac{\underbrace{\frac{X}{K_R}\left(1 + \frac{X}{K_R}\right)^3}_{\text{R State}} + \underbrace{L_0 \frac{X}{K_T}\left(1 + \frac{X}{K_T}\right)^3}_{\text{T State}}}{\underbrace{\left(1 + \frac{X}{K_R}\right)^4}_{\text{R State}} + \underbrace{L_0 \left(1 + \frac{X}{K_T}\right)^4}_{\text{T State}}}$$

$$\bar{R} = \frac{\left(1 + \frac{X}{K_R}\right)^4}{\underbrace{\left(1 + \frac{X}{K_R}\right)^4}_{\text{R State}} + \underbrace{L_0 \left(1 + \frac{X}{K_T}\right)^4}_{\text{T State}}}.$$

FIGURE A.1 Mathematical approaches to multiple ligand binding by an oligomeric protein. (A) The sequential model of Adair (1925) formulated for the four steps of oxygen binding by hemoglobin (Hb_0 to Hb_4), with these species represented by P_0 to P_4 in the binding equations defined with respect to ($\bar{Y}$). (B) Transformation of the Adair equation into a simple binding isotherm in the case of identical sites. (C) Multiple equilibria for the Monod-Wyman-Changeux (MWC) model and equations for the binding ($\bar{Y}$) and state ($\bar{R}$) functions.

be described by the transition between a low-affinity, relatively stable T (tense) state and a high-affinity ligand-stabilized R (relaxed) state. In the absence of oxygen, the T state is predominant ($L_0 = [T_0][R_0] \gg 1$). As oxygen is added to the system, causing bonding to occur, the R state is progressively stabilized and the T-R distribution shifts in favor of the R state, which predominates at full saturation with oxygen. As shown in the $\bar{Y}$ equation applying the MWC model to hemoglobin (Figure A.1C), the two halves of the equation contain terms for the R and T states, respectively, which individually have the same form as the noncooperative tetramer shown in Figure A.1B. A further distinction between the Adair and MWC models concerns the fractional response f_R, or $\bar{R}$. Early on it was recognized that this "state function" does not necessarily coincide in the MWC model with the "binding function" ($\bar{Y}$). The implications of this difference were worked out with respect to the enzyme aspartate transcarbamylase (Changeux and Rubin, 1968): since the T-R distribution only slightly favors T in the absence of substrate, the curve for $\bar{R}$ displays higher values than the curve for ($\bar{Y}$) as a function of substrate concentration, as we noted in Chapter 1. The equation for $\bar{R}$ is given at the bottom of Figure A.1C.

The application of these principles to nicotinic receptors and other membrane proteins had to be carried out indirectly, mainly because ligand binding and response (channel opening) were less amenable to experimental measurement than in the case of soluble proteins such as hemoglobin or aspartate transcarbamylase. There is yet another fundamental difference between the Adair and MWC models, this one having to do with the occurrence of desensitized states for receptors. Applying the MWC model to such systems, as we saw in Chapter 6, requires a minimum of four conformational states (B, A, I, and D). Some measure of simplification is possible, however, since the full interaction scheme (Figure A.2A) reduces to a linear cascade (Figure A.2B) because of the progression to significantly slower transition rates. In this case, the system is fully defined by the individual ligand-binding on and off rates for each state and the state-interconversion rate constants (Edelstein et al., 1996). Each allosteric equilibrium constant governing the interconversion between two states (Figure A.1B) can be determined from the ratio of the appropriate rate constants. For example, the B $\rightleftharpoons$ A equilibrium is defined by $^{BA}L_0$, where $^{BA}L_0 = [B_0]/[A_0] = {}^{AB}k_0/{}^{BA}k_0$. The full complement of rate constants permits a complete description to be given for the B, A, I, and D states, in this case with two ligand-binding sites for each state (Figure A.2D). The interconversion rates are constrained by ligand binding according to linear free-energy relations (Edelstein et al., 1996), as shown in Figure A.2C for the B and A states (see Figure 6.2 for a more schematic representation of the B, A, I, and D states).

For any degree of ligand binding, the probability of channel opening, P_{open}, is determined by the equilibrium ligand-binding constants for each state (as defined above) and the intrinsic interconversion equilibrium (allosteric) constants, or L values. With two identical ligand-binding sites, the equilibrium distribution of all forms is given by the partition function, F:

$$F = \Sigma B_i + \Sigma A_i + \Sigma D_i$$
$$= {}^{BA}L_0 \, {}^{AD}L_0 (1 + X/{}^BK)^2 + {}^{AD}L_0 (1 + X/{}^AK)^2 + (1 + X/{}^DK)^2, \quad (A.4)$$

where X is the concentration of ligand, $^{BA}L_0 = [B_0]/[A_0]$ and $^{AD}L_0 = [A_0]/[D_0]$, and ^BK, ^AK, and ^DK are equilibrium dissociation constants. The value of P_{open} with respect to the A and B states for the

Acetylcholine Receptor Dynamics

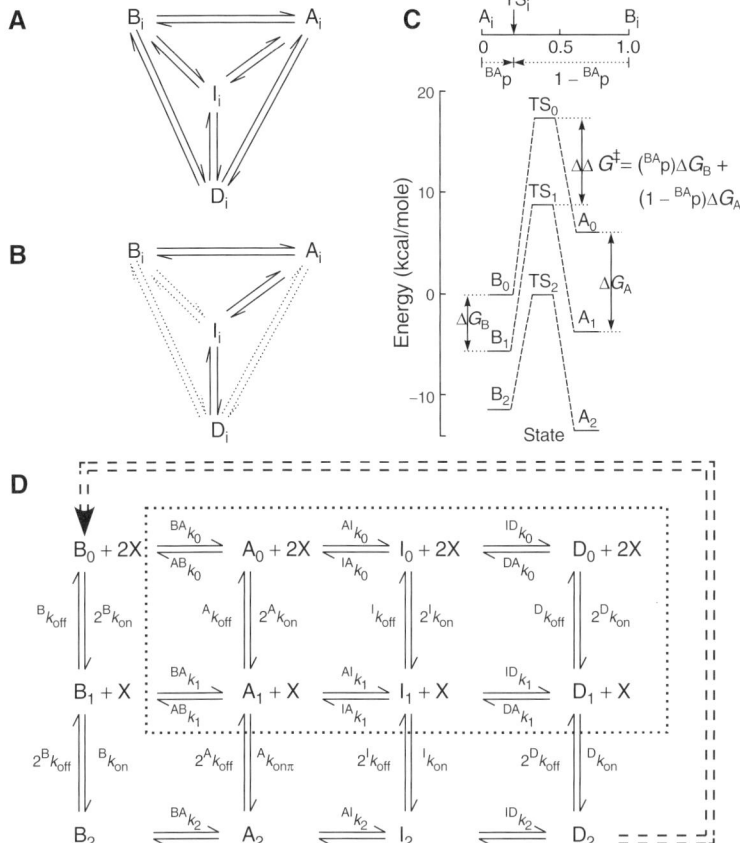

FIGURE A.2 Interconversion pathways and models. (A) Tetrahedral model for interactions between four conformational states, B, A, I, and D. (B) Linear model resulting from kinetic pathway selection on the basis of the hierarchy of reaction rates (Edelstein et al., 1996). The kinetic pathways that can be ignored are indicated by dotted lines. (C) Linear free-energy relations permitting interconversion rates for the B and A states to be fixed by a transition-state parameter, ^{BA}p. The B and A states each bind ligand with a characteristic intrinsic affinity that yields energy ladders for molecules with 0, 1, or 2 ligands bound (indicated by the subscript), the step sizes being indicated by ΔG_B and ΔG_A, respectively. The two rate constants for interconversion of the fully liganded forms, $B_2 \rightarrow A_2$ and $A_2 \rightarrow B_2$, fix the height of the transition state barrier, TS_2. The position ^{BA}p, on a hypothetical scale from 0 to 1 (shown at the top of the panel), characterizes the transition state such that the more closely the transition state resembles A, the nearer ^{BA}p is to zero. The rates for the unliganded and monoliganded forms are then determined by the heights of the transition state barriers TS_1 and TS_0, which are at steps above TS_2 fixed by ΔG in terms of ^{BA}p, ΔG_B, and ΔG_A according to the relationship indicated (Edelstein and Changeux, 1998). Applying these relations to detailed kinetic studies of muscle receptors, a value of $^{BA}p = 0.2$ was obtained (Edelstein et al., 1996), indicating that with each ligand-binding step the increase in the B → A rate is considerably larger (~250-fold) than the decrease in the A → B rate (~4-fold). (D) Ligand-binding and interconversion reactions for a linear MWC-type model. The states within the dashed rectangle are not considered by the purely sequential model. The "cyclic pathway" posited to explain recovery from a strong, desensitizing agonist pulse without passage via the open state (Katz and Thesleff, 1957; Franke et al., 1993) is indicated by the dashed double line. (From Changeux and Edelstein, 1998.)

MWC model—i.e., $\Sigma A_i/(\Sigma A_i + \Sigma B_i)$, which applies to dose-response curves where desensitization can be neglected or for single channel measurements within bursts—is given by:

$$P_{open}(MWC) = \frac{\left(1 + \dfrac{X}{K_A}\right)^2}{\left(1 + \dfrac{X}{K_A}\right)^2 + {}^{BA}L_0\left(1 + \dfrac{X}{K_B}\right)^2}. \qquad (A.5)$$

This equation depends on the equilibrium of the unliganded forms ($B_0 \rightleftharpoons A_0$) defined by ${}^{BA}L_0$. An equivalent equation based on the equilibrium between the fully liganded forms ($B_2 \rightleftharpoons A_2$) can be derived with respect to K_{open}, which has the advantage of being directly comparable to the relevant equation in the "sequential" model widely employed by electrophysiologists (see below). In the sequential model, channel opening is assumed to occur only for fully liganded receptors (i.e., in the case of reactions outside the dashed rectangle in (Figure A.2D). The MWC equation for P_{open} takes the form:

$$P_{open}(MWC) = \frac{\left(\dfrac{K_A}{X} + 1\right)^2}{\left(\dfrac{K_A}{X} + 1\right)^2 + K_{open}\left(\dfrac{K_B}{X} + 1\right)^2}, \qquad (A.6)$$

where $K_{open} = {}^{BA}L_2 = {}^{BA}L_0(K_A/K_B)^2$.

The corresponding equation for P_{open} in the sequential model—see, for example, Colquhoun and Sakmann (1985), and many subsequent authors—is essentially a modified version of (A.1) that is intended to handle two sites:

$$2X + R \underset{}{\overset{K_1}{\rightleftharpoons}} X + XR \underset{}{\overset{K_2}{\rightleftharpoons}} X_2R \underset{}{\overset{K_{open}}{\rightleftharpoons}} X_2R^*. \qquad (A.7)$$

In this case, where $K_1 = K_2 = \mathbf{K_B}$, the equation for $P_{open}(Seq.)$ takes the form:

$$P_{open}(Seq.) = \frac{1}{1 + K_{open}\left(\dfrac{K_B}{X} + 1\right)^2}, \qquad (A.8)$$

which corresponds to the form of the equation for the MWC model (A.5), but with no binding to the A state.

An alternative, but equivalent, way of comparing the MWC and sequential models is to present the basic form of the MWC equation (A.5) with the terms of the numerator and the terms on the left of the denominator expanded to explicitly show all A states (unliganded:

1; monoliganded: $2X/K_A$; and diliganded: $[X/K_A]^2$):

$$P_{open}(MWC) = \frac{1 + 2\dfrac{X}{K_A} + \left(\dfrac{X}{K_A}\right)^2}{1 + 2\dfrac{X}{K_A} + \left(\dfrac{X}{K_A}\right)^2 + {}^{BA}L_0\left(1 + \dfrac{X}{K_B}\right)^2}. \quad (A.9)$$

Assuming that the only A state species presented is the diliganded form A_2 (the only form of the A state that figures in the sequential model), an equation can be derived that fully describes the sequential model:

$$P_{open}(Seq.) = \frac{\left(\dfrac{X}{K_A}\right)^2}{\left(\dfrac{X}{K_A}\right)^2 + {}^{BA}L_0\left(1 + \dfrac{X}{K_B}\right)^2}. \quad (A.10)$$

Where receptors have very large values for ${}^{BA}L_0$, both models can satisfactorily represent experimental data. In the case of human muscle nicotinic receptors (Edelstein et al., 1997), for example, where ${}^{BA}L_0 = 9 \times 10^8$, diliganded molecules account for virtually all of the open-channel activity, with maximal contributions to P_{open} by unliganded molecules in the open state (A_0) and monoliganded molecules in the open state (A_1) of only 1.1×10^{-9} and 2.6×10^{-5}, respectively—well within the margin of error for experimental observation. However, in the case of certain mutant receptors—for example, ε T264P (Edelstein et al., 1997a)—the estimated value of ${}^{BA}L_0$ is greatly reduced (dropping to 100), and the contributions of A_0 and A_1 are no longer negligible, with maximal contributions to P_{open} of 1% and 8%, respectively. Moreover, with respect to single channel measurements, the contributions of A_0 and A_1 in terms of numbers of events are superior to A_2 (see Chapter 8, in particular Figure 8.2). In this case, the sequential model cannot satisfactorily account for the experimental data since it neglects the strong contributions of A_0 and A_1. Generally speaking, as we noted in Chapter 1, this comparison of the two mechanistic approaches brings out the Darwinian features of the allosteric model, according to which different conformations freely coexist and the ligand "selects" the conformation whose relative stability will be increased the most by binding of the ligand. The sequential model, by contrast, implies a kind of molecular Lamarckism, whereby the ligand "instructs," or induces, the protein to assume a conformation complementary to the ligand's structure.

APPENDIX

An important aspect of the recent development of the allosteric model involves its success in explaining the striking mutational changes in α7-nicotinic receptor molecules that convert competitive antagonists into partial agonists (Bertrand et al., 1992), as well as the opposite effect exerted by mutant forms of the glycine receptor that convert partial agonists into competitive antagonists (Langosh et al., 1994; Rajendra et al., 1994). In contrast to rather convoluted hypotheses that make it necessary to posit the existence of an inhibitory subsite (Laube et al., 1995), the allosteric model provides a simple and widely applicable explanation of these phenomena based on changes in the intrinsic equilibrium between basal and active states (Galzi et al., 1996a). The key concept here is that so-called competitive antagonists may in some cases actually be very weak agonists. Although they bind somewhat more tightly to the A state than to the B state, the small advantage enjoyed by the A state in this respect is not sufficient to overcome the large advantage of intrinsic stability enjoyed by the B state. As a result, compounds that exert virtually no agonistic effect act in the presence of a strong agonist as competitive antagonists. On the other hand, a mutation that diminishes the intrinsic stability of the B state compared to the A state (corresponding to a decrease in the value of the allosteric constant, L, where L is defined as $L = [B]/[A]$) will cause the small binding advantage of the A state to appear as a significant partial agonist effect. This effect, constituting what is known as an "L phenotype," is presented in quantitative form in Figure A.3 for simulations of data involving mutant and wild-type α7-nicotinic receptors for which the mutation leads to a strong decrease in L. Its corresponding "mirror image" effect involves mutations in the glycine receptor that increase L and convert partial antagonists to agonists (Galzi et al., 1996a).

The allosteric model has also been applied to the problem of explaining more complex changes observed in connection with the "γ phenotype" (Galzi et al., 1966). This phenotype is assumed to result from changes in the state of activity of the ion channel (for example, from nonconducting to conducting) in one (or possibly more) conformation(s), with no alteration of the intrinsic binding parameters for each state—that is, of the receptor's pharmacological specificity—nor of the equilibria or kinetics of interconversion. Assume, for example, that a desensitized conformation exhibiting high affinity for agonists, but which has a closed channel, becomes conducting after a mutation. In a three-state model such as this, having one activatable state and two conducting states, the expected changes in physiological response as compared to wild-type response are fourfold: (1) desensitization of the response to agonists is reduced, since isomerization to a desensitized conformation is no longer accompanied by closing of the ion channel; (2) the sensitivity for activation is now greater, since

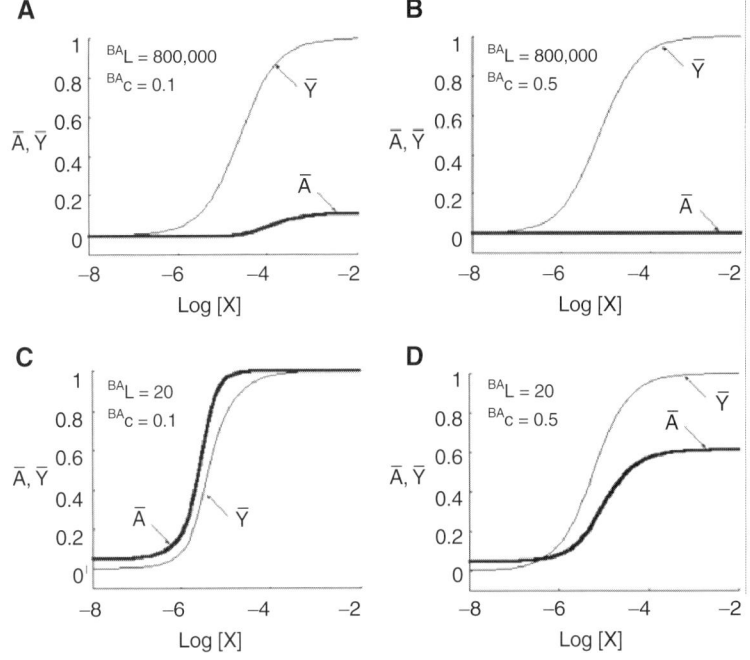

FIGURE A.3 The L phenotype as illustrated by the binding function and state function, $\bar{Y}$ and $\bar{A}$, respectively, in curves for four combinations of $^{BA}L_0$ and ^{BA}c (Edelstein and Changeux, 1996). (A) High $^{BA}L_0$ and low ^{BA}c: complete binding results in a partial agonist response. (B) High $^{BA}L_0$ and high ^{BA}c: complete binding results in a competitive antagonist response. (C) Low $^{BA}L_0$ and low ^{BA}c: the same intrinsic ligand affinities as in (A) result in a full agonist response. (D) Low $^{BA}L_0$ and high ^{BA}c: the same intrinsic ligand affinities as in (B) result in a partial agonist response. The values of $^{BA}L_0$ and ^{BA}c correspond to modeling results obtained using a two-state model (Galzi et al., 1996a) for the nAChR α7-subunit, wild-type ($^{BA}L_0$ = 800,000), and the α7-receptor-channel mutant V251T ($^{BA}L_0$ = 20), with respect to the agonist ACh (^{BA}c = 0.1) and the partial agonist dihydro-β-erythroidine (^{BA}c = 0.5) on the basis of published experimental data (Devillers-Thiéry et al., 1992; Galzi et al., 1992). In addition, the absolute values of the binding affinities are fixed by K_A = 2.5 × 10^{-6} M for ACh, and K_A = 3.5 × 10^{-6} M for DHβE. (From Edelstein and Changeux, 1998.)

desensitized conformations exhibit higher affinity for agonists; (3) a new conducting state, distinct from the wild-type conducting state, may be observed; and (4) the pharmacological drug profiles of the two conducting states now differ. In low concentrations, agonists bring about a high-affinity desensitized (but nonetheless conducting) state and, in high concentrations, two conducting states. Stabilization of the desensitized conformation by competitive antagonists, on the other hand, will activate only a new conducting state, regardless of their concentration.

Analogies may be drawn between the phenotypes of the M2 mutant receptors L247T and T244Q or V251T. Were the L247T mutant assumed to be an instance of an L phenotype, a change in the L value alone would not fully account for the experimentally determined ACh and DHβE dose-response curves (Revah et al., 1991; Bertrand et al., 1992; Devillers-Thiéry et al., 1992). Indeed, where the L value yields an appropriate EC$_{50}$ for ACh, DHβE will behave not as a full agonist but rather as a partial agonist, as in the case of the T244Q or V251T mutants; moreover, under no circumstances will the apparent affinity for DHβE be higher than for ACh, contrary to observation. These facts, together with the occurrence in L247T of two conducting states with distinct pharmacological profiles (Revah et al., 1991; Bertrand et al., 1992; Devillers-Thiéry et al., 1992), favor an interpretation in terms of the γ-phenotype scheme. In this case, simulated dose-response curves satisfactorily fit the experimental data, as shown in Figure A.4 for both ACh (ligand 2) and DHβE (ligand 1), assuming that one of the con-

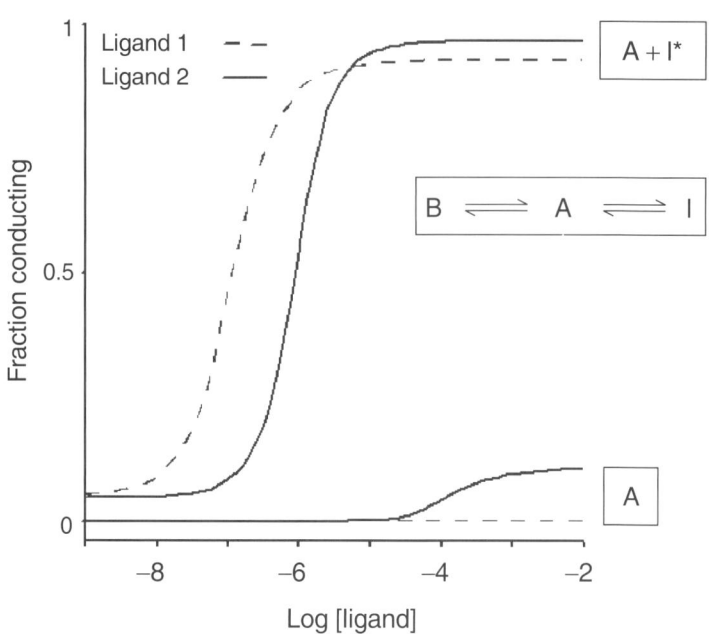

FIGURE A.4 Theoretical dose-response relationships describing the γ phenotype. The curves are generated with a three-state model, B ⇌ A ⇌ I, assuming either that only the A conformation or that both the A and I conformations (I*) contribute to the physiological response, for the curves as indicated. The L value for the B ⇌ A transition is $^{BA}L_0 = 8 \times 10^5$, and for the A ⇌ I transition $^{AI}L_0 = 1.2 \times 10^{-5}$. The affinity values for ligand 1 are $K_B = 2.5 \times 10^{-5}$ M, $K_A = 2.5 \times 10^{-6}$ M, $K_I = 10^{-6}$ M; and for ligand 2, $K_B = 7 \times 10^{-6}$ M, $K_A = 3.5 \times 10^{-6}$ M, $K_I = 3 \times 10^{-7}$ M. Intrinsic affinities increase from state B to A to I for both ligands. Ligand 1 is a competitive antagonist when the I state is characterized by a closed channel and becomes an agonist when it is characterized by an open channel. Ligand 2, which is an agonist in both cases, stabilizes one or two conducting states depending on the biological activity of the I conformation (Galzi et al., 1996a). (From Edelstein and Changeux, 1998.)

ducting states is identical to the wild-type conducting states (not stabilized by DHβE), while the other (assumed to correspond to a wild-type desensitized conformation) shows greater affinity for DHβE binding than for ACh binding (Galzi et al., 1996a).

Application of the fundamental concepts of allosteric equilibria to proteins such as hemoglobin, enzymes, and ligand-gated channels can also be extended to other membrane proteins, notably G-protein coupled receptors (GPCR) and receptor tyrosine kinases. In the case of GPCR, a large family of receptor proteins that had long been considered to be monomeric in form, recent evidence strongly suggests they are in fact dimeric (see review in Milligan, 2004). These proteins, together with their kinetic properties, can be readily accommodated within the allosteric framework (Palanche et al., 2001). It should be noted, however, that a different nomenclature is generally employed in referring to allosteric states: the resting state is designated R and the active state R* (or R0 and R1), with R2, etc., applied to successive states (Palanche et al., 2001). GPCR exhibit "constitutive" mutations at various loci within the receptor molecule that result in dramatically enhanced G-protein-linked biological activity in the absence of an agonist (Lefkowitz et al., 1993), along the same lines observed for nicotinic receptors. Similarly, the allosteric equilibrium between resting and active states can be shifted in either direction: fully or partially from the resting to the active state in the presence of full or partial agonists, or from active to resting under the influence of "inverse" agonists. True "neutral" antagonists—molecules that are neutral with respect to the allosteric equilibrium—have also been described.

WORKS CITED

Abramson, S. N., Li, Y., Culver, P., and Taylor, P. (1989). An analog of lophotoxin reacts covalently with Tyr190 in the alpha-subunit of the nicotinic acetylcholine receptor. *J. Biol. Chem.* 264:12666–12672.

Ackermann, E. J. and Taylor, P. (1997). Nonidentity of the alpha-neurotoxin binding sites on the nicotinic acetylcholine receptor revealed by modification in alpha-neurotoxin and receptor structures. *Biochemistry* 36:12836–12844.

Ackers, G. K., Perrella, M., Holt, J. M., Denisov, I., and Huang, Y. (1997). Thermodynamic stability of the asymmetric doubly-ligated hemoglobin tetramer (alpha+CNbeta+CN) (alphabeta): methodological and mechanistic issues. *Biochemistry* 36:10822–10829.

Ackers, G. K., Holt, J. M., Huang, Y., Grinkove, Y., Klinger, A. L., and Denisov, L. (2000). Confirmation of a unique intra-dimer cooperativity in the human hemoglobin alpha(1)beta(1)half-oxygenated intermediate supports the symmetry rule model of allosteric regulation. *Proteins* 4:23–43.

Adair, G. S. (1925). The hemoglobin system. VI. The oxygen dissociation curve of hemoglobin. *J. Biol. Chem.* 63:529–545.

Agnati, L. F., Zoli, M., Stromberg, I., and Fuxe, K. (1995). Intercellular communication in the brain: wiring versus volume transmission. *Neuroscience* 69:711–726.

Akaaboune, M., Culican, S. M., Turney, S. G., and Lichtman, J. W. (1999). Rapid and reversible effects of activity on acetylcholine receptor density at the neuromuscular junction *in vivo*. *Science* 286:503–507.

Akaaboune, M., Grady, R. M., Turney, S., Sanes, J. R., and Lichtman, J. W. (2002). Neurotransmitter receptor dynamics studied *in vivo* by reversible photo-unbinding of fluorescent ligands. *Neuron* 34:865–876.

Albuquerque, E. X., Barnard, E. A., Chiu, T. H., Lapa, A. J., Dolly, J. O., Jansson, S. E., Daly, J., and Witkop, B. (1973). Acetylcholine receptor and ion conductance modulator sites at the murine neuromuscular junction: evidence from specific toxin reactions. *Proc. Natl. Acad. Sci. USA* 70:949–953.

Albuquerque, E. X., Kuba, K., and Daly, J. (1974). Effect of histrionicotoxin on the ionic conductance modulator of the cholinergic receptor: a quantitative analysis of the endplate current. *J. Pharmacol. Exp. Ther.* 189:513–524.

Albuquerque, E. X., Tsai, M. C., Aronstam, R. S., Witkop, B., Eldefrawi, A. T., and Eldefrawi, M. E. (1980). Phencyclidine interactions with the ionic channel of the acetylcholine receptor and electrogenic membrane. *Proc. Natl. Acad. Sci. USA* 77:1224–1228.

Alkondon, M., Pereira, E. F., Cortes, W. S., Maelicke, A., and Albuquerque, E. X. (1997). Choline is a selective agonist of alpha7 nicotinic acetylcholine receptors in the rat brain neurons. *Eur. J. Neurosci.* 9:2734–2742.

Altiok, N., Bessereau, J. L., and Changeux, J.-P. (1995). ErbB3 and ErbB2/neu mediate the effect of heregulin on acetylcholine receptor gene expression in muscle: differential expression at the endplate. *EMBO J.* 14:4258–4266.

Altiok, N., Altiok, S., and Changeux, J.-P. (1997). Heregulin-stimulated acetylcholine receptor gene expression in muscle: requirement for MAP kinase and evidence for a parallel inhibitory pathway independent of electrical activity. *EMBO J.* 16:717–725.

Works Cited

Altiok, N. and Changeux, J.-P. (2001). Electrical activity regulates AChR gene expression via JNK, PKCzeta and Sp1 in skeletal chick muscle. *FEBS Lett.* 487:333–338.

Amador, M. and Dani, J. A. (1995). Mechanism for modulation of nicotinic acetylcholine receptors that can influence synaptic transmission. *J. Neurosci.* 15:4525–4532.

Anand, R., Conroy, W. G., Schoepfer, R., Whiting, P., and Lindstrom, J. (1991). Neuronal nicotinic acetylcholine receptors expressed in *Xenopus* oocytes have a pentameric quaternary structure. *J. Biol. Chem.* 266:11192–11198.

Anderson, C. R. and Stevens, C. F. (1973). Voltage clamp analysis of acetylcholine produced end-plate current fluctuations at frog neuromuscular junction. *J. Physiol.* 235:655–691.

Anderson, M. J., Cohen, M. W., and Zorychta, E. (1977). Effects of innervation on the distribution of acetylcholine receptors on cultured muscle cells. *J. Physiol.* 268:731–756.

Angelini, G., Sparapani, C., and Speranza, M. (1984). Reactions of phynylium ions with gaseous hydrocarbons. I. Methane, ethane and propane. *Tetrahedron* 40:4865–4871.

Anholt, R., Lindstrom, J., and Montal, M. (1980). Functional equivalence of monomeric and dimeric forms of purified acetylcholine receptors from *Torpedo californica* in reconstructed lipid vesicles. *Eur. J. Biochem.* 109:481–487.

Anholt, R., Lindstrom, J., and Montal, M. (1985). The Molecular Basis of Neurotransmission: Structure and Function of the Nicotinic Acetylcholine Receptor. In A. N. Martonosi, ed. *The Enzymes of Biological Membranes,* 335–401. New York: Plenum Press.

Aoshima, H., Cash, D. J., and Hess, G. P. (1980). Acetylcholine receptor-controlled ion flux in electroplax membrane vesicles: a minimal mechanism based on rate measurements in the millisecond to minute time region. *Biochem. Biophys. Res. Commun.* 92:896–904.

Aoshima, H. (1984). A second, slower inactivation process in acetylcholine receptor-rich membrane vesicles prepared from *Electrophorus electricus. Arch. Biochem. Biophys.* 235:312–318.

Apel, E. D., Roberds, S. L., Campbell, K. P., and Merlie, J.-P. (1995). Rapsyn may function as a link between the acetylcholine receptor and the agrin-binding dystrophin-associated glycoprotein complex. *Neuron* 15:115–126.

Apel, E. D., Shah, S., Bowen, D. C., DeChiara, T. M., Stitt, T. N., Sanes, J. R., and Yancopoulos, G. D. (1997). Rapsyn is required for MuSK signaling and recruits synaptic components to a MuSK-containing scaffold. *Proc. Natl. Acad. Sci. USA* 94:8848–8853.

Arias, H. R. (2000). Localization of agonist and competitive antagonist binding sites on nicotinic acetylcholine receptors. *Neurochem. Int.* 36:595–645.

Arias, H. R. and Blanton, M. P. (2002). Molecular and physicochemical aspects of local anesthetics acting on nicotinic acetylcholine receptor-containing membranes. *Mini. Rev. Med. Chem.* 2:385–410.

Ariëns, E. J. (1954). Affinity and intrinsic activity in the theory of competitive inhibition. *Arch. Int. Pharmacodyn.* 99:32–49.

Armstrong, C. M. (1966). Time course of TEA(+)-induced anomalous rectification in squid giant axons. *J. Gen. Physiol.* 50:491–503.

Armstrong, C. M. (1969). Inactivation of the potassium conductance and related phenomena caused by quaternary ammonium ion injection in squid axons. *J. Gen. Physiol.* 54:553–575.

Armstrong, N., Sun, Y., Chen, G. Q., and Gouaux, E. (1998). Structure of a glutamate-receptor ligand-binding core in complex with kainate. *Nature* 395:913–917.

Arnone, A. (1972). X-ray diffraction study of binding of 2,3-diphosphoglycerate to human deoxyhemoglobin. *Nature* 237:146–149.

Aubert, I., Araujo, D. M., Cecyre, D., Robitaille, Y., Gauthier, S., and Quirion, R. (1992). Comparative alterations of nicotinic and muscarinic binding sites in Alzheimer's and Parkinson's diseases. *J. Neurochem.* 58:529–541.

Augustinsson, K. B. (1948). Cholinesterases: a study in comparative enzymology. *Acta Physiol. Scand.* 182 (Suppl.):50–53.

Augustinsson, K. B. (1950). Acetylcholine esterase and cholinesterase. In *The Enzymes,* 443–472. New York: Academic Press.

Avila, O. L., Drachman, D. B., and Pestronk, A. (1989). Neurotransmission regulates stability of acetylcholine receptors at the neuromuscular junction. *J. Neurosci.* 9:2902–2906.

Axelrod, D., Ravdin, P., Koppel, D. E., Schlessinger, J., Webb, W. W., Elson, E. L., and Podleski, T. R. (1976). Lateral motion of fluorescently labeled acetylcholine receptors in membranes of developing muscle fibers. *Proc. Natl. Acad. Sci. USA* 73:4594–4598.

Axelsson, J. and Thesleff, S. (1959). A study of super-sensitivity in denervated mammalian skeletal muscle. *J. Physiol.* 147:178–193.

Aylwin, M. L. and White, M. M. (1994). Gating properties of mutant acetylcholine receptors. *Mol. Pharmacol.* 46:1149–1155.

Baenziger, J. E. and Chew, J. P. (1997). Desensitization of the nicotinic acetylcholine receptor mainly involves a structural change in solvent-accessible regions of the polypeptide backbone. *Biochemistry* 36:3617–3624.

Baldwin, J. and Chothia, C. (1979). Haemoglobin: the structural changes related to ligand binding and its allosteric mechanism. *J. Mol. Biol.* 129:175–220.

Baldwin, T. J., Theriot, J. A., Yoshihara, C. M., and Burden, S. J. (1988). Regulation of transcript encoding the 43K subsynaptic protein during development and after denervation. *Development* 104:557–564.

Bali, M. and Akabas, M. H. (2004). Defining the propofol binding site location on the GABAA receptor. *Mol. Pharmacol.* 65:68–76.

Ballivet, M., Patrick, J., Lee, J., and Heinemann, S. (1982). Molecular cloning of cDNA coding for the gamma subunit of *Torpedo* acetylcholine receptor. *Proc. Natl. Acad. Sci. USA* 79:4466–4470.

Ballivet, M., Alliod, C., Bertrand, S., and Bertrand, D. (1996). Nicotinic acetylcholine receptors in the nematode *Caenorhabditis elegans*. *J. Mol. Biol.* 258:261–269.

Bansal, A., Singer, J. H., Hwang, B. J., Xu, W., Beaudet, A., and Feller, M. B. (2000). Mice lacking specific nicotinic acetylcholine receptor subunits exhibit dramatically altered spontaneous activity patterns and reveal a limited role for retinal waves in forming ON and OFF circuits in the inner retina. *J. Neurosci.* 20:7672–7681.

Bargmann, C. I. (1998). Neurobiology of the *Caenorhabditis elegans* genome. *Science* 282:2028–2033.

Barkas, T., Mauron, A., Roth, B., Alliod, C., Tzartos, S. J., and Ballivet, M. (1987). Mapping the main immunogenic region and toxin-binding site of the nicotinic acetylcholine receptor. *Science* 235:77–80.

Barlow, R. B. and Ing, H. R. (1948). Curare-like action of polymethylene bis-quaternary ammonium salts. *J. Pharm. Chemother.* 3:298–304.

Barnard, E. A., Wieckowski, J., and Chiu, T. H. (1971). Cholinergic receptor molecules and cholinesterase molecules at mouse skeletal muscle junctions. *Nature* 234:207–209.

Barnard, E. A., Miledi, R., and Sumikawa, K. (1982). Translation of exogenous messenger RNA coding for nicotinic acetylcholine receptors produces functional receptors in *Xenopus* oocytes. *Proc. R. Soc. Lond. B Biol. Sci.* 215:241–246.

Barrantes, F. J., Changeux, J.-P., Lunt, G. G., and Sobel, A. (1975). Differences between detergent-extracted acetylcholine receptor and "cholinergic proteolipid." *Nature* 256:325–327.

Barrantes, F. J. (1976). Intrinsic fluorescence of the membrane-bound acetylcholine receptor: its quenching by suberyldicholine. *Biochem. Biophys. Res. Commun.* 72:479–488.

Barrantes, F. J. (1980). Modulation of acetylcholine receptor states by thiol modification. *Biochemistry* 19:2957–2965.

Barrantes, F. J. (1997). The acetylcholine receptor ligand-gated channel as a molecular target of disease and therapeutic agents. *Neurochem. Res.* 22:391–400.

Barria, A., Derkach, V., and Soderling, T. (1997). Identification of the Ca2+/calmodulin-dependent protein kinase II regulatory phosphorylation site in the alpha-amino-3-hydroxyl-5-methyl-4-isoxazole-propionate-type glutamate receptor. *J. Biol. Chem.* 272:32727–32730.

Barto, A. G., Sutton, R. S., and Anderson, C. W. (1983). Neuron-like adaptive elements that can solve difficult learning and control problems. *IEEE Trans. Syst. Man. Cybern* 13:835.

Bartoli, M., Ramarao, M. K., and Cohen, J. B. (2001). Interactions of the rapsyn RING-H2 domain with dystroglycan. *J. Biol. Chem.* 276:24911–24917.

Bartus, R. T., Dean, R. L., 3rd, Beer, B., and Lippa, A. S. (1982). The cholinergic hypothesis of geriatric memory dysfunction. *Science* 217:408–414.

Batchelor, A. H., Piper, D. E., De la Brousse, F. C., McKnight, S. L., and Wolberger, C. (1998). The structure of GABPalpha/beta: an ETS domain-ankyrin repeat heterodimer bound to DNA. *Science* 279:1037–1041.

Baulac, S., Huberfeld, G., Gourfinkel-An, I., Mitropoulou, G., Beranger, A., Prud'homme, J. F., Baulac, M., Brice, A., Bruzzone, R., and LeGuern, E. (2001). First genetic evidence of GABA(A) receptor dysfunction in epilepsy: a mutation in the gamma2-subunit gene. *Nat. Genet.* 28:46–48.

Beaudet, A. and Descarries, L. (1978). The monoamine innervation of rat cerebral cortex: synaptic and nonsynaptic axon terminals. *Neuroscience* 3:851–860.

Behr, J.-P., Lehn, J.-M., and Vierling, P. (1976). Stable ammonium cryptates of chiral macrocyclic receptor molecules bearing amino acid side-chains. *JCS Chem. Comm.* 16:621–623.

Behr, J.-P., Lehn, J.-M., and Vierling, P. (1982). Molecular receptors. Structural effects and substrate recognition in binding of organic and biogenic ammonium ions by chiral polyfunctional macrocyclic polyethers bearing amino acid and other side-chains. *Helv. Chim. Acta* 65:1853–1867.

Belleau, B. (1964). A molecular theory of drug action based on induced conformational perturbations of receptors. *J. Med. Chem.* 7:776–784.

Benatar, M., Blaes, F., Johnston, I., Wilson, K., Vincent, A., Beeson, D., and Lang, B. (2001). Presynaptic neuronal antigens expressed by a small cell lung carcinoma cell line. *J. Neuroimmunol.* 113:153–162.

Benke, T. A., Lüthi, A., Isaas, J. T. R., and Collingridge, G. L. (1998). Modulation of AMPA receptor unitary conductance by synaptic activity. *Nature* 393:793–797.

Bennett, M. R., Pettigrew, A. G., and Taylor, R. S. (1973). The formation of synapses in reinnervated and cross-reinnervated adult avian muscle. *J. Physiol.* 230:331–357.

Bennett, M. V. L., Wurtzel, M., and Grundfest, H. (1961). The electrophysiology of electric organs of marine electric fishes. I. Properties of electroplaques of *Torpedo nobiliana*. *J. Gen. Physiol.* 44:757–804.

Benoit, P. and Changeux, J.-P. (1993). Voltage dependencies of the effects of chlorpromazine on the nicotinic receptor channel from mouse muscle cell line S018. *Neurosci. Lett.* 160:81–84.

Benwell, M. E., Balfour, D. J., and Anderson, J. M. (1988). Evidence that tobacco smoking increases the density of (-)-[3H]nicotine binding sites in human brain. *J. Neurochem.* 50:1243–1247.

Berg, D. K., Kelly, R. B., Sargent, P. B., Williamson, P., and Hall, Z. W. (1972). Binding of α-bungarotoxin to acetylcholine receptors in mammalian muscle. *Proc. Natl. Acad. Sci. USA* 69:147–151.

Berg, D. K. and Hall, Z. W. (1975). Loss of alpha-bungarotoxin from junctional and extrajunctional acetylcholine receptors in rat diaphragm muscle *in vivo* and in organ culture. *J. Physiol.* 252:771–789.

Bergstrom, R. G., Landells, R. G., Wahl, G. H., and Zollinger, H. (1976). Dediazoniation of arenediazonium ions in homogeneous solution. VII. On the intermediacy of the phynyl cation. *J. Am. Chem. Soc.* 98:3301–3305.

Bernhardt, J. and Neumann, E. (1978). Kinetic analysis of receptor-controlled tracer efflux from sealed membrane fragments. *Proc. Natl. Acad. Sci. USA* 75:3756–3760.

Bernhardt, J. and Neumann, E. (1981). Single channel gating events in tracer flux. *Biophys. Chem.* 14:303–316.

Bernstein, J. (1902). Untersuchungen zur Thermodynamik der bioelektrischen Ströme. *Pfluegers Arch. Ges. Physiol.* 92:521–562.

Bernstein, J. and Tschermak, A. (1904). Über die Frage: Praexistenz Theorie oder Alterationstheorie des Muskelstromes. *Pfluegers Arch. Ges. Physiol.* 103:67–83.

Beroukhim, R. and Unwin, N. (1995). Three-dimensional location of the main immunogenic region of the acetylcholine receptor. *Neuron* 15:323–331.

Bertrand, D., Devillers-Thiéry, A., Revah, F., Galzi, J. L., Hussy, N., Mulle, C., Bertrand, S., Ballivet, M., and Changeux, J.-P. (1992). Unconventional pharmacology of a neuronal nicotinic receptor mutated in the channel domain. *Proc. Natl. Acad. Sci. USA* 89:1261–1265.

Bertrand, D., Galzi, J.-L., Devillers-Thiéry, A., Bertrand, S., and Changeux, J.-P. (1993). Mutations at two distinct sites within the channel domain M2 alter calcium permeability of neuronal α7 nicotinic receptor. *Proc. Natl. Acad. Sci. USA* 90:6971–6975.

Bertrand, D., Ballivet, M., Gomez, M., Bertrand, S., Phannavong, B., and Gundelfinger, E. D. (1994). Physiological properties of neuronal nicotinic receptors reconstituted from the vertebrate beta-2 subunit and *Drosophila* alpha subunits. *Eur. J. Neurosci.* 6:869–875.

Bertrand, D., Picard, F., Le Hellard, S., Weiland, S., Favre, I., Phillips, H., Bertrand, S., Berkovic, S. F., Malafosse, A., and Mulley, J. (2002). How mutations in the nAChRs can cause ADNFLE epilepsy. *Epilepsia* 43:112–122.

Bertrand, S., Devillers-Thiéry, A., Palma, E., Buisson, B., Edelstein, S. J., Corringer, P.-J., Changeux, J.-P., and Bertrand, D. (1997). Paradoxical allosteric effects of competitive inhibitors on neuronal alpha7 nicotinic receptor mutants. *Neuroreport* 8:3591–3596.

Bessereau, J. L., Stratford-Perricaudet, L. D., Piette, J., Le Poupon, C., and Changeux, J.-P.

(1994). *In vivo* and *in vitro* analysis of electrical activity-dependent expression of muscle acetylcholine receptor genes using adenovirus. *Proc. Natl. Acad. Sci. USA* 91: 1304–1308.

Bessis, A., Simon-Chazottes, D., Devillers-Thiéry, A., Guenet, J. L., and Changeux, J.-P. (1990). Chromosomal localization of the mouse genes coding for alpha2, alpha3, alpha4 and beta2 subunits of neuronal nicotinic acetylcholine receptor. *FEBS Lett.* 264:48–52.

Bessis, A., Savatier, N., Devillers-Thiéry, A., Bejanin, S., and Changeux, J.-P. (1993). Negative regulatory elements upstream of a novel exon of the neuronal nicotinic acetylcholine receptor alpha2 subunit gene. *Nucl. Acids Res.* 21:2185–2192.

Bessis, A., Salmon, A. M., Zoli, M., Le, N. N., Picciotto, M., and Changeux, J.-P. (1995). Promoter elements conferring neuron-specific expression of the beta2-subunit of the neuronal nicotinic acetylcholine receptor studied *in vitro* and in transgenic mice. *Neuroscience* 69:807–819.

Bessis, A., Champtiaux, N., Chatelin, L., and Changeux, J.-P. (1997). The neuron-restrictive silencer element: a dual enhancer/silencer crucial for patterned expression of a nicotinic receptor gene in the brain. *Proc. Natl. Acad. Sci. USA* 94:5906–5911.

Betz, H., Bourgeois, J.-P., and Changeux, J.-P. (1977). Evidence for degradation of the acetylcholine (nicotinic) receptor in skeletal muscle during the development of the chick embryo. *FEBS Lett.* 77:219–224.

Betz, H. and Changeux, J.-P. (1979). Regulation of muscle acetylcholine receptor synthesis *in vitro* by cyclic nucleotide derivatives. *Nature* 278:749–752.

Betz, H., Bourgeois, J. P., and Changeux, J.-P. (1980). Evolution of cholinergic proteins in developing slow and fast skeletal muscles in chick embryo. *J. Physiol.* 302:197–218.

Betz, H. (1987). Biology and structure of the mammalian glycine receptor. *Trends Neurosci.* 10:113–117.

Bezakova, G. and Ruegg, M. A. (2003). New insights into the roles of agrin. *Nat. Rev. Mol. Cell Biol.* 4:295–308.

Bhushan, A. and McNamee, M. G. (1993). Correlation of phospholipid structure with functional effects on the nicotinic acetylcholine receptor: a modulatory role for phosphatidic acid. *Biophys. J.* 64:716–723.

Biesecker, G. (1973). Molecular properties of the cholinergic receptor purified from *Electrophorus electricus*. *Biochemistry* 12:4403–4409.

Birnbaumer, L., Pohl, S. L., Michiel, H., Krans, M. J., and Rodbell, M. (1970). The actions of hormones on the adenyl cyclase system. *Adv. Biochem. Psychopharmacol.* 3:185–208.

Blake, C. C., Koenig, D. F., Mair, G. A., North, A. C., Phillips, D. C., and Sarma, V. R. (1965). Structure of hen egg-white lysozyme: a three-dimensional Fourier synthesis at 2 Angstrom resolution. *Nature* 206:757–761.

Blanchard, S. G., Elliott, J., and Raftery, M. A. (1979a). Interaction of local anesthetics with *Torpedo californica* membrane-bound acetylcholine receptor. *Biochemistry* 18:5880–5885.

Blanchard, S. G., Quast, U., Reed, K., Lee, T., Schimerlik, M. I., Vandlen, R., Claudio, T., Strader, C. D., Moore, H. P., and Raftery, M. A. (1979b). Interaction of [125I]-alpha-bungarotoxin with acetylcholine receptor from *Torpedo californica*. *Biochemistry* 18:1875–1883.

Blanton, M. P. and Cohen, J. B. (1994). Identifying the lipid-protein interface of the *Torpedo* nicotinic acetylcholine receptor: secondary structure implications. *Biochemistry* 33:2859–2872.

Blanton, M. P., Dangott, L. J., Raja, S. K., Lala, A. K., and Cohen, J. B. (1998a). Probing the structure of the nicotinic acetylcholine receptor ion channel with the uncharged photoactivable compound [^{3}H]-diazofluorene. *J. Biol. Chem.* 273:8659–8668.

Blanton, M. P., McCardy, E. A., Huggins, A., and Parikh, D. (1998b). Probing the structure of the nicotinic acetylcholine receptor with the hydrophobic photoreactive probes [125I]TID-BE and [125I]TIDPC/16. *Biochemistry* 37:14545–14555.

Blatt, Y., Montal, M. S., Lindstrom, J. M., and Montal, M. (1986). Monoclonal antibodies specific to the beta and gamma subunits of the *Torpedo* acetylcholine receptor inhibit single-channel activity. *J. Neurosci.* 6:481–486.

Bliss, T. V., Collingridge, G. L., and Morris, R. G. (2003). Introduction: long-term potentiation and structure of the issue. *Philos. Trans. R. Soc. Lond. B Biol. Sci.* 358:607–611.

Bloch, R. J. and Hall, Z. W. (1983). Cytoskeletal components of the vertebrate neuromuscular junction: vinculin, alpha-actinin, and filamin. *J. Cell Biol.* 97:217–223.

Bloch, R. J. and Froehner, S. C. (1987). The relationship of the postsynaptic 43K protein to

Works Cited

acetylcholine receptors in receptor clusters isolated from cultured rat myotubes. *J. Cell Biol.* 104:645–654.

Blount, P. and Merlie, J.-P. (1989). Molecular basis of the two nonequivalent ligand binding sites of the muscle nicotinic acetylcholine receptor. *Neuron* 3:349–357.

Blount, P. and Merlie, J.-P. (1991). Characterization of an adult muscle acetylcholine receptor subunit by expression in fibroblasts. *J. Biol. Chem.* 266:14692–14696.

Boheim, G., Hanke, W., Barrantes, F. J., Eibl, H., Sakmann, B., Fels, G., and Maelicke, A. (1981). Agonist-activated ionic channels in acetylcholine receptor reconstituted into planar lipid bilayers. *Proc. Natl. Acad. Sci. USA* 78:3586–3590.

Bohler, S., Gay, S., Bertrand, S., Corringer, P.-J., Edelstein, S. J., Changeux, J.-P., and Bertrand, D. (2001). Desensitization of neuronal nicotinic acetylcholine receptors conferred by N-terminal segments of the β2 subunit. *Biochemistry* 201:2066–2074.

Bolger, M. B., Dionne, V., Chrivia, J., Johnson, D. A., and Taylor, P. (1984). Interaction of a fluorescent acyldicholine with the nicotinic acetylcholine receptor and acetylcholinesterase. *Mol. Pharmacol.* 26:57–69.

Bon, F., Lebrun, E., Gomel, J., van Rapenbusch, R., Cartaud, J., Popot, J. L., and Changeux, J.-P. (1982). [Relative orientation of the two oligomers composing the heavy form of the acetylcholine receptor from *Torpedo marmorata*.] *C. R. Séances Acad. Sci. III* 295:199–204.

Bon, F., Lebrun, E., Gomel, J., Van Rapenbusch, R., Cartaud, J., Popot, J. L., and Changeux, J.-P. (1984). Image analysis of the heavy form of the acetylcholine receptor from *Torpedo marmorata*. *J. Mol. Biol.* 176:205–237.

Bormann, J., Rundström, N., Betz, H., and Langosh, D. (1993). Residues within transmembrane segment M2 determine chloride conductance of glycine receptor homo- and hetero-oligomers. *EMBO J.* 12:3729–3737.

Boulter, J., Evans, K., Goldman, D., Martin, G., Treco, D., Heinemann, S., and Patrick, J. (1986a). Isolation of a cDNA clone coding for a possible neural nicotinic acetylcholine receptor α-subunit. *Nature* 319:368–374.

Boulter, J., Evans, K., Martin, G., Mason, P., Stengelin, S., Goldman, D., Heinemann, S., and Patrick, J. (1986b). Isolation and sequence of cDNA clones coding for the precursor to the gamma subunit of mouse muscle nicotinic acetylcholine receptor. *J. Neurosci. Res.* 16:37–49.

Boulter, J., O'Shea-Greenfield, A., Duvoisin, R. M., Connolly, J. G., Wada, E., Jensen, A., Gardner, P. D., Ballivet, M., Deneris, E. S., McKinnon, D., and et al. (1990). Alpha3, alpha5, and beta4: three members of the rat neuronal nicotinic acetylcholine receptor-related gene family form a gene cluster. *J. Biol. Chem.* 265:4472–4482.

Bourgeois, J.-P., Tsuji, S., Boquet, P., Pillot, J., Ryter, A. M., and Changeux, J.-P. (1971). Localization of the cholinergic receptor protein by immunofluorescence in eel electroplax. *FEBS Lett.* 16:92–94.

Bourgeois, J.-P., Ryter, A. M., Menez, P., Fromageot, P., Boquet, P., and Changeux, J.-P. (1972). Localization of the cholinergic receptor protein in eel electroplax by high resolution autoradiography. *FEBS Lett.* 25:127–133.

Bourgeois, J.-P., Popot, J. L., Ryter, A., and Changeux, J.-P. (1973). Consequences of denervation on the distribution of the cholinergic (nicotinic) receptor sites from *Electrophorus electricus* revealed by high resolution autoradiography. *Brain Res.* 62:557–563.

Bourgeois, J.-P., Betz, H., and Changeux, J. P. (1978a). Effets de la paralysie chronique de l'embryon de poulet par le flaxedil sur le developpement de la jonction neuro-musculaire. *C. R. Hebd. Séances Acad. Sci. D Sci. Nat.* 286:773–776.

Bourgeois, J.-P., Popot, J. L., Ryter, A., and Changeux, J.-P. (1978b). Quantitative studies on the localization of the cholinergic receptor protein in the normal and denervated electroplaque from *Electrophorus electricus*. *J. Cell Biol.* 79:200–216.

Bourgeois, J.-P., Toutant, M., Gouze, J. L., and Changeux, J.-P. (1986). Effect of activity on the selective stabilization of the motor innervation of fast muscle posterior latissimus dorsi from chick embryo. *Int. J. Dev. Neurosci.* 4:415–429.

Bovet, D., Depierre, F., Courvoisier, S., and de Lestrange, Y. (1949). [Synthetic curarizing drugs. II. Phenolic ethers with quaternary ammonium groups. The action of tris(diethylamino-ethoxy)benzene triiodoethylate.] *Arch. Int. Pharmacodyn. Ther.* 80:172–188.

Boyd, N. D. and Cohen, J. B. (1980). Kinetics of binding of [3H]acetylcholine and [3H]carbamoylcholine to *Torpedo* postsynaptic membranes: slow conformational transitions of the cholinergic receptor. *Biochemistry* 19:5344–5353.

Boyd, R. T. (1994). Sequencing and promoter analysis of the genomic region between the rat neuronal nicotinic acetylcholine receptor beta4 and alpha3 genes. *J. Neurobiol.* 25:960–973.

Brejc, K., van Dijk, W. J., Klaassen, R. V., Schuurmans, M., van Der Oost, J., Smit, A. B., and Sixma, T. K. (2001). Crystal structure of an ACh-binding protein reveals the ligand-binding domain of nicotinic receptors. *Nature* 411:269–276.

Bren, N. and Sine, S. M. (1997). Identification of residues in the adult nicotinic acetylcholine receptor that confer selectivity for curariform antagonists. *J. Biol. Chem.* 272:30793–30798.

Brenner, H. R. and Rudin, W. (1989). On the effect of muscle activity on the end-plate membrane in denervated mouse muscle. *J. Physiol.* 410:501–512.

Brenner, H. R., Witzemann, V., and Sakmann, B. (1990). Imprinting of acetylcholine receptor messenger RNA accumulation in mammalian neuromuscular synapses. *Nature* 344:544–547.

Bridgman, P. C., Carr, C., Pedersen, S. E., and Cohen, J. B. (1987). Visualization of the cytoplasmic surface of *Torpedo* postsynaptic membranes by freeze-etch and immunoelectron microscopy. *J. Cell Biol.* 105:1829–1846.

Brisson, A. D. (1978). Acetylcholine Receptor: Protein Structure. *Proceedings of the Ninth International Conference on Electron Microscopy*, 2:180–181.

Brisson, A. D. (1980). Étude structurale de protéines membranaires au moyen des méthodes optiques et numériques d'analyse d'images de microscopie électronique. Ph.D. thesis, University of Grenoble.

Brisson, A. and Unwin, P. N. (1984). Tubular crystals of acetylcholine receptor. *J. Cell Biol.* 99:1202–1211.

Brisson, A. and Unwin, P. N. (1985). Quaternary structure of the acetylcholine receptor. *Nature* 315:474–477.

Broide, R. S., Orr-Urtreger, A., and Patrick, J. W. (2001). Normal apoptosis levels in mice expressing one alpha7 nicotinic receptor null and one L250T mutant allele. *Neuroreport* 12:1643–1648.

Bruses, J. L. and Rutishauser, U. (2001). Membrane lipid rafts are necessary for the maintenance of the (alpha)7 nicotinic acetylcholine receptor in somatic spines of ciliary neurons. *J. Biol. Chem.* 276:31745–31751.

Buc, M. H. and Buc, H. (1968). Allosteric interactions between AMP and orthophosphate sites on phosphorylase b from rabbit muscle. *Fed. Eur. Biochem. Soc.* 4:109–130.

Buisson, B., Gopalakrishnan, M., Arneric, S. P., Sullivan, J. P., and Bertrand, D. (1996). Human α4β2 neuronal nicotinic acetylcholine receptor in HEK 293 cells: a patch-clamp study. *J. Neurosci.* 16:7880–7891.

Buisson, B. and Bertrand, D. (1998). Steroid Modulation of the Nicotinic Receptor. In E. E. Baulieu, P. Robel, and M. Schumacher, eds. *Neurosteroids: A New Regulatory Function in the Nervous System*, 207–223. Totowa, N.J.: Humana Press.

Buisson, B. and Bertrand, D. (2001). Chronic exposure to nicotine upregulates the human alpha4-beta2 nicotinic acetylcholine receptor function. *J. Neurosci.* 21:1819–1829.

Burden, S. (1977a). Development of the neuromuscular junction in the chick embryo: the number, distribution, and stability of acetylcholine receptors. *Dev. Biol.* 57:317–329.

Burden, S. (1977b). Acetylcholine receptors at the neuromuscular junction: developmental change in receptor turnover. *Dev. Biol.* 61:79–85.

Burden, S. J., Sargent, P. B., and McMahan, U. J. (1979). Acetylcholine receptors in regenerating muscle accumulate at original synaptic sites in the absence of the nerve. *J. Cell Biol.* 82:412–425.

Burden, S. J., DePalma, R. L., and Gottesman, G. S. (1983). Crosslinking of proteins in acetylcholine receptor-rich membranes: association between the beta-subunit and the 43 kd subsynaptic protein. *Cell* 35:687–692.

Burden, S. J. (1985). The subsynaptic 43-kDa protein is concentrated at developing nerve-muscle synapses *in vitro*. *Proc. Natl. Acad. Sci. USA* 82:8270–8273.

Burden, S. J. (1998). The formation of neuromuscular synapses. *Genes Dev.* 12:133–148.

Burden, S. J. (2002). Building the vertebrate neuromuscular synapse. *J. Neurobiol.* 53:501–511.

Burger, K., Gimpl, G., and Fahrenholz, F. (2000). Regulation of receptor function by cholesterol. *Cell Mol. Life Sci.* 57:1577–1592.

Burgermeister, W., Catterall, W. A., and Witkop, B. (1977). Histrionicotoxin enhances agonist-induced desensitization of acetylcholine receptor. *Proc. Natl. Acad. Sci. USA* 74:5754–5758.

Bursztajn, S., Berman, S. A., and Gilbert, W. (1989). Differential expression of acetylcholine receptor mRNA in nuclei of cultured muscle cells. *Proc. Natl. Acad. Sci. USA* 86:2928–2932.

Buss, R. R. and Oppenheim, R. W. (2004). Role of programmed cell death in normal neuronal development and function. *Anat. Sci. Int.* 79:191–197.

Caldarone, B. J., Duman, C. H., and Picciotto, M. R. (2000). Fear conditioning and latent inhibition in mice lacking the high-affinity subclass of nicotinic acetylcholine receptors in the brain. *Neuropharmacology* 39:2779–2784.

Campos-Caro, A., Sala, S., Ballesta, J. J., Vicente-Agullo, F., Criado, M., and Sala, F. (1996). A single residue in the M2-M3 loop is a major determinant of coupling between binding and gating in neuronal nicotinic receptors. *Proc. Natl. Acad. Sci. USA* 93:6118–6123.

Camus, G., Jasmin, B. J., and Cartaud, J. (1998). Polarized sorting of nicotinic acetylcholine receptors to the postsynaptic membrane in *Torpedo* electrocyte. *Eur. J. Neurosci.* 10: 839–852.

Carlson, J., Noguchi, K., and Ellison, G. (2001). Nicotine produces selective degeneration in the medial habenula and fasciculus retroflexus. *Brain Res.* 906:127–134.

Caroni, P., Rotzler, S., Britt, J. C., and Brenner, H. R. (1993). Calcium influx and protein phosphorylation mediate the metabolic stabilization of synaptic acetylcholine receptors in muscle. *J. Neurosci.* 13:1315–1325.

Carr, C., McCourt, D., and Cohen, J. B. (1987). The 43-kilodalton protein of *Torpedo* nicotinic postsynaptic membranes: purification and determination of primary structure. *Biochemistry* 26:7090–7102.

Cartaud, J., Benedetti, E. L., Kasai, M., and Changeux, J.-P. (1971). *In vitro* excitation of purified membrane fragments by cholinergic agonists. IV: Ultrastructure, at high resolution, of the AcChE-rich and ATP-ase-rich microsacs. *J. Membr. Biol.* 6:81–88.

Cartaud, J., Benedetti, E. L., Cohen, J. B., Meunier, J. C., and Changeux, J.-P. (1973). Presence of a lattice structure in membrane fragments rich in nicotinic receptor protein from the electric organ of *Torpedo marmorata*. *FEBS Lett.* 33:109–113.

Cartaud, J., Benedetti, L., Sobel, A., and Changeux, J.-P. (1978). A morphological study of the cholinergic receptor protein from *Torpedo marmorata* in its membrane environment and in its detergent extracted purified form. *J. Cell Sci.* 29:313–337.

Cartaud, J., Sobel, A., Rousselet, A., Devaux, P. F., and Changeux, J.-P. (1981). Consequences of alkaline treatment for the ultrastructure of the acetylcholine-receptor-rich membranes from *Torpedo marmorata* electric organ. *J. Cell Biol.* 90:418–426.

Cartaud, J. and Changeux, J.-P. (1993). Post-transcriptional compartmentalization of acetylcholine receptor biosynthesis in the subneural domain of muscle and electrocyte junctions. *Eur. J. Neurosci.* 5:191–202.

Cartaud, A., Coutant, S., Petrucci, T. C., and Cartaud, J. (1998). Evidence for *in situ* and *in vitro* association between beta-dystroglycan and the subsynaptic 43K rapsyn protein: consequence for acetylcholine receptor clustering at the synapse. *J. Biol. Chem.* 273: 11321–11326.

Cartier, G. E., Yoshikami, D., Gray, W. R., Luo, S., Olivera, B. M., and McIntosh, J. M. (1996). A new alpha-conotoxin which targets alpha3beta2 nicotinic acetylcholine receptors. *J. Biol. Chem.* 271:7522–7528.

Cascio, M., Shenkel, S., Grodzicki, R. L., Sigworth, F. J., and Fox, R. O. (2001). Functional reconstitution and characterization of recombinant human alpha1-glycine receptors. *J. Biol. Chem.* 276:20981–20988.

Celesia, G. G. (2001). Disorders of membrane channels or channelopathies. *Clin. Neurophysiol.* 112:2–18.

Celie, P. H., van Rossum-Fikkert, S. E., van Dijk, W. J., Brejc, K., Smit, A. B., and Sixma, T. K. (2004). Nicotine and carbamylcholine binding to nicotinic acetylcholine receptors as studied in AChBP crystal structures. *Neuron* 41:907–914.

Chagas, C., Penna-Franca, E., Hasson, A., Crocker, C., Nishie, K., and Garcia, E. (1957). Studies of the mechanisms of curarisation. *An. Acad. Brasil Ci.* 29:53–62.

Champtiaux, N. (2002). Inactivation génique de la sous-unité α6 du récepteur nicotinique de l'acétylcholine chez la souris. Étude de la composition des récepteurs nictoniques exprimés dans les neurones dopaminergique. Ph.D. thesis, University of Paris-6.

Champtiaux, N. and Changeux, J.-P. (2002). Knock-out and knock-in mice to investigate the role of nicotinic receptors in the central nervous system. *Curr. Drug Targets CNS Neurol. Disord.* 1:319–330.

Champtiaux, N., Han, Z. Y., Bessis, A., Rossi, F. M., Zoli, M., Marubio, L., McIntosh, J. M., and Changeux, J.-P. (2002). Distribution and pharmacology of alpha 6-containing nicotinic acetylcholine receptors analyzed with mutant mice. *J. Neurosci.* 22:1208–1217.

Champtiaux, N., Gotti, C., Cordero-Erausquin, M., David, D. J., Przybylski, C., Léna, C.,

Clementi, F., Moretti, M., Rossi, F., Le Novère, N., et al. (2003). Subunit composition of functional nicotinic receptors in dopaminergic neurons investigated with knock-out mice. *J. Neurosci.* 23:7820–7829.

Chan, R. Y., Boudreau-Lariviere, C., Angus, L. M., Mankal, F. A., and Jasmin, B. J. (1999). An intronic enhancer containing an N-box motif is required for synapse- and tissue-specific expression of the acetylcholinesterase gene in skeletal muscle fibers. *Proc. Natl. Acad. Sci. USA* 96:4627–4632.

Chang, C. C. and Lee, C. Y. (1963). Isolation of neurotoxins from the venom of *Bungarus multicinctus* and their modes of neuromuscular blocking action. *Arch. Int. Pharmacodyn.* 144:316–332.

Chang, C. C. and Huang, M. C. (1975). Turnover of junctional and extrajunctional acetylcholine receptors of the rat diaphragm. *Nature* 253:643–644.

Chang, G., Spencer, R. H., Lee, A. T., Barclay, M. T., and Rees, D. C. (1998). Structure of the MscL homolog from *Mycobacterium tuberculosis:* a gated mechanosensitive ion channel. *Science* 282:2220–2226.

Chang, H. W. (1974). Purification and characterization of acetylcholine receptor. I. From *Electrophorus electricus. Proc. Natl. Acad. Sci. USA* 71:2113–2117.

Chang, H. W. and Bock, E. (1977). Molecular forms of acetylcholine receptor: effects of calcium ions and a sulfhydryl reagent on the occurrence of oligomers. *Biochemistry* 16: 4513–4520.

Changeux, J.-P. (1961). The feedback control mechanism of biosynthetic L-threonine deaminase by L-isoleucine. *Cold Spring Harb. Symp. Quant. Biol.* 26:313–318.

Changeux, J.-P. (1965). Sur les propriétés allostériques de la L-thréonine et de la L-leucine sur la L-thréonine désaminase. VI. Discussion générale. *Bull. Soc. Chim. Biol.* 47:281–300.

Changeux, J.-P. (1966). Responses of acetylcholinesterase from *Torpedo marmorata* to salts and curarizing drugs. *Mol. Pharmacol.* 2:369–392.

Changeux, J.-P., Podleski, T. R., and Wofsy, L. (1967a). Affinity labeling of the acetylcholine-receptor. *Proc. Natl. Acad. Sci. USA* 58:2063–2070.

Changeux, J.-P., Thiéry, J.-P., Tung, T., and Kittel, C. (1967b). On the cooperativity of biological membranes. *Proc. Natl. Acad. Sci. USA* 57:335–341.

Changeux, J.-P. and Podleski, T. (1968). On the excitability and cooperativity of the electroplax membrane. *Proc. Natl. Acad. Sci. USA* 59:944–950.

Changeux, J.-P. and Rubin, M. M. (1968). Allosteric interactions in aspartate transcarbamylase. III. Interpretations of experimental data in terms of the model of Monod, Wyman, and Changeux. *Biochemistry* 7:553–561.

Changeux, J.-P. (1969). Remarks on the Symmetry and Cooperative Properties of Biological Membranes. In A. Engström and B. Stranberg, eds. *Nobel Symposium: Symmetry and Functions in Biological Systems at the Molecular Level,* 235–256. New York: Wiley.

Changeux, J.-P., Gautron, J., Israel, M., and Podleski, T. (1969). [Separation of excitable membranes from the electric organ of *Electrophorus electricus.*] *C. R. Acad. Sci. Hebd. Séances Acad. Sci. D Sci. Nat.* 269:1791.

Changeux, J.-P., Kasai, M., Huchet, M., and Meunier, J. C. (1970a). Extraction à partir du tissu électrique de gymnote d'une protéine présentant plusieurs propriétés caractéristiques du récepteur physiologique de l'acétylcholine. *C. R. Acad. Sci. Hebd. Séances Acad. Sci. D Sci. Nat.* 270:2864–2867.

Changeux, J.-P., Kasai, M., and Lee, C. Y. (1970b). Use of a snake venom toxin to characterize the cholinergic receptor protein. *Proc. Natl. Acad. Sci. USA* 67:1241–1247.

Changeux, J.-P., Meunier, J. C., and Huchet, M. (1971). Studies on the cholinergic receptor protein of *Electrophorus electricus.* I. An assay *in vitro* for the cholinergic receptor site and solubilization of the receptor protein from electric tissue. *Mol. Pharmacol.* 7:538–553.

Changeux, J.-P., Huchet, M., and Cartaud, J. (1972). [Partial reconstitution of an excitable membrane after dissolution by sodium deoxycholate.] *C. R. Acad. Sci. Hebd. Séances Acad. Sci. D Sci. Nat.* 274:122–125.

Changeux, J.-P., Courrège, P., and Danchin, A. (1973). A theory of the epigenesis of neural networks by selective stabilization of synapses. *Proc. Natl. Acad. Sci. USA* 70:2974–2978.

Changeux, J.-P. (1975). The Cholinergic Receptor Protein from Fish Electric Organ. In L. L. Iversen, S. D. Iversen, and S. H. Snyder, eds. *Handbook of Psychopharmacology,* 235–301. New York: Plenum.

Works Cited

Changeux, J.-P., Benedetti, L., Bourgeois, J.-P., Brisson, A., Cartaud, J., Devaux, P., Grunhagen, H., Moreau, M., Popot, J. L., Sobel, A., and Weber, M. (1976). Some structural properties of the cholinergic receptor protein in its membrane environmental relevant to its function as a pharmacological receptor. *Cold Spring Harb. Symp. Quant. Biol.* 40:211–230.

Changeux, J.-P. and Danchin, A. (1976). Selective stabilization of developing synapses as a mechanism for the specification of neuronal networks. *Nature* 264:705–712.

Changeux, J.-P. (1979). Claude Bernard's Experiments on Curare and Motor Endplate Synaptic Transmission. In E. D. Robin, ed. *Claude Bernard and the Internal Environment: A Memorial Symposium,* 73–95. New York: Marcel Dekker.

Changeux, J.-P., Heidmann, T., Popot, J. L., and Sobel, A. (1979). Reconstitution of a functional acetylcholine regulator under defined conditions. *FEBS Lett.* 105:181–187.

Changeux, J.-P. (1981). The acetylcholine receptor: an "allosteric" membrane protein. *Harvey Lectures* 75:85–254.

Changeux, J.-P. (1983a). *L'homme neuronal.* Paris: Fayard.

Changeux, J.-P. (1983b). Concluding remarks: on the "singularity" of nerve cells and its ontogenesis. *Prog. Brain Res.* 58:465–478.

Changeux, J.-P., Devillers-Thiéry, A., and Chemouilli, P. (1984). Acetylcholine receptor: an allosteric protein. *Science* 225:1335–1345.

Changeux, J.-P. (1985). *Neuronal Man: The Biology of Mind.* Trans. Laurence Garey. New York: Pantheon. [Translation of Changeux, 1983a.]

Changeux, J.-P., Pinset, C., and Ribera, A. B. (1986). Effects of chlorpromazine and phencyclidine on mouse C2 acetylcholine receptor kinetics. *J. Physiol.* 378:497–513.

Changeux, J.-P., Devillers-Thiéry, A., Giraudat, J., Dennis, M., Heidmann, T., Revah, F., Mulle C., Heidmann, O., Klarsfeld, A., Fontaine, B., et al. (1987a). The Acetylcholine Receptor: Functional Organization and Evolution during Synapse Formation. In O. Hayaishi, ed. *Strategy and Prospects in Neuroscience,* 29–76. Tokyo/Utrecht: Japan Scientific Societies Press/VNU Science Press.

Changeux, J.-P., Klarsfeld, A. Z., and T., H. (1987b). The Acetylcholine Receptor and Molecular Models for Short- and Long-Term Learning. In J.-P. Changeux and M. Konishi, eds. *Dahlem Konferenzen: The Cellular and Molecular Bases of Learning,* 31–83. London: Wiley.

Changeux, J.-P. and Connes, A. (1989). *Matière à pensée.* Paris: Odile Jacob.

Changeux, J.-P. (1990). Functional Architecture and Dynamics of the Nicotinic Acetylcholine Receptor: An Allosteric Ligand-Gated Channel. In J.-P. Changeux, R. R. Llinás, D. Purves, and F. F. Bloom, eds. *Fidia Research Foundation Neuroscience Award Lectures,* vol. 4, 17–168. New York: Raven Press.

Changeux, J.-P. and Connes, A. (1995). *Conversations on Mind, Matter, and Mathematics.* Trans. M. B. DeBevoise. Princeton: Princeton University Press. [Translation of Changeux and Connes, 1989.]

Changeux, J.-P., Bertrand, D., Corringer, P.-J., Dehaene, S., Edelstein, S., Léna, C., Le Novère, N., Marubio, L., Picciotto, M., and Zoli, M. (1998). Brain nicotinic receptors: structure and regulation, role in learning and reinforcement. *Brain Res. Rev.* 26:198–216.

Changeux, J.-P. and Edelstein, S. J. (1998). Allosteric receptors after 30 years. *Neuron* 21:959–980.

Changeux, J.-P. and Edelstein, S. J. (2001). Allosteric mechanisms in normal and pathological nicotinic acetylcholine receptors. *Curr. Opin. Neurobiol.* 11:369–377.

Changeux, J.-P. (2002). *L'homme de vérité.* Paris: Odile Jacob.

Changeux, J.-P. (2004). *The Physiology of Truth.* Trans. M. B. DeBevoise. Cambridge, Mass.: Harvard University Press. [Translation of Changeux, 2002.]

Chapman, M. L., VanDongen, H. M., and VanDongen, A. M. (1997). Activation-dependent subconductance levels in the drk1 K channel suggest a subunit basis for ion permeation and gating. *Biophys. J.* 72:708–719.

Chappell, J., McMahan, R., Chiba, A., and Gallagher, M. (1998). A re-examination of the role of basal forebrain cholinergic neurons in spatial working memory. *Neuropharmacology* 37:481–487.

Chatrenet, B., Tremeau, O., Bontems, F., Goeldner, M. P., Hirth, C. G., and Menez, A. (1990). Topography of toxin-acetylcholine receptor complexes by using photoactivatable toxin derivatives. *Proc. Natl. Acad. Sci. USA* 87:3378–3382.

Chavez-Noriega, L. E., Crona, J. H., Washburn, M. S., Urrutia, A., Elliott, K. J., and Johnson, E. C. (1997). Pharmacological characterization of recombinant human neuronal nicotinic acetylcholine receptors hα2β2, hα2β4, hα3β2, hα3β4, hα4β2, hα4β4, and hα7 expressed in *Xenopus* oocytes. *J. Pharmacol. Exp. Ther.* 280:346–356.

Chiara, D. C. and Cohen, J. B. (1997). Identification of amino acids contributing to high and low affinity d-tubocurarine sites in the *Torpedo* nicotinic acetylcholine receptor. *J. Biol. Chem.* 272:32940–32950.

Chiara, D. C., Middleton, R. E., and Cohen, J. B. (1998). Identification of tryptophan 55 as the primary site of [3H]nicotine photoincorporation in the gamma-subunit of the *Torpedo* nicotinic acetylcholine receptor. *FEBS. Lett.* 423:223–226.

Chiara, D. C., Xie, Y., and Cohen, J. B. (1999). Structure of the agonist-binding sites of the *Torpedo* nicotinic acetylcholine receptor: affinity-labeling and mutational analyses identify gamma Tyr-111/delta Arg-113 as antagonist affinity determinants. *Biochemistry* 38:6689–6698.

Choquet, D. and Triller, A. (2003). The role of receptor diffusion in the organization of the postsynaptic membrane. *Nat. Rev. Neurosci.* 4:251–265.

Clark, A. J. (1926). The antagonism of acetylcholine by atropine. *J. Physiol. (London)* 61:547–556.

Clarke, P. B. (1993). Nicotinic receptors in mammalian brain: localization and relation to cholinergic innervation. *Prog. Brain Res.* 98:77–83.

Claudio, T., Ballivet, M., Patrick, J., and Heinemann, S. (1983). Nucleotide and deduced amino acid sequences of *Torpedo californica* acetylcholine receptor gamma subunit. *Proc. Natl. Acad. Sci. USA* 80:1111–1115.

Cohen, B. N., Labarca, C., Czyzyk, L., Davidson, N., and Lester, H. A. (1992). Tris+/Na+ permeability ratios of nicotinic acetylcholine receptors are reduced by mutations near the intracellular end of the M2 region. *J. Gen. Physiol.* 99:545–572.

Cohen, G., Han, Z. Y., Grailhe, R., Gallego, J., Gaultier, C., Changeux, J.-P., and Lagercrantz, H. (2002). Beta2 nicotinic acetylcholine receptor subunit modulates protective responses to stress: a receptor basis for sleep-disordered breathing after nicotine exposure. *Proc. Natl. Acad. Sci. USA* 99:13272–13277.

Cohen, J. B., Weber, M., Huchet, M., and Changeux, J.-P. (1972). Purification from *Torpedo marmorata* electric tissue of membrane fragments particularly rich in cholinergic receptor protein. *FEBS Lett.* 26:43–47.

Cohen, J. B. and Changeux, J.-P. (1973a). Interaction of a fluorescent ligand with membrane-bound cholinergic receptor from *Torpedo marmorata*. *Biochemistry* 12:4855–4864.

Cohen, J. B. and Changeux, J.-P. (1973b). [Interaction of a fluorescent ligand with the protein receptor of acetylcholine present in *Torpedo marmorata* membrane fragments]. *C. R. Acad. Sci. Hebd. Séances Acad. Sci. D Sci. Nat.* 277:603–606.

Cohen, J. B., Weber, M., and Changeux, J.-P. (1974). Effects of local anesthetics and calcium on the interaction of cholinergic ligands with the nicotinic receptor protein from *Torpedo marmorata*. *Mol. Pharmacol.* 10:904–932.

Cohen, J. B. (1978). Ligand Binding Properties of Membrane-Bound Cholinergic Receptor of *Torpedo marmorata*. In A. K. Solomon, and M. Karnovsky, eds. *Molecular Specialization and Symmetry in Membrane Function*, 99–127. Cambridge, Mass.: Harvard University Press.

Cohen, J. B., Sharp, S. D., and Liu, W. S. (1991). Structure of the agonist-binding site of the nicotinic acetylcholine receptor: [3H]acetylcholine mustard identifies residues in the cation-binding subsite. *J. Biol. Chem.* 266:23354–23364.

Cohen, M. W. (1972). The development of neuromuscular connexions in the presence of D-tubocurarine. *Brain Res.* 41:457–463.

Cole, N. B. and Lippincott-Schwartz, J. (1995). Organization of organelles and membrane traffic by microtubules. *Curr. Opin. Cell Biol* 7:55–64.

Collingridge, G. L. (2003). The induction of N-methyl-D-aspartate receptor-dependent long-term potentiation. *Philos. Trans. R. Soc. Lond. B Biol. Sci.* 358:635–641.

Colquhoun, D. and Rang, H. P. (1976). Effects of inhibitors of the binding of iodinated alpha-bungarotoxin to acetylcholine receptors in rat muscle. *Mol. Pharmacol.* 12:519–535.

Colquhoun, D. (1979). The Link between Drug Binding and Response: Theories and Observations. In R. D. O'Brien, ed. *The Receptors: General Principles and Procedures*, 93–142. New York: Plenum Press.

Colquhoun, D. and Sakmann, B. (1981). Fluctuations in the microsecond time range of the current through single acetylcholine receptor ion channels. *Nature* 294:464–466.

Colquhoun, D. and Sakmann, B. (1985). Fast events in single-channel currents activated by acetylcholine and its analogues at the frog muscle end-plate. *J. Physiol.* 369:501–557.

Colquhoun, D. (1998). Binding, gating, affinity and efficacy: the interpretation of structure-activity relationships for agonists and of the effects of mutating receptors. *Br. J. Pharmacol.* 125:924–947.

Colquhoun, D. and Sakmann, B. (1998). From muscle endplate to brain synapses: a short history of synapses and agonist-activated ion channels. *Neuron* 20:381–387.

Colquhoun, L. M. and Patrick, J. W. (1997). Alpha3, beta2, and beta4 form heterotrimeric neuronal nicotinic acetylcholine receptors in *Xenopus* oocytes. *J. Neurochem.* 69:2355–2362.

Connolly, J. G., Gibb, A. J., and Colquhoun, D. (1995). Heterogeneity of neuronal nicotinic acetylcholine receptors in thin slices of rat medial habenula. *J. Physiol.* 484:87–105.

Conti-Tronconi, B. M. and Raftery, M. A. (1982). The nicotinic cholinergic receptor: correlation of molecular structure with functional properties. *Ann. Rev. Biochem.* 51:491–530.

Coombs, S. E. and Hess, G. (1984). Acetylcholine Receptor: Some Methods Developed to Study a Membrane-Bound Regulatory Protein. In E. Elson, W. Frazier, and L. Glaser, eds. *Cell Membranes,* 295–364. New York: Plenum Press.

Cooper, D. and Reich, E. (1972). Neurotoxin from venom of the cobra *Naja-Naja siamensis.* *J. Biol. Chem.* 247:3008–3013.

Corbin, J., Methot, N., Wang, H. H., Baenziger, J. E., and Blanton, M. P. (1998a). Secondary structure analysis of individual transmembrane segments of the nicotinic acetylcholine receptor by circular dichroism and Fourier transform infrared spectroscopy. *J. Biol. Chem.* 273:771–777.

Corbin, J., Wang, H. H., and Blanton, M. P. (1998b). Identifying the cholesterol binding domain in the nicotinic acetylcholine receptor with [125I]azido-cholesterol. *Biochem. Biophys. Acta* 1414:65–74.

Cordero-Erausquin, M., and Changeux, J.-P. (2001). Tonic nicotinic modulation of serotoninergic transmission in the spinal cord. *Proc. Natl. Acad. Sci. USA* 98:2803–2807.

Cormier, A., Paas, Y., Zini, R., Tillement, J.-P., Lagrue, G., Changeux, J.-P., and Grailhe, R. (2004). Long-term exposure to nicotine modulates the level and activity of acetylcholine receptors in white blood cells of smokers and model mice. *Mol. Pharmacol.* 66:1712–1718.

Corrigall, W. A., Coen, K. M., and Adamson, K. L. (1994). Self-administered nicotine activates the mesolimbic dopamine system through the ventral tegmental area. *Brain Res.* 653:278–284.

Corrigall, W. A., Coen, K. M., Zhang, J., and Adamson, K. L. (2001). GABA mechanisms in the pedunculopontine tegmental nucleus influence particular aspects of nicotine self-administration selectively in the rat. *Psychopharmacology (Berl.)* 158:190–197.

Corringer, P.-J., Galzi, J. L., Eisele, J. L., Bertrand, S., Changeux, J.-P., and Bertrand, D. (1995). Identification of a new component of the agonist binding site of the nicotinic alpha 7 homooligomeric receptor. *J. Biol. Chem.* 270:11749–11752.

Corringer, P.-J., Bertrand, S., Bohler, S., Edelstein, S. J., Changeux, J.-P., and Bertrand, D. (1998). Critical elements determining diversity in agonist binding and desensitization of neuronal nicotinic acetylcholine receptors. *J. Neurosci.* 18:648–657.

Corringer, P.-J., Bertrand, S., Galzi, J. L., Devillers-Thiéry, A., Changeux, J.-P., and Bertrand, D. (1999). Mutational analysis of the charge selectivity filter of the alpha7 nicotinic acetylcholine receptor. *Neuron* 22:831–843.

Corringer, P.-J., Le Novère, N., and Changeux, J.-P. (2000). Nicotinic receptors at the amino acid level. *Ann. Rev. Pharmacol. Toxicol.* 40:431–458.

Corriveau, R., Romano, S., Conroy, W., Olivia, L., and Berg, D. K. (1995). Expression of neuronal acetylcholine receptor genes in vertebrate skeletal muscle during development. *J. Neurosci.* 15:1372–1383.

Corsi, P. (1988). Luigi Galvani: Biographical Sketch. In A. G. Gilman et al., eds. *Fidia Research Foundation Neuroscience Award Lectures,* vol. 2, 3–6. New York: Raven Press.

Couteaux, R. (1958). Morphological and cyotchemical observations on the post-synaptic membrane at motor endplates and ganglionic synapses. *Exp. Cell Res.* 5:294–322.

Couteaux, R. (1972). Structure and Cytochemical Characteristics of the Neuromuscular Junction. In J. Cheymol, ed. *Neuromuscular Blocking and Stimulating Agents,* 7–56. Oxford: Pergamon.

Couteaux, R. (1978). *Recherches morphologiques et cytochimiques sur l'organisation des tissus excitables.* Paris: Robin et Mareuge.

Couteaux, R. (1981). Structure of the subsynaptic sarcoplasm in the interfolds of the frog neuromuscular junction. *J. Neurocytol.* 10:947–962.

Couturier, S., Bertrand, D., Matter, J.-M., Hernandez, M.-C., Bertrand, S., Millar, N., Valera, S., Barkas, T., and Ballivet, M. (1990a). A neuronal nicotinic acetylcholine receptor subunit (α7) is developmentally regulated and forms a homo-oligomeric channel blocked by α-BTX. *Neuron* 5:847–856.

Couturier, S., Erkman, L., Valera, S., Rungger, D., Bertrand, S., Boulter, J., Ballivet, M., and Bertrand, D. (1990b). Alpha5, alpha3, and non-alpha3: three clustered avian genes encoding neuronal nicotinic acetylcholine receptor-related subunits. *J. Biol. Chem.* 265:560–17567.

Cox, D. H., Cui, J., and Aldrich, R. W. (1997). Allosteric gating of a large conductance Ca-activated K^+ channel. *J. Gen. Physiol.* 110:257–281.

Cox, R. N., Kaldany, R. R., Brandt, P. W., Ferren, B., Hudson, R. A., and Karlin, A. (1984). A continuous-flow, rapid-mixing, photolabeling technique applied to the acetylcholine receptor. *Analyt. Biochem.* 136:476–486.

Cox, R. N., Kaldany, R. R., DiPaola, M., and Karlin, A. (1985). Time-resolved photolabeling by quinacrine azide of a noncompetitive inhibitor site of the nicotinic acetylcholine receptor in a transient, agonist-induced state. *J. Biol. Chem.* 260:7186–7193.

Criado, M., Dominguez del Toro, E., Carrasco-Serrano, C., Smillie, F. I., Juiz, J. M., Viniegra, S., and Ballesta, J. J. (1997). Differential expression of alpha-bungarotoxin-sensitive neuronal nicotinic receptors in adrenergic chromaffin cells: a role for transcription factor Egr-1. *J. Neurosci.* 17:6554–6564.

Croxen, R., Newland, C., Beeson, D., Oosterhuis, H., Chauplannaz, G., Vincent, A., and Newsom-Davis, J. (1997). Mutations in different functional domains of the human muscle acetylcholine receptor alpha subunit in patients with the slow-channel congenital myasthenic syndrome. *Hum. Mol. Genet.* 6:767–774.

Cuatrecasas, P. (1971). Affinity chromotography. *Ann. Rev. Biochem.* 40:259–278.

Cuatrecasas, P. (1972). Affinity chromatography and purification of the insulin receptor of liver cell membranes. *Proc. Natl. Acad. Sci. USA* 69:1277–1281.

Cui, C., Booker, T. K., Allen, R. S., Grady, S. R., Whiteaker, P., Marks, M. J., Salminen, O., Tritto, T., Butt, C. M., Allen, W. R., et al. (2003). The beta3 nicotinic receptor subunit: a component of alpha-conotoxin MII-binding nicotinic acetylcholine receptors that modulate dopamine release and related behaviors. *J. Neurosci.* 23:11045–11053.

Cull-Candy, S. G. and Usowicz, M. M. (1987). Multiple-conductance channels activated by excitatory amino acids in cerebellar neurons. *Nature* 325:525–528.

Culver, P., Fenical, W., and Taylor, P. (1984). Lophotoxin irreversibly inactivates the nicotinic acetylcholine receptor by preferential association at one of the two primary agonist sites. *J. Biol. Chem.* 259:3763–3770.

Czajkowski, C. and Karlin, A. (1995). Structure of the nicotinic acetylcholine-binding site: identification of acidic residues in the δ subunit within 0.9 nm of the δ subunit binding site disulfide. *J. Biol. Chem.* 270:3160–3164.

Dale, H. (1914). The actions of certain esters and esters of choline, and their relation to muscarine. *J. Pharmacol. Exp. Ther.* 6:147–190.

Dale, H. (1943). Modes of drug action. General introductory address. *Trans. Faraday Soc.* 39:319–322.

Daly, J. W., Karle, I., Myers, C. W., Tokuyama, T., Waters, J. A., and Witkop, B. (1971). Histrionicotoxins: roentgen-ray analysis of the novel allenic and acetylenic spiro-alkaloids isolated from a Colombian frog, *Dendrobates histrionicus*. *Proc. Natl. Acad. Sci. USA* 68:1870–1875.

Damle, V. N., Hamilton, S., Valderrama, R., and Karlin, A. (1976). The binding properties of acetylcholine receptor in membrane from *Torpedo* electric tissue. *Pharmacologist* 18:146–185.

Damle, V. N. and Karlin, A. (1978). Affinity labeling of one of the two alpha-neurotoxin binding sites in acetylcholine receptor from *Torpedo californica*. *Biochemistry* 17:2039–2045.

Damle, V. N. and Karlin, A. (1980). Effects of agonists and antagonists on the reactivity of the binding site disulfide in acetylcholine receptor from *Torpedo californica*. *Biochemistry* 19:3924–3932.

Daubas, P., Devillers-Thiéry, A., Geoffroy, B., Martinez, S., Bessis, A., and Changeux, J.-P. (1990). Differential expression of the neuronal acetylcholine receptor alpha2 subunit gene during chick brain development. *Neuron* 5:49–60.

Works Cited

Davidson, E. H., Rast, J. P., Oliveri, P., Ransick, A., Calestani, C., Yuh, C. H., Minokawa, T., Amore, G., Hinman, V., Arenas-Mena, C., et al. (2002). A genomic regulatory network for development. *Science* 295:1669–1678.

De Kerchove D'Exaerde, A., Cartaud, J., Ravel-Chapuis, A., Seroz, T., Pasteau, F., Angus, L. M., Jasmin, B. J., Changeux, J.-P., and Schaeffer, L. (2002). Expression of mutant Ets protein at the neuromuscular synapse causes alterations in morphology and gene expression. *EMBO Rep.* 22:22.

De Robertis, E. (1971). Molecular biology of synaptic receptors. *Science* 171:963–971.

Debru, C. (1983). *L'esprit des protéines*. Paris: Hermann.

Decker, M. W., Majchrzak, M. J., and Anderson, D. J. (1992). Effects of nicotine on spatial memory deficits in rats with septal lesions. *Brain Res.* 572:281–285.

Dehaene, S. and Changeux, J.-P. (1989). A simple model of prefrontal cortex function in delayed-response tasks. *J. Cognitive Neurosci.* 1:244–261.

Dehaene, S. and Changeux, J.-P. (1991). The Wisconsin card sorting test: theoretical analysis and modeling in a neuronal network. *Cerebral Cortex* 1:62–79.

Dehaene, S., Kerszberg, M., and Changeux, J.-P. (1998). A neuronal model of a global workspace in effortful cognitive tasks. *Proc. Natl. Acad. Sci. USA* 95:14529–14534.

Dehaene, S. and Changeux, J.-P. (2000). Reward-dependent learning in neuronal networks for planning and decision making. *Prog. Brain Res.* 126:217–229.

Dehaene, S., Sergent, C., and Changeux, J.-P. (2003). A neuronal network model linking subjective reports and objective physiological data during conscious perception. *Proc. Natl. Acad. Sci. USA* 100:8520–8525.

Dehaene, S. and Changeux, J.-P. (2005). Neural Mechanisms for Access to Consciousness. In M. Gazzaniga, ed. *The New Cognitive Neuroscience*, vol. 3. Cambridge, Mass.: The MIT Press.

Del Castillo, J. and Katz, B. (1955a). On the localization of acetylcholine receptors. *J. Physiol.* 128:157–181.

Del Castillo, J. and Katz, B. (1955b). Local activity at a depolarized nerve-muscle junction. *J. Physiol.* 128:396–411.

Del Castillo, J. and Katz, B. (1956). Biophysical aspects of neuro-muscular transmission. *Prog. Biophys. Biochem.* 6:121–170.

Del Castillo, J. and Katz, B. (1957a). Interaction at endplate receptors between different choline derivatives. *Proc. Roy. Soc. Ser. B* 146:369–381.

Del Castillo, J. and Katz, B. (1957b). A study of curare action with an electrical micromethod. *Proc. Roy. Soc. Ser. B* 146:369–381.

Deleglane, A. M. and McNamee, M. G. (1980). Independent activation of the acetylcholine receptor from *Torpedo californica* at two sites. *Biochemistry* 19:890–895.

Deneris, E. S., Boulter, J., Swanson, L. W., Patrick, J., and Heinemann, S. (1989). Beta-3: a new member of nicotinic acetylcholine receptor gene family is expressed in brain. *J. Biol. Chem.* 264:6268–6272.

Dennis, M., Giraudat, J., Kotzyba-Hibert, F., Goeldner, M., Hirth, C., Chang, J. Y., Lazure, C., Chretien, M., and Changeux, J.-P. (1986). A photoaffinity ligand of the acetylcholine binding site predominantly labels the region 179–207 of the alpha-subunit on native acetylcholine receptor from *Torpedo marmorata*. *FEBS Lett.* 207:243–249.

Dennis, M., Giraudat, J., Kotzyba-Hibert, F., Goeldner, M., Hirth, C., Chang, J. Y., Lazure, C., Chretien, M., and Changeux, J.-P. (1988). Amino acids of the *Torpedo mamorata* acetylcholine receptor alpha subunit labeled by a photoaffinity ligand for the acetylcholine binding site. *Biochemistry* 27:2346–2357.

Descarries, L., Gisiger, V., and Steriade, M. (1997). Diffuse transmission by acetylcholine in the CNS. *Prog. Neurobiol.* 53:603–625.

Descarries, L. and Mechawar, N. (2000). Ultrastructural evidence for diffuse transmission by monoamine and acetylcholine neurons of the central nervous system. *Prog. Brain Res.* 125:27–47.

Descarries, L., Mechawar, N., Aznavour, N., and Watkins, K. C. (2004). Structural determinants of the roles of acetylcholine in cerebral cortex. *Prog. Brain Res.* 145:45–58.

Devillers-Thiéry, A., Changeux, J.-P., Paroutaud, P., and Strosberg, A.-D. (1979). The amino-terminal sequence of the 40,000 molecular weight subunit of the acetylcholine receptor protein from *Torpedo marmorata*. *FEBS Lett.* 104:99–105.

Devillers-Thiéry, A., Giraudat, J., Bentaboulet, M., and Changeux, J.-P. (1983). Complete mRNA coding sequence of the acetylcholine binding alpha-subunit of *Torpedo mamorata* acetylcholine receptor: a model for the transmembrane organization of the polypeptide chain. *Proc. Natl. Acad. Sci. USA* 80:2067–2071.

Devillers-Thiéry, A., Galzi, J. L., Bertrand, S., Changeux, J.-P., and Bertrand, D. (1992). Stratified organization of the nicotinic acetylcholine receptor channel. *Neuroreport* 3: 1001–1004.

Devreotes, P. N. and Fambrough, D. M. (1975). Acetylcholine receptor turnover in membranes of developing muscle fibers. *J. Cell Biol.* 65:335–358.

Devreotes, P. N. and Fambrough, D. M. (1976). Synthesis of acetylcholine receptors by cultured chick myotubes and denervated mouse extensor digitorum longus muscles. *Proc. Natl. Acad. Sci. USA* 73:161–164.

Devreotes, P. N., Gardner, J. M., and Fambrough, D. M. (1977). Kinetics of biosynthesis of acetylcholine receptor and subsequent incorporation into plasma membrane of cultured chick skeletal muscle. *Cell* 10:365–373.

Di Chiara, G. (2000). Role of dopamine in the behavioural actions of nicotine related to addiction. *Eur. J. Pharmacol.* 393:295–314.

Di Chiara, G. (2002). Nucleus accumbens shell and core dopamine: differential role in behavior and addiction. *Behav. Brain Res.* 137:75–114.

Dias, R., Thomas, K. L., Augood, S. J., Everitt, B. J., Robbins, T. W., and Roberts, A. C. (1997). Primate analogue of the Wisconsin Card Sorting Test: effects of excitotoxic lesions of the prefrontal cortex in the marmoset. *Neuroscience* 80:717–730.

DiPaola, M., Czajkowski, C., and Karlin, A. (1989). The sidedness of the COOH terminus of the acetylcholine receptor delta subunit. *J. Biol. Chem.* 264:15457–15463.

Dolly, J. O., Albuquerque, E. X., Sarvey, J., Mallick, B., and Barnard, E. A. (1977). Binding of perhydro-histrionicotoxin to the postsynaptic membrane of skeletal muscle in relation to its blockage of acetylcholine-induced depolarization. *Mol. Pharmacol.* 13:1–14.

Donger, C., Krejci, E., Serradell, A. P., Eymard, B., Bon, S., Nicole, S., Chateau, D., Gary, F., Fardeau, M., Massoulie, J., and Guicheney, P. (1998). Mutation in the human acetylcholinesterase-associated collagen gene, COLQ, is responsible for congenital myasthenic syndrome with end-plate acetylcholinesterase deficiency (Type Ic). *Am. J. Hum. Genet.* 63:967–975.

Dorus, S., Vallender, E. J., Evans, P. D., Anderson, J. R., Gilbert, S. L., Mahowald, M., Wyckoff, G. J., Malcom, C. M., and Lahn, B. T. (2004). Accelerated evolution of nervous system genes in the origin of *Homo sapiens. Cell* 119:1027–1040

Doster, W., Hess, B., Watters, D., and Maelicke, A. (1980). Translational diffusion coefficient and molecular weight of the acetylcholine receptor from *Torpedo marmorata. FEBS Lett.* 113:312–314.

Dowding, A. J. and Hall, Z. W. (1987). Monoclonal antibodies specific for each of the two toxin-binding sites of Torpedo acetylcholine receptor. *Biochemistry* 26:6372–6381.

Doyère, M. (1840). Mémoire sur les tardigrades. *Ann. Sci. Nat.* 14:269–361.

Doyle, D. A., Cabral, J. M., Pfuetzner, R. A., Kuo, A., Gulbis, J. M., Cohen, S. L., Chait, B. T., and MacKinnon, R. (1998). The structure of the potassium channel: molecular basis of K^+ conduction and selectivity. *Science* 280:69–77.

Drouin, C., Darracq, L., Trovero, F., Blanc, G., Glowinski, J., Cotecchia, S., and Tassin, J. P. (2002). Alpha1b-adrenergic receptors control locomotor and rewarding effects of psychostimulants and opiates. *J. Neurosci.* 22:2873–2884.

Duclert, A., Piette, J., and Changeux, J.-P. (1990). Induction of acetylcholine receptor alpha-subunit gene expression in chicken myotubes by blocking electrical activity requires on-going protein synthesis. *Proc. Natl. Acad. Sci. USA* 87:1391–1395.

Duclert, A., Piette, J., and Changeux, J.-P. (1991). Influence of innervation of myogenic factors and acetylcholine receptor alpha-subunit mRNAs. *Neuroreport* 2:25–28.

Duclert, A., Savatier, N., and Changeux, J.-P. (1993). An 83-nucleotide promoter of the acetylcholine receptor epsilon-subunit gene confers preferential synaptic expression in mouse muscle. *Proc. Natl. Acad. Sci. USA* 90:3043–3047.

Duclert, A. and Changeux, J.-P. (1995). Acetylcholine receptor gene expression at the developing neuromuscular junction. *Physiol. Rev.* 75:339–368.

Duclert, A., Savatier, N., Schaeffer, L., and Changeux, J.-P. (1996). Identification of an element crucial for the sub-synaptic expression of the acetylcholine receptor epsilon-subunit gene. *J. Biol. Chem.* 271:17433–17438.

Dunbrack, R. L. J. and Karplus, M. (1994). Conformational analysis of the backbone-dependent rotamer preferences of protein sidechains. *Nat. Struct. Biol.* 1:334–340.

Dunn, S. M., Blanchard, S. G., and Raftery, M. A. (1980). Kinetics of carbamylcholine binding to membrane-bound acetylcholine receptor monitored by fluorescence changes of a covalently bound probe. *Biochemistry* 19:5645–5652.

Dunn, S. M. and Raftery, M. A. (1982a). Multiple binding sites for agonists on *Torpedo californica* acetylcholine receptor. *Biochemistry* 21:6264–6272.

Dunn, S. M. and Raftery, M. A. (1982b). Activation and desensitization of *Torpedo* acetylcholine receptor: evidence for separate binding sites. *Proc. Natl. Acad. Sci. USA* 79:6757–6761.

Eccles, J. (1964). *The Physiology of Synapses*. Berlin: Springer-Verlag.

Eccles, J. C., Fatt, P., and Kuffler, S. W. (1941). Nature of the "endplate potential" in curarized muscle. *J. Neurophysiol.* 4:362–387.

Edelstein, S. J. (1971). Extensions of the allosteric model for hemoglobin. *Nature* 230:224–227.

Edelstein, S. J. (1972). An allosteric mechanism for the acetylcholine receptor. *Biochem. Biophys. Res. Commun.* 48:1160–1165.

Edelstein, S. J. (1975). Cooperative interactions of hemoglobin. *Ann. Rev. Biochem.* 44:209–232.

Edelstein, S. J. (1996). An allosteric theory for hemoglobin incorporating asymmetric states to test the putative molecular model for cooperativity. *J. Mol. Biol.* 257:737–744.

Edelstein, S. J., Schaad, O., Henry, E., Bertrand, D., and Changeux, J.-P. (1996). A kinetic mechanism for nicotinic acetylcholine receptors based on multiple allosteric transitions. *Biol. Cybern.* 75:361–380.

Edelstein, S. J., Schaad, O., and Changeux, J.-P. (1997a). Myasthenic nicotinic receptor mutant interpreted in terms of the allosteric model. *C. R. Acad. Sci. Paris* 320:953–961.

Edelstein, S. J., Schaad, O., and Changeux, J.-P. (1997b). Single bindings versus single channel recordings: a new approach to ionotropic receptors. *Biochemistry* 36:13755–13760.

Edelstein, S. J. and Changeux, J.-P. (1998). Allosteric transitions of the acetylcholine receptor. *Adv. Prot. Chem.* 51:121–184.

Edman, L., Mets, U., and Rigler, R. (1996). Conformational transitions monitored by single molecules in solution. *Proc. Natl. Acad. Sci. USA* 93:6710–6715.

Edmonds, B., Gibb, A. J., and Colquhoun, D. (1995). Mechanisms of activation of muscle nicotinic acetylcholine receptors and the time course of endplate currents. *Ann. Rev. Physiol.* 57:469–493.

Edsall, J. T. (1980). Hemoglobin and the origins of the concept of allosterism. *Fed. Proc.* 39:226–235.

Eftimie, R., Brenner, H. R., and Buonanno, A. (1991). Myogenin and MyoD join a family of skeletal muscle genes regulated by electrical activity. *Proc. Natl. Acad. Sci. USA* 88:349–1353.

Eglen, S. J., Demas, J., and Wong, R. O. (2003). Mapping by waves: patterned spontaneous activity regulates retinotopic map refinement. *Neuron* 40:1053–1055.

Ehrenpreis, S. (1960). Isolation and identification of the acetylcholine receptor protein of electric tissue. *Biochem. Biophys. Acta* 44:561–577.

Ehrlich, P. (1957). *Collected Papers*. Ed. H. Dale and F. Himmelweit. Vol 2. London: Pergamon Press.

Eigen, M. (1967). Kinetics of reaction control and information transfer in enyzmes and nucleic acids. *Nobel Symp.* 5:333–369.

Eigen, M. and Rigler, R. (1994). Sorting single molecules: application to diagnostics and evolutionary biotechnology. *Proc. Natl. Acad. Sci. USA* 91:5740–5747.

Eiselé, J. L., Bertrand, S., Galzi, J. L., Devillers-Thiéry, A., Changeux, J.-P., and Bertrand, D. (1993). Chimaeric nicotinic-serotonergic receptor combines distinct ligand binding and channel specificities. *Nature* 366:479–483.

Eldefrawi, A. T., Eldefrawi, M. E., Albuquerque, E. X., Oliveira, A. C., Mansour, N., Adler, M., Daly, J. W., Brown, G. B., Burgermeister, W., and Witkop, B. (1977). Perhydrohistrionicotoxin: a potential ligand for the ion conductance modulator of the acetylcholine receptor. *Proc. Natl. Acad. Sci. USA* 74:2172–2176.

Eldefrawi, M. E. and Eldefrawi, A. T. (1973). Purification and molecular properties of the acetylcholine receptor from *Torpedo* electroplax. *Arch. Biochem. Biophys.* 159:362–373.

Eldefrawi, M. E., Aronstam, R. S., Bakry, N. M., Eldefrawi, A. T., and Albuquerque, E. X. (1980). Activation, inactivation, and desensitization of acetylcholine receptor channel complex detected by binding of perhydrohistrionicotoxin. *Proc. Natl. Acad. Sci. USA* 77:2309–2313.

Elgoyhen, A. B., Johnson, D. S., Boulter, J., Vetter, D. E., and Heinemann, S. (1994). Alpha9: an acetylcholine receptor with novel pharmacological properties expressed in rat cochlear hair cells. *Cell* 79:705–715.

Elgoyhen, A. B., Vetter, D. E., Katz, E., Rothlin, C. V., Heinemann, S. F., and Boulter, J. (2001). Alpha10: a determinant of nicotinic cholinergic receptor function in mammalian

vestibular and cochlear mechanosensory hair cells. *Proc. Natl. Acad. Sci. USA* 98:3501–3506.
Elliott, J., Blanchard, S. G., Wu, W., Miller, J., Strader, C. D., Hartig, P., Moore, H. P., Racs, J., and Raftery, M. A. (1980). Purification of *Torpedo californica* post-synaptic membranes and fractionation of their constituent proteins. *Biochem. J.* 185:667–677.
Elliott, T. R. (1904). On the action of adrenalin. *Proc. Physiol. Soc., X–XI.*
Elliott, T. R. (1905). The action of adrenalin. *J. Physiol.* 32:401–467.
Engel, A. G., Lambert, E. H., and Gomez, M. R. (1977). A new myasthenic syndrome with end-plate acetylcholinesterase deficiency, small nerve terminals, and reduced acetylcholine release. *Ann. Neurol.* 1:315–330.
Engel, A. G. (1993). The investigation of congenital myasthenic syndromes. *Ann. N.Y. Acad. Sci.* 681:425–434.
Engel, A. G., Ohno, K., Milone, M., Wang, H. L., Nakano, S., Bouzat, C., Pruitt, J. N., 2nd, Hutchinson, D. O., Brengman, J. M., Bren, N. et al. (1996). New mutations in acetylcholine receptor subunit genes reveal heterogeneity in the slow-channel congenital myasthenic syndrome. *Hum. Mol. Genet.* 5:1217–1227.
Engel, A. G., Ohno, K., and Sine, S. M. (1999). Congenital myasthenic syndromes: recent advances. *Arch. Neurol.* 56:163–167.
Engel, A. G., Ohno, K., and Sine, S. M. (2003a). Sleuthing molecular targets for neurological diseases at the neuromuscular junction. *Nat. Rev. Neurosci.* 4:339–352.
Engel, A. G., Ohno, K., and Sine, S. M. (2003b). Congenital myasthenic syndromes: a diverse array of molecular targets. *J. Neurocytol.* 32:1017–1037.
Epstein, M. and Racker, E. (1978). Reconstitution of carbamylcholine-dependent sodium ion flux and desensitization of the acetylcholine receptor from *Torpedo californica*. *J. Biol. Chem.* 253:6660–6662.
Eunson, L. H., Rea, R., Zuberi, S. M., Youroukos, S., Panayiotopoulos, C. P., Liguori, R., Avoni, P., McWilliam, R. C., Stephenson, J. B., Hanna, M. G., et al. (2000). Clinical, genetic, and expression studies of mutations in the potassium channel gene KCNA1 reveal new phenotypic variability. *Ann. Neurol.* 48:647–656.
Fahr, A., Lauffer, L., Schmidt, D., Heyn, M. P., and Hucho, F. (1985). Covalent labeling of functional states of the acetylcholine receptor. Effects of antagonists on the receptor conformation. *Eur. J. Biochem.* 147:483–487.
Fairclough, R. H., Finer-Moore, J., Love, R. A., Kristofferson, D., Desmeules, P. J., and Stroud, R. M. (1983). Subunit organization and structure of an acetylcholine receptor. *Cold Spring Harb. Symp. Quant. Biol.* 48 (Pt. 1):9–20.
Fallon, J. R., Nitkin, R. M., Reist, N. E., Wallace, B. G., and McMahan, U. J. (1985). Acetylcholine receptor-aggregating factor is similar to molecules concentrated at neuromuscular junctions. *Nature* 315:571–574.
Falls, D. L., Rosen, K. M., Corfas, G., Lane, W. S., and Fischbach, G. D. (1993). ARIA, a protein that stimulates acetylcholine receptor synthesis, is a member of the neu ligand family. *Cell* 72:801–815.
Fambrough, D. (1979). Control of acetylcholine receptors in skeletal muscle. *Physiol. Rev.* 59:165–227.
Fatt, P. (1950). Membrane potential changes at the motor endplate. *J. Physiol.* 111:408–442.
Fatt, P. and Katz, B. (1950). Membrane potential changes at the motor endplate. *J. Physiol.* 111:46–47.
Fatt, P. and Katz, B. (1951). An analysis of the endplate potential recorded with an intracellular electrode. *J. Physiol.* 115:320–369.
Fatt, P. and Katz, B. (1952). Spontaneous subthreshold activity at motor nerve-endings. *J. Physiol.* 117:109–128.
Faure, P., Neumeister, H., Faber, D. S., and Korn, H. (2003). Symbolic analysis of swimming trajectories reveals scale invariance and provides a model for fish locomotion. *Fractals* 11:233–243.
Feldberg, W., Fessard, A., and Nachmansohn, D. (1940). The cholinergic nature of the nervous supply to the electric organ of the Torpedo (*Torpedo marmorata*). *J. Physiol.* 97:3–4.
Feller, M. B., Wellis, D. P., Stellwagen, D., Werblin, F. S., and Shatz, C. J. (1996). Requirement for cholinergic synaptic transmission in the propagation of spontaneous retinal waves. *Science* 272:1182–1187.
Feltz, A. and Trautmann, A. (1982). Desensitization at the frog neuro-muscular junction: a biphasic process. *J. Physiol.* 322:257–272.

Fenton, J. W. and Singer, S. J. (1965). Affinity labeling of antibodies to the p-azophenyltrimethylammonium hapten and a structural relationship among antibody active sites of different specificities. *Biochem. Biophys. Res. Commun.* 20:315–320.

Fertuck, H. C. and Salpeter, M. M. (1976). Quantitation of junctional and extrajunctional acetylcholine receptors by electron microscope autoradiography after 125I-alpha-bungarotoxin binding at mouse neuromuscular junctions. *J. Cell Biol.* 69:144–158.

Fessard, A. (1958). Les organes électriques. In P. Grassé, ed. *Traité de zoologie*, 1143–1238. Paris: Masson.

Figl, A., Cohen, B. N., Quick, M. W., Davidson, N., and Lester, H. A. (1992). Regions of beta4/beta2 subunit chimeras that contribute to the agonist selectivity of neuronal nicotinic receptors. *FEBS Lett.* 308:245–248.

Finer-Moore, J. and Stroud, R. M. (1984). Amphipathic analysis and possible formation of the ion channel in an acetylcholine receptor. *Proc. Natl. Acad. Sci. USA* 81:155–159.

Fischbach, G. D. and Schuetze, S. M. (1980). A post-natal decrease in acetylcholine channel open time at rat end-plates. *J. Physiol.* 303:125–137.

Fischer, E. (1894). Einfluss der Configuration auf die Wirkung der Enzyme. *Ber. Dtsch. Chem. Ges.* 27:2985–2986.

Fletterick, R. J. and Madsen, N. B. (1980). The structures and related functions of phosphorylase a. *Ann. Rev. Biochem.* 49:31–61.

Flores, C. M., Rogers, S. W., Pabreza, L. A., Wolfe, B. B., and Keller, K. J. (1992). A subtype of nicotinic cholinergic receptor in rat brain is composed of α4β2 subunits and is up-regulated by chronic nicotine treatment. *Mol. Pharmacol.* 41:31–37.

Flucher, B. E. and Daniels, M. P. (1989). Distribution of Na^+ channels and ankyrin in neuromuscular junctions is complementary to that of acetylcholine receptors and the 43 kd protein. *Neuron* 3:163–175.

Fontaine, B., Klarsfeld, A., Hokfelt, T., and Changeux, J.-P. (1986). Calcitonin gene-related peptide, a peptide present in spinal cord motoneurons, increases the number of acetylcholine receptors in primary cultures of chick embryo myotubes. *Neurosci. Lett.* 71:59–65.

Fontaine, B., Klarsfeld, A., and Changeux, J.-P. (1987). Calcitonin gene-related peptide and muscle activity regulate acetylcholine receptor alpha-subunit mRNA levels by distinct intracellular pathways. *J. Cell Biol.* 105:1337–1342.

Fontaine, B., Sassoon, D., Buckingham, M., and Changeux, J.-P. (1988). Detection of the nicotinic acetylcholine receptor alpha-subunit mRNA by *in situ* hybridization at neuromuscular junctions of 15-day-old chick striated muscles. *EMBO J.* 7:603–609.

Fontaine, B. and Changeux, J.-P. (1989). Localization of nicotinic acetylcholine receptor alpha-subunit transcripts during myogenesis and motor endplate development in the chick. *J. Cell Biol.* 108:1025–1037.

Forman, S. A., Yellen, G., and Thiele, E. A. (1996). Alternative mechanism for pathogenesis of an inherited epilepsy by a nicotinic AChR mutation. *Nat. Genet.* 13:396–397.

Frail, D. E., Mudd, J., Shah, V., Carr, C., Cohen, J. B., and Merlie, J.-P. (1987). cDNAs for the postsynaptic 43-kDa protein of *Torpedo* electric organ encode two proteins with different carboxyl termini. *Proc. Natl. Acad. Sci. USA* 84:6302–6306.

Frail, D. E., McLaughlin, L. L., Mudd, J., and Merlie, J.-P. (1988). Identification of the mouse muscle 43,000-dalton acetylcholine receptor-associated protein (RAPsyn) by cDNA cloning. *J. Biol. Chem.* 263:15602–15607.

Frank, E., Gautvik, K., and Sommerschild, H. (1975). Cholinergic receptors at denervated mammalian motor end-plates. *Acta Physiol. Scand.* 95:66–76.

Frank, E. and Fischbach, G. D. (1979). Early events in neuromuscular junction formation *in vitro*: induction of acetylcholine receptor clusters in the postsynaptic membrane and morphology of newly formed synapses. *J. Cell Biol.* 83:143–158.

Franke, C., Parnas, H., Hovav, G., and Dudel, J. (1993). A molecular scheme for the reaction between acetylcholine and nicotinic channels. *Biophys. J.* 64:339–356.

Fratiglioni, L. and Wang, H. X. (2000). Smoking and Parkinson's and Alzheimer's disease: review of the epidemiological studies. *Behav. Brain Res.* 113:117–120.

Froehner, S. C. (1981). Identification of exposed and buried determinants of the membrane-bound acetylcholine receptor from *Torpedo californica*. *Biochemistry* 20:4905–4915.

Froehner, S. C. (1984). Peripheral proteins of postsynaptic membranes from *Torpedo* electric organ identified with monoclonal antibodies. *J. Cell Biol.* 99:88–96.

Froehner, S. C., Luetje, C. W., Scotland, P. B., and Patrick, J. (1990). The postsynaptic 43K protein clusters muscle nicotinic acetylcholine receptors in *Xenopus* oocytes. *Neuron* 5:403–410.

Fromm, L. and Burden, S. J. (1998). Synapse-specific and neuregulin-induced transcription require an ets site that binds GABPalpha/GABPbeta. *Genes. Dev.* 12:3074–3083.

Fruchart-Gaillard, C., Gilquin, B., Antil-Delbeke, S., Le Novère, N., Tamiya, T., Corringer, P.-J., Changeux, J.-P., Menez, A., and Servent, D. (2002). Experimentally based model of a complex between a snake toxin and the alpha 7 nicotinic receptor. *Proc. Natl. Acad. Sci. USA* 99:3216–3221.

Fu, A. K. Y., Fu, W. Y., Cheung, J., Tsim, K. W. K., Ip, F. C. F., Wang, J. H., and Ip, N. Y. (2001). Cdk5 is involved in neuregulin-induced AChR expression at the neuromuscular junction. *Nat. Neurosci.* 4:374–381.

Fuchs, S. (1979). Immunological analysis of acetylcholine receptor. *Adv. Cytopharmacol.* 3:279–286.

Fucile, S., Matter, J. M., Erkman, L., Ragozzino, D., Barabino, B., Grassi, F., Alema, S., Ballivet, M., and Eusebi, F. (1998). The neuronal alpha6 subunit forms functional heteromeric acetylcholine receptors in human transfected cells. *Eur. J. Neurosci.* 10:172–178.

Fuhrer, C., Gautam, M., Sugiyama, J. E., and Hall, Z. W. (1999). Roles of rapsyn and agrin in interaction of postsynaptic proteins with acetylcholine receptors. *J. Neurosci.* 19:6405–6416.

Fumagalli, G., Balbi, S., Cangiano, A., and Lomo, T. (1990). Regulation of turnover and number of acetylcholine receptors at neuromuscular junctions. *Neuron* 4:563–569.

Furchgott, R. F. (1966). The use of beta-haloalkylamines in the differentiation of receptors and in the determination of dissociation constants of receptor-agonist complexes. *Adv. Drug Res.* 3:21–55.

Furukawa, T. (1957). Properties of the procaine endplate potential. *Jpn. J. Physiol.* 7:199–212.

Fusco, M. D., Becchetti, A., Patrignani, A., Annesi, G., Gambardella, A., Quattrone, A., Ballabio, A., Wanke, E., and Casari, G. (2000). The nicotinic receptor beta2 subunit is mutant in nocturnal frontal lobe epilepsy. *Nat. Genet.* 26:275–276.

Fuster, J. M. (2001). The prefrontal cortex—an update: time is of the essence. *Neuron* 30:319–333.

Fuxe, K. and Agnati, L. (1991). Two Principal Modes of Electro-Chemical Communication in the Brain: Volume versus Wiring Transmission. In K. Fuxe and L. Agnati, eds. *Volume Transmission in the Brain: Novel Mechanisms for Neural Transmission,* 1–9. New York: Raven Press.

Gaddum, J. H. (1957). Symposium on drug antagonism. Theories of drug antagonism. *Pharmacol. Rev.* 9:211–218.

Gallagher, D. T., Gilliland, G. L., Xiao, G., Zondlo, J., Fisher, K. E., Chinchilla, D., and Eisenstein, E. (1998). Structure and control of pyridoxal phosphate dependent allosteric threonine deaminase. *Structure* 6:465–475.

Gallagher, M. and Colombo, P. J. (1995). Ageing: the cholinergic hypothesis of cognitive decline. *Curr. Opin. Neurobiol.* 5:161–168.

Gallagher, M. J., Chiara, D. C., and Cohen, J. B. (2001). Interactions between 3-(Trifluoromethyl)-3-(m- [(125)I]iodophenyl)diazirine and tetracaine, phencyclidine, or histrionicotoxin in the *Torpedo* series nicotinic acetylcholine receptor ion channel. *Mol. Pharmacol.* 59:1514–1522.

Galzi, J.-L., Revah, F., Black, D., Goeldner, M., Hirth, C., and Changeux, J.-P. (1990). Identification of a novel amino acid alpha-tyrosine 93 within the cholinergic ligand-binding sites of the acetylcholine receptor by photoaffinity labeling: additional evidence for a three-loop model of the cholinergic ligands-binding sites. *J. Biol. Chem.* 265:10430–10437.

Galzi, J.-L., Revah, F., Bouet, F., Menez, A., Goeldner, M., Hirth, C., and Changeux, J.-P. (1991a). Allosteric transitions of the acetylcholine receptor probed at the amino acid level with a photolabile cholinergic ligand. *Proc. Natl. Acad. Sci. USA* 88:5051–5055.

Galzi, J.-L., Bertrand, D., Devillers-Thiéry, A., Revah, F., Bertrand, S., and Changeux, J.-P. (1991b). Functional significance of aromatic amino acids from three peptide loops of the α7 neuronal nicotinic receptor site investigated by site-directed mutagenesis. *FEBS Lett.* 294:198–202.

Galzi, J.-L., Devillers-Thiéry, A., Hussy, N., Bertrand, S., Changeux, J.-P., and Bertrand, D. (1992). Mutations in the channel domain of a neuronal nicotinic receptor convert ion selectivity from cationic to anionic. *Nature* 359:500–505.

Galzi, J.-L. and Changeux, J.-P. (1994). Neurotransmitter-gated ion channels as unconventional allosteric proteins. *Curr. Opin. Struct. Biol.* 4:554–565.

Galzi, J.-L., Edelstein, S. J., and Changeux, J.-P. (1996a). The multiple phenotypes of allosteric receptor mutants. *Proc. Natl. Acad. Sci. USA* 93:1853–1858.

Works Cited

Galzi, J.-L., Bertrand, S., Corringer, P.-J., Changeux, J.-P., and Bertrand, D. (1996b). Identification of calcium binding sites that regulate potentiation of a neuronal nicotinic acetylcholine receptor. *EMBO J.* 15:5824–5832.

Gault, J., Robinson, M., Berger, R., Drebing, C., Logel, J., Hopkins, J., Moore, T., Jacobs, S., Meriwether, J., Choi, M. J., et al. (1998). Genomic organization and partial duplication of the human alpha7 neuronal nicotinic acetylcholine receptor gene (CHRNA7). *Genomics* 52:173–185.

Gautam, M., DeChiara, T. M., Glass, D. J., Yancopoulos, G. D., and J.R., S. (1999). Distinct phenotypes of mutant mice lacking agrin, MuSK, or rapsyn. *Brain Res. Dev. Brain Res.* 14:171–178.

Gehring, W. J. and Knight, R. T. (2000). Prefrontal-cingulate interactions in action monitoring. *Nat. Neurosci.* 3:516–520.

Gerhart, J. C. and Pardee, A. B. (1962). The enzymology of control by feedback inhibition. *J. Biol. Chem.* 237:891–896.

Gerhart, J. C. and Schachman, H. K. (1965). Distinct subunits for the regulation and catalytic activity of aspartate transcarbamylase. *Biochemistry* 4:1054–1062.

Gerhart, J. C. and Schachman, H. K. (1968). Allosteric interactions in aspartate transcarbamylase. II. Evidence for different conformational states of the protein in the presence and absence of specific ligands. *Biochemistry* 7:538–552.

Gershoni, J. M., Hawrot, E., and Lentz, T. L. (1983). Binding of alpha-bungarotoxin to isolated alpha subunit of the acetylcholine receptor of *Torpedo californica*: quantitative analysis with protein blots. *Proc. Natl. Acad. Sci. USA* 80:4973–4977.

Giacobini, G., Filogamo, G., Weber, M., Boquet, P., and Changeux, J.-P. (1973). Effects of a snake alpha-neurotoxin on the development of innervated skeletal muscles in chick embryo. *Proc. Natl. Acad. Sci. USA* 70:1708–1712.

Giacobini-Robecchi, M. G., Giacobini, G., Filogamo, G., and Changeux, J.-P. (1975). Effect of the type A toxin from *C. botulinum* on the development of skeletal muscles and of their innervation in chick embryo. *Brain Res.* 83:107–121.

Gill, E. W. and Rang, H. P. (1966). An alkylating derivative of benzilylcholine with specific and long-lasting parasympatholytic activity. *Mol. Pharmacol.* 2:284–297.

Giraudat, J., Devillers-Thiéry, A., Auffray, C., Rougeon, F., and Changeux, J.-P. (1982). Identification of a cDNA clone coding for the acetylcholine binding subunit of *Torpedo marmorata* acetylcholine receptor. *EMBO J.* 1:713–717.

Giraudat, J., Montecucco, C., Bisson, R., and Changeux, J.-P. (1985). Transmembrane topology of acetylcholine receptor subunits probed with photoreactive phospholipids. *Biochemistry* 24:3121–3127.

Giraudat, J., Dennis, M., Heidmann, T., Chang, J. Y., and Changeux, J.-P. (1986). Structure of the high-affinity binding site for noncompetitive blockers of the acetylcholine receptor: serine-262 of the delta subunit is labeled by [3H]chlorpromazine. *Proc. Natl. Acad. Sci. USA* 83:2719–2723.

Giraudat, J., Dennis, M., Heidmann, T., Haumont, P. Y., Lederer, F., and Changeux, J.-P. (1987). Structure of the high-affinity binding site for noncompetitive blockers of the acetylcholine receptor: [3H]chlorpromazine labels homologous residues in the beta and delta chains. *Biochemistry* 26:2410–2418.

Giraudat, J., Gali, J., Revah, F., Changeux, J.-P., Haumont, P., and Lederer, F. (1989). The noncompetitive blocker [(3)H]chlorpromazine labels segment M2 but not segment M1 of the nicotinic acetylcholine receptor alpha-subunit. *FEBS Lett.* 253:190–198.

Gisiger, T., Dehaene, S., and Changeux, J.-P. (2000). Computational models of association cortex. *Curr. Opin. Neurobiol.* 10:250–259.

Goeldner, M. P. and Hirth, C. G. (1980). Specific photoaffinity labeling induced by energy transfer: application to irreversible inhibition of acetylcholinesterase. *Proc. Natl. Acad. Sci. USA* 77:6439–6442.

Goeldner, M. P., Hirth, C. G., Kieffer, B., and Ourisson, G. (1982). Photosuicide inhibition: a step towards specific photoaffinity labelling. *Trends Biochem. Sci.* 7:310–312.

Goldin, A. L., Snutch, T., Lubbert, H., Dowsett, A., Marshall, J., Auld, V., Downey, W., Fritz, L. C., Lester, H. A., and Dunn, R. (1986). Messenger RNA coding for only the alpha subunit of the rat brain Na channel is sufficient for expression of functional channels in *Xenopus* oocytes. *Proc. Natl. Acad. Sci. USA* 83:7503–7507.

Goldman, D. and Staple, J. (1989). Spatial and temporal expression of acetylcholine receptor RNAs in innervated and denervated rat soleus muscle. *Neuron* 3:219–228.

Goldner, F. M., Dineley, K. T., and Patrick, J. W. (1997). Immunohistochemical localization of the nicotinic acetylcholine receptor subunit alpha6 to dopaminergic neurons in the substantia nigra and ventral tegmental area. *Neuroreport* 8:2739–2742.

Gomez, C. M. and Gammack, J. T. (1995). A leucine-to-phenylalanine substitution in the acetylcholine receptor ion channel in a family with the slow-channel syndrome. *Neurology* 45:982–985.

Gomez, C. M., Maselli, R., Gammack, J., Lasalde, J., Tamamizu, S., Cornblath, D. R., Lehar, M., McNamee, M., and Kuncl, R. W. (1996). A beta-subunit mutation in the acetylcholine receptor channel gate causes severe slow-channel syndrome. *Ann. Neurol.* 39:712–723.

Göpfert, H. and Schaefer, H. (1938). Über den direkt und indirekt erregten Aktionsstorm und die Funktion der motorischen Endplatte. *Pfluegers Arch.* 239:597–619.

Gordon, A. S., Davis, C. G., and Diamond, I. (1977). Phosphorylation of membrane proteins at a cholinergic synapse. *Proc. Natl. Acad. Sci. USA* 74:263–267.

Gorne-Tschelnokow, U., Strecker, A., Kaduk, C., Naumann, D., and Hucho, F. (1994). The transmembrane domains of the nicotinic acetylcholine receptor contain alpha-helical and beta structures. *EMBO J.* 13:338–341.

Grady, S. R., Meinerz, N. M., Cao, J., Reynolds, A. M., Picciotto, M. R., Changeux, J.-P., McIntosh, J. M., Marks, M. J., and Collins, A. C. (2001). Nicotinic agonists stimulate acetylcholine release from mouse interpeduncular nucleus: a function mediated by a different nAChR than dopamine release from striatum. *J. Neurochem.* 76:258–268.

Gramolini, A. O., Angus, L. M., Schaeffer, L., Burton, E. A., Tinsley, J. M., Davies, K. E., Changeux, J.-P., and Jasmin, B. J. (1999). Induction of utrophin gene expression by heregulin in skeletal muscle cells: role of the N-box motif and GA binding protein. *Proc. Natl. Acad. Sci. USA* 96:3223–3227.

Granon, S., Vidal, C., Thinus-Blanc, C., Changeux, J.-P., and Poucet, B. (1994). Working memory, response selection, and effortful processing in rats with medial prefrontal lesions. *Behav. Neurosci.* 108:883–891.

Granon, S., Poucet, B., Thinus-Blanc, C., Changeux, J.-P., and Vidal, C. (1995). Nicotinic and muscarinic receptors in the rat prefrontal cortex: differential roles in working memory, response selection and effortful processing. *Psychopharmacology (Berl.)* 119:139–144.

Granon, S., Faure, P., and Changeux, J.-P. (2003). Executive and social behaviors under nicotinic receptor regulation. *Proc. Natl. Acad. Sci. USA* 100:9596–9601.

Gready, J. E., Ranganathan, S., Schofield, P. R., Matsuo, Y., and Nishikawa, K. (1997). Predicted structure of the extracellular region of ligand-gated ion-channel receptors shows SH2-like and SH3-like domains forming the ligand-binding site. *Protein Sci.* 6:983–998.

Grenhoff, J., Aston-Jones, G., and Svensson, T. H. (1986). Nicotinic effects on the firing pattern of midbrain dopamine neurons. *Acta Physiol. Scand.* 128:351–358.

Groot-Kormelink, P. J., Luyten, W., Colquhoun, D., and Sivilotti, L. G. (1998). A reporter mtation approach shows incorporation of the "orphan" subunit beta3 into a functional nicotinic receptor. *J. Biol. Chem.* 273:15317–15320.

Grosman, C. and Auerbach, A. (2000). Kinetic, mechanistic, and structural aspects of unliganded gating of acetylcholine receptor channels: a single-channel study of second transmembrane segment 12′ mutants. *J. Gen. Physiol.* 115:621–635.

Grosman, C., Zhou, M., and Auerbach, A. (2000). Mapping the conformational wave of acetylcholine receptor channel gating. *Nature* 403:773–776.

Grubb, M. S., Rossi, F. M., Changeux, J.-P., and Thompson, I. D. (2003). Abnormal functional organization in the dorsal lateral geniculate nucleus of mice lacking the beta2 subunit of the nicotinic acetylcholine receptor. *Neuron* 40:1161–1172.

Grunhagen, H. and Changeux, J.-P. (1975). [Structural transitions of the *Torpedo* cholinergic receptor in its membrane state as demonstrated by a fluorescent local anesthetic: quinacrine]. *C. R. Acad. Sci. Hebd. Séances Acad. Sci. D Sci. Nat.* 281:1047–1050.

Grunhagen, H. H. and Changeux, J.-P. (1976). Studies on the electrogenic action of acetylcholine with *Torpedo marmorata* electric organ. V. Qualitative correlation between pharmacological effects and equilibration processes of the cholinergic receptor protein as revealed by the structural probe quinacrine. *J. Mol. Biol.* 106:517–535.

Grunhagen, H. H., Iwatsubo, M., and Changeux, J. (1976). [Rapid changes in the intensity of fluorescence observed in the presence of cholinergic agonists with *Torpedo marmorata* membrane fragments rich in cholinergic receptors and labelled with quinacrine]. *C. R. Acad. Sci. Hebd. Séances Acad. Sci. D Sci. Nat.* 283:1105–1108.

Grunhagen, H. H., Iwatsubo, M., and Changeux, J.-P. (1977). Fast kinetic studies on the interaction of cholinergic agonists with the membrane-bound acetylcholine receptor from *Torpedo marmorata* as revealed by quinacrine fluorescence. *Eur. J. Biochem.* 80:225–242.

Works Cited

Grutter, T. and Changeux, J.-P. (2001). Nicotinic receptors in wonderland. *Trends Biochem. Sci.* 26:459–463.

Grutter, T., Prado de Carvalho, L., Le Novere, N., Corringer, P.-J., Edelstein, S., and Changeux, J.-P. (2003). An H-bond between two residues from different loops of the acetylcholine binding site contributes to the activation mechanism of nicotinic receptors. *EMBO J.* 22:1990–2003.

Gu, Y., Silberstein, L., and Hall, Z. W. (1985). The effects of a myasthenic serum on the acetylcholine receptors of C2 myotubes. I. Immunological distinction between the two toxin-binding sites of the receptor. *J. Neurosci.* 5:1909–1916.

Gunthorpe, M. J. and Lummis, S. C. (2001). Conversion of the ion selectivity of the 5-HT (3a) receptor from cationic to anionic reveals a conserved feature of the ligand-gated ion channel superfamily. *J. Biol. Chem.* 276:10977–10983.

Guo, X. and Wecker, L. (2002). Identification of three cAMP-dependent protein kinase (PKA) phosphorylation sites within the major intracellular domain of neuronal nicotinic receptor alpha4 subunits. *J. Neurochem.* 82:439–447.

Haggerty, J. G. and Froehner, S. C. (1981). Restoration of 125I-alpha-bungarotoxin binding activity to the alpha subunit of *Torpedo* acetylcholine receptor isolated by gel electrophoresis in sodium dodecyl sulfate. *J. Biol. Chem* 256:8294–8297.

Haldane, J. B. S. (1930). *Enzymes.* New York: Longmans, Green.

Hamburger, V. (1970). Embryonic Motility in Vertebrates. In F. O. Schmitt, ed. *The Neurosciences,* 141–151. New York: Rockefeller University Press.

Hamill, O. P. and Sakmann, B. (1981). Multiple conductance states of single acetylcholine receptor channels in embryonic muscle cells. *Nature* 294:462–464.

Hamilton, S. L., McLaughlin, M., and Karlin, A. (1977). Disulfide bond cross-linked dimer in acetylcholine receptor from *Torpedo californica. Biochem. Biophys. Res. Commun.* 79:692–699.

Hammes, G. G. and Wu, C.-W. (1974). Kinetics of allosteric enzymes. *Ann. Rev. Biophys. Bioeng.* 3:1–33.

Han, Z. Y., Le Novère, N., Zoli, M., Hill, J. A., Jr., Champtiaux, N., and Changeux, J.-P. (2000). Localization of nAChR subunit mRNAs in the brain of *Macaca mulatta. Eur. J. Neurosci.* 12:3664–3674.

Han, Z. Y., Zoli, M., Cardona, A., Bourgeois, J. P., Changeux, J.-P., and Le Novère, N. (2003). Localization of [3H]nicotine, [3H]cytisine, [3H]epibatidine, and [125I]alpha-bungarotoxin binding sites in the brain of *Macaca mulatta. J. Comp. Neurol.* 461:49–60.

Hardison, R. C., Chui, D. H., Riemer, C. R., Miller, W., Carver, M. F., Molchanova, T. P., Efremov, G. D., and Huisman, T. H. (1998). Access to a syllabus of human hemoglobin variants (1996) via the World Wide Web. *Hemoglobin* 22:113–127.

Harel, M., Kasher, R., Nicolas, A., Guss, J. M., Balass, M., Fridkin, M., Smit, A. B., Brejc, K., Sixma, T. K., Katchalski-Katzir, E., et al. (2001). The binding site of acetylcholine receptor as visualized in the X-Ray structure of a complex between alpha-bungarotoxin and a mimotope peptide. *Neuron* 32:265–275.

Harper, C. M. and Engel, A. G. (1998). Quinidine sulfate therapy for the slow-channel congenital myasthenic syndrome. *Ann. Neurol.* 43:480–484.

Harris, D. A., Falls, D. L., and Fischbach, G. D. (1989). Differential activation of myotube nuclei following exposure to an acetylcholine receptor-inducing factor. *Nature* 337:173–176.

Harris, W. A. (1981). Neural activity and development. *Ann. Rev. Physiol.* 43:689–710.

Hart, I. K. (2000). Acquired neuromyotonia: a new autoantibody-mediated neuronal potassium channelopathy. *Am. J. Med. Sci.* 319:209–216.

Hartig, P. R. and Raftery, M. A. (1977). Lactoperoxidase catalyzed membrane surface labeling of the acetylcholine receptor from *Torpedo californica. Biochem. Biophys. Res. Commun.* 78:16–22.

Harvey, S. C. and Luetje, C. W. (1996). Determinants of competitive antagonist sensitivity on neuronal nicotinic receptor beta subunits. *J. Neurosci.* 16:3798–3806.

Hayward, L. J., Sandoval, G. M., and Cannon, S. C. (1999). Defective slow inactivation of sodium channels contributes to familial periodic paralysis. *Neurology* 52:1447–1453.

Hazelbauer, G. L. and Changeux, J.-P. (1974). Reconstitution of a chemically excitable membrane. *Proc. Natl. Acad. Sci. USA* 71:1479–1483.

He, X. P., Patel, M., Whitney, K. D., Janumpalli, S., Tenner, A., and McNamara, J. O. (1998). Glutamate receptor GluR3 antibodies and death of cortical cells. *Neuron* 20:153–163.

Heidmann, O., Buonanno, A., Geoffroy, B., Robert, B., Guenet, J.-L., Merlie, J.-P., and Changeux, J.-P. (1986). Chromosomal localization of muscle nicotinic acetylcholine receptor genes in the mouse. *Science* 234:866–868.

Heidmann, T., Iwatsubo, M., and Changeux, J.-P. (1977). [Study by a rapid mixing method of the interaction between a fluorescent cholinergic agonist and membrane fragments rich in cholinergic receptors from *Torpedo marmorata*]. *C. R. Acad. Sci. Hebd. Séances Acad. Sci. D Sci. Nat.* 284:771–774.

Heidmann, T. and Changeux, J.-P. (1978). Structural and functional properties of the acetylcholine receptor protein in its purified and membrane-bound states. *Ann. Rev. Biochem.* 47:317–357.

Heidmann, T. and Changeux, J.-P. (1979). Fast kinetic studies on the interaction of a fluorescent agonist with the membrane-bound acetylcholine receptor from *Torpedo marmorata*. *Eur. J. Biochem.* 94:255–279.

Heidmann, T. and Changeux, J.-P. (1980). Interaction of a fluorescent agonist with the membrane-bound acetylcholine receptor from *Torpedo marmorata* in the millisecond time range: resolution of an "intermediate" conformational transition and evidence for positive cooperative effects. *Biochem. Biophys. Res. Commun.* 97:889–896.

Heidmann, T., Sobel, A., and Changeux, J.-P. (1980). Conservation of the kinetic and allosteric properties of the acetylcholine receptor in its Na cholate soluble 9 S form: effect of lipids. *Biochem. Biophys. Res. Commun.* 93:127–133.

Heidmann, T. and Changeux, J.-P. (1981). Stabilization of the high affinity state of the membrane-bound acetylcholine receptor from *Torpedo marmorata* by noncompetitive-blockers: evidence for dual interaction and pharmacological selectivity. *FEBS Lett.* 131: 239–244.

Heidmann, T. and Changeux, J.-P. (1982). Un modèle moléculaire de régulation d'efficacité au niveau postsynaptique d'une synapse chimique. *C. R. Acad. Sci. Paris* 295:665–670.

Heidmann, T., Oswald, R., and Changeux, J.-P. (1982). [The high affinity binding site for chlorpromazine is present only as a single copy per cholinergic receptor molecule and is shared by four polypeptide chains]. *C. R. Séances Acad. Sci. III* 295:345–349.

Heidmann, T., Bernhardt, J., Neumann, E., and Changeux, J.-P. (1983a). Rapid kinetics of agonist binding and permeability response analyzed in parallel on acetylcholine receptor-rich membranes from *Torpedo marmorata*. *Biochemistry* 22:5452–5459.

Heidmann, T., Oswald, R. E., and Changeux, J.-P. (1983b). Multiple sites of action for noncompetitive blockers on acetylcholine receptor-rich membrane fragments from *Torpedo marmorata*. *Biochemistry* 22:3112–3127.

Heidmann, T. and Changeux, J.-P. (1984). Time-resolved photolabeling by the noncompetitive blocker chlorpromazine of the acetylcholine receptor in its transiently open and closed ion channel conformations. *Proc. Natl. Acad. Sci. USA* 81:1897–1901.

Heidmann, T. and Changeux, J.-P. (1986). Characterization of the transient agonist-triggered state of the acetylcholine receptor rapidly labeled by the noncompetitive blocker [3H]chlorpromazine: additional evidence for the open channel conformation. *Biochemistry* 25:6109–6113.

Heilbronn, E., Karlsson, E., and Widlund, L. (1973). Purification of a Membrane Constituent, Possibly the Nicotinic Cholinergic Receptor, from the Electric Organ of *Torpedo marmorata*. In *Proceedings of the Symposium on Cholinergic Transmission of the Nerve Impulse*, vol. 15, 151–158. Paris: INSERM.

Heilbronn, E. and Mattson, C. (1974). The nicotinic cholinergic receptor protein: improved purification method, preliminary amino acid composition and observed autoimmune response. *J. Neurochem.* 22:315–317.

Heilbronn, E. (1975). Current research on the nature of cholinergic receptors. *Croat. Chem. Acta* 47:395–408.

Heinemann, S., Boulter, J., Deneris, E., Connolly, J., Gardner, P., Wada, E., Wada, K. D. R., Ballivet, M., Swanson, L., and Patrick, J. (1989). Brain and Muscle Nicotinic Acetylcholine Receptor: A Gene Family. In A. Maelicke, ed. *Molecular Biology of Neuroreceptors and Ion Channels*, 13–30. Berlin: Springer-Verlag.

Heinemann, S., Boulter, J., Deneris, E., Connolly, J., Duvoisin, R., Papke, R., and Patrick, J. (1990). The brain nicotinic acetylcholine receptor gene family. *Prog. Brain Res.* 86: 195–203.

Heinemann, S. H., Terlau, H., Stuhmer, W., Imoto, K., and Numa, S. (1992). Calcium channel characteristics conferred on the sodium channel by single mutations. *Nature* 356: 441–443.

Helmholtz, H. (1850). Messungen über den zeitlichen Verlauf der Zwicking animalischer

Works Cited

Muskeln und die Fortpflanzungsgeschwindigkeit der Reizung in den Nerven. *Müllers Arch. Anat. Physiol.*:276–374.

Henis, Y. I., Levitzki, A., and Gafni, A. (1979). Evidence for ligand-induced conformational changes in rabbit-muscle glyceraldehyde-3-phosphate dehydrogenase. *Eur. J. Biochem.* 97:519–528.

Henri, V. (1903). *Lois générales de l'action des diastases.* Paris: Hermann.

Henry, E. R., Jones, C. M., Hofrichter, J., and Eaton, W. A. (1997). Can a two-state MWC allosteric model explain hemoglobin kinetics? *Biochemistry* 36:6511–6528.

Henry, E. R., Bettati, S., Hofrichter, J., and Eaton, W. A. (2002). A tertiary two-state allosteric model for hemoglobin. *Biophys. Chem.* 98:149–164.

Hensley, P. and Schachman, H. K. (1979). Communication between dissimilar subunits in aspartate transcarbamoylase: effect of inhibitor and activator on the conformation of the catalytic polypeptide chains. *Proc. Natl. Acad. Sci. USA* 76:3732–3736.

Hernandez, M. C., Erkman, L., Matter-Sadzinski, L., Roztocil, T., Ballivet, M., and Matter, J. M. (1995). Characterization of the nicotinic acetylcholine receptor beta3 gene. Its regulation within the avian nervous system is effected by a promoter 143 base pairs in length. *J. Biol. Chem.* 270:3224–3233.

Herz, J. M., Johnson, D. A., and Taylor, P. (1987). Interaction of noncompetitive inhibitors with the acetylcholine receptor: the site specificity and spectroscopic properties of ethidium binding. *J. Biol. Chem.* 262:7238–7247.

Herz, J. M., Johnson, D. A., and Taylor, P. (1989). Distance between the agonist and noncompetitive inhibitor sites on the nicotinic acetylcholine receptor. *J. Biol. Chem.* 264:12439–12448.

Hess, G. P., Cash, D. J., and Aoshima, H. (1979). Acetylcholine receptor controlled ion fluxes in membrane vesicles investigated by fast reaction techniques. *Nature* 282:329–331.

Hess, G. P., Pasquale, E. B., Walker, J. W., and McNamee, M. G. (1982). Comparison of acetylcholine receptor-controlled cation flux in membrane vesicles from *Torpedo californica* and *Electrophorus electricus:* chemical kinetic measurements in the millisecond region. *Proc. Natl. Acad. Sci. USA* 79:963–967.

Hess, G. P., Cash, D. J., and Aoshima, H. (1983). Acetylcholine receptor-controlled ion translocation: chemical kinetic investigation of the mechanism. *Ann. Rev. Biophys. Bioeng.* 12:443–473.

Heuser, J. E. and Salpeter, S. R. (1979). Organization of acetylcholine receptors in quick-frozen, deep-etched, and rotary-replicated *Torpedo* postsynaptic membrane. *J. Cell Biol.* 82:150–173.

Higman, H., Podleski, T. R., and Bartels, E. (1963). Apparent dissociation constants between carbamylcholine, d-tubocurarine and the receptor. *Biochem. Biophys. Acta* 75:187–193.

Hill, A. V. (1910). The mode of action of nicotine and curari, determined by the form of the contraction curve and the method of temperature coefficients. *J. Physiol.* 39:361–373.

Hill, J. A., Jr., Nghiem, H. O., and Changeux, J.-P. (1991). Serine-specific phosphorylation of nicotinic receptor associated 43K protein. *Biochemistry* 30:5579–5585.

Hill, J. A., Jr., Zoli, M., Bourgeois, J.-P., and Changeux, J.-P. (1993). Immunocytochemical localization of a neuronal nicotinic receptor: the β2-subunit. *J. Neurosci.* 13:1551–1568.

Hille, B. (1992). *Ionic Channels of Excitable Membranes.* 2nd ed. Sunderland, Mass.: Sinauer.

Hilmas, C., Pereira, E. F., Alkondon, M., Rassoulpour, A., Schwarcz, R., and Albuquerque, E. X. (2001). The brain metabolite kynurenic acid inhibits alpha7 nicotinic receptor activity and increases non-alpha7 nicotinic receptor expression: physiopathological implications. *J. Neurosci.* 21:7463–7473.

Hirokawa, N. and Heuser, J. E. (1982). Internal and external differentiations of the postsynaptic membrane at the neuromuscular junction. *J. Neurocytol.* 11:487–510.

Hoch, W., McConville, J., Helms, S., Newsom-Davis, J., Melms, A., and Vincent, A. (2001). Auto-antibodies to the receptor tyrosine kinase MuSK in patients with myasthenia gravis without acetylcholine receptor antibodies. *Nat. Med.* 7:365–368.

Hodgkin, A. L. and Huxley, A. F. (1952). A quantitative description of membrane current and its application to conduction and excitation in nerve. *J. Physiol.* 117:500–544.

Hökfelt, T., Fried, G., Hansen, S., Holets, V., Lundberg, J. M., and Skirboll, L. (1986). Neurons with multiple messengers—distribution and possible functional significance. *Prog. Brain Res.* 65:115–137.

Holmes, W. E., Sliwkowski, M. X., Akita, R. W., Henzel, W. J., Lee, J., Park, J. W., Yansura, D., Abadi, N., Raab, H., and Lewis, G. D. (1992). Identification of heregulin, a specific activator of p185erbB2. *Science* 256:1205–1210.

Holtzman, E., Wise, D., Wall, J., and Karlin, A. (1982). Electron microscopy of complexes of

isolated acetylcholine receptor, biotinyl-toxin, and avidin. *Proc. Natl. Acad. Sci. USA* 79: 310–314.
Hoover, F. and Goldman, D. (1992). Temporally correlated expression of nAChR genes during development of the mammalian retina. *Exp. Eye Res.* 54:561–571.
Hucho, F. and Changeux, J.-P. (1973). Molecular weight and quaternary structure of the cholinergic receptor protein extracted by detergents from *Electrophorus electricus* electric tissue. *FEBS Lett.* 38:11–15.
Hucho, F., Oberthur, W., and Lottspeich, F. (1986). The ion channel of the nicotinic acetylcholine receptor is formed by the homologous helices M2 of the receptor subunits. *FEBS Lett.* 205:137–142.
Hucho, F., Tsetlin, V. I., and Machold, J. (1996). The emerging three-dimensional structure of a receptor: the nicotinic acetylcholine receptor. *Eur. J. Biochem.* 239:539–557.
Huganir, R. L., Schell, M. A., and Racker, E. (1979). Reconstitution of the purified acetylcholine receptor from *Torpedo californica*. *FEBS Lett.* 108:155–160.
Huganir, R. L. and Greengard, P. (1990). Regulation of neurotransmitter receptor desensitization by protein phosphorylation. *Neuron* 5:555–567.
Huh, K. H. and Fuhrer, C. (2002). Clustering of nicotinic acetylcholine receptors: from the neuromuscular junction to interneuronal synapses. *Mol. Neurobiol.* 25:79–112.
Hull, C. L. (1943). *Principles of Behavior: An Introduction to Behavior Theory.* New York: D. Appleton-Century.
Hunkapiller, M. W., Strader, C. D., Hood, L., and Raftery, M. A. (1979). Amino terminal amino acid sequence of the major polypeptide subunit of *Torpedo californica* acetylcholine receptor. *Biochem. Biophys. Res. Comm.* 91:164–169.
Husain, S. S., Ziebell, M. R., Ruesch, D., Hong, F., Arevalo, E., Kosterlitz, J. A., Olsen, R. W., Forman, S. A., Cohen, J. B., and Miller, K. W. (2003). 2-(3-Methyl-3H-diaziren-3-yl)ethyl 1-(1-phenylethyl)-1H-imidazole-5-carboxylate: a derivative of the stereoselective general anesthetic etomidate for photolabeling ligand-gated ion channels. *J. Med. Chem.* 46: 1257–1265.
Hyman, S. E. (1996). Addiction to cocaine and amphetamine. *Neuron* 16:901–904.
Ikeda, S. R., Aronstam, R. S., Daly, J. W., Aracava, Y., and Albuquerque, E. X. (1984). Interactions of bupivacaine with ionic channels of the nicotinic receptor: electrophysiological and biochemical studies. *Mol. Pharmacol.* 26:293–303.
Ikonen, E. (2001). Roles of lipid rafts in membrane transport. *Curr. Opin. Cell Biol.* 13: 470–477.
Imoto, K., Methfessel, C., Sakmann, B., Mishina, M., Mori, Y., Konno, T., Fukuda, K., Kurasaki, M., Bujo, H., Fujita, Y., and et al. (1986). Location of a delta-subunit region determining ion transport through the acetylcholine receptor channel. *Nature* 324:670–674.
Imoto, K., Busch, C., Sakmann, B., Mishina, M., Konno, T., Nakai, J., Bujo, H., Mori, Y., Fukuda, K., and Numa, S. (1988). Rings of negatively charged amino acids determine the acetylcholine receptor channel conductance. *Nature* 335:645–648.
Iwata, S., Kamata, K., Yoshida, S., Minowa, T., and Ohta, T. (1994). T and R states in the crystals of bacterial L-lactate dehydrogenase reveal the mechanism of allosteric control. *Nat. Struct. Biol.* 1:176–185.
Jackson, M. B. (1986). Kinetics of unliganded acetylcholine receptor channel gating. *Biophys. J.* 49:663–672.
Jackson, M. B., Imoto, K., Mishina, M., Konno, T., Numa, S., and Sakmann, B. (1990). Spontaneous and agonist-induced openings of an acetylcholine receptor channel composed of bovine α-, β- and δ-subunits. *Pfluegers Arch. Eur. J. Physiol.* 417:129–135.
Jahr, C. E. and Stevens, C. F. (1987). Glutamate activates multiple single channel conductances in hippocampal neurons. *Nature* 325:522–525.
Jardetzky, O. (1996). Protein dynamics and conformational transitions in allosteric proteins. *Prog. Biophys. Mol. Biol.* 65:171–219.
Jasmin, B. J., Cartaud, J., Bornens, M., and Changeux, J.-P. (1989). Golgi apparatus in chick skeletal muscle: changes in its distribution during end plate development and after denervation. *Proc. Natl. Acad. Sci. USA* 86:7218–7222.
Jasmin, B. J., Changeux, J.-P., and Cartaud, J. (1990). Compartmentalization of cold-stable and acetylated microtubules in the subsynaptic domain of chick skeletal muscle fibre. *Nature* 344:673–675.
Jasmin, B. J., Changeux, J.-P., and Cartaud, J. (1991). Organization and dynamics of microtubules in *Torpedo marmorata* electrocyte: selective association with specialized domains of the postsynaptic membrane. *Neuroscience* 43:151–162.

Works Cited

Jasmin, B. J., Antony, C., Changeux, J.-P., and Cartaud, J. (1995). Nerve-dependent plasticity of the Golgi complex in skeletal muscle fibres: compartmentalization within the subneural sarcoplasm. *Eur. J. Neurosci.* 7:470–479.

Jen, J., Wan, J., Graves, M., Yu, H., Mock, A. F., Coulin, C. J., Kim, G., Yue, Q., Papazian, D. M., and Baloh, R. W. (2001). Loss-of-function EA2 mutations are associated with impaired neuromuscular transmission. *Neurology* 57:1843–1848.

Jenkinson, D. H. (1960). The antagonism between tubocurarine and substances which depolarize the motor end-plate. *J. Physiol. (London)* 152:309–324.

Johnson, D. A., Voet, J. G., and Taylor, P. (1984). Fluorescence energy transfer between cobra alpha-toxin molecules bound to the acetylcholine receptor. *J. Biol. Chem* 259:5717–5725.

Johnson, D. A. and Ayres, S. (1996). Quinacrine noncompetitive inhibitor binding site localized on the *Torpedo* acetylcholine receptor in the open state. *Biochemistry* 35:6330–6336.

Johnson, J. W. and Ascher, P. (1987). Glycine potentiates the NMDA response in cultured mouse brain neurons. *Nature* 325:529–531.

Johnson, L. N. and Barford, D. (1990). Glycogen phosphorylase: the structural basis of the allosteric response and comparison with other allosteric proteins. *J. Biol. Chem.* 265:2409–2412.

Jonas, P., Baumann, A., Merz, B., and Gundelfinger, E. D. (1990). Structure and developmental expression of the D alpha2 gene encoding a novel nicotinic acetylcholine receptor protein of *Drosophila melanogaster*. *FEBS Lett.* 269:264–268.

Jonas, P. and Sakmann, B. (1992). Glutamate receptor channels in isolated patches from CA1 and CA3 pyramidal cells of rat hippocampal slices. *J. Physiol.* 455:143–171.

Jones, B. E. (2003). Arousal systems. *Front. Biosci.* 8:438–451.

Jones, G., Moore, C., Hashemolhosseini, S., and Brenner, H. R. (1999). Constitutively active MuSK is clustered in the absence of agrin and induces ectopic postsynaptic-like membranes in skeletal muscle fibers. *J. Neurosci.* 19:3376–3383.

Jones, R. and Vrbova, G. (1974). Two factors responsible for the development of denervation hypersensitivity. *J. Physiol.* 236:517–538.

Kagawa, Y. and Racker, E. (1971). Partial resolution of the enzymes catalyzing oxidative phosphorylation. *J. Biol. Chem.* 246:5477–5487.

Kalbaugh, T. L., VanDongen, H. M., and VanDongen, A. M. (2004). Ligand-binding residues integrate affinity and efficacy in the NMDA receptor. *Mol. Pharmacol.* 66:209–219.

Kaldany, R. R. J. and Karlin, A. (1983). Reaction of quinacrine mustard with the acetylcholine receptor from *Torpedo californica*: functional consequences and sites of labeling. *J. Biol. Chem.* 258:6232–6202.

Kandel, E. (1976). *The Cellular Basis of Behavior*. San Francisco: Freeman.

Kao, P. N., Dwork, A. J., Kaldany, R. R., Silver, M. L., Wideman, J., Stein, S., and Karlin, A. (1984). Identification of the alpha subunit half-cystine specifically labeled by an affinity reagent for the acetylcholine receptor binding site. *J. Biol. Chem.* 259:11662–11665.

Kao, P. N. and Karlin, A. (1986). Acetylcholine receptor binding site contains a disulfide cross-link between adjacent half-cystinyl residues. *J. Biol. Chem.* 261:8085–8088.

Karlin, A. and Bartels, E. (1966a). Effects of blocking sulfhydryl groups and of reducing disulfide bonds on the acetylcholine-activated permeability system of the electroplax. *Biochem. Biophys. Acta* 126:525–535.

Karlin, A. and Bartels, E. (1966b). Effects of blocking sulfhydryl groups and of reducing disulfide bonds on the acetylcholine-activated permeability systems of the electroplax. *Biochem. Biophys. Acta* 126:525–535.

Karlin, A. (1967a). On the application of "a plausible model" of allosteric proteins to the receptor of acetylcholine. *J. Theor. Biol.* 16:306–320.

Karlin, A. (1967b). Chemical distinctions between acetylcholinesterase and the acetylcholine receptor. *Biochem. Biophys. Acta* 139:358–362.

Karlin, A. and Winnik, M. (1968). Reduction and specific alkylation of the receptor for acetylcholine. *Proc. Natl. Acad. Sci. USA* 60:668–674.

Karlin, A. (1969). Chemical modification of the active site of the acetylcholine receptor. *J. Gen. Physiol.* 54:265–264.

Karlin, A. and Cowburn, D. A. (1973). The affinity labeling of partially purified acetylcholine receptor from electric tissue of *Electrophorus*. *Proc. Natl. Acad. Sci. USA* 70:3636–3640.

Karlin, A., Weill, C. L., McNamee, M. G., and Valderrama, R. (1976). Facets of the structures of acetylcholine receptors from *Electrophorus* and *Torpedo*. *Cold Spring Harb. Symp. Quant. Biol.* 40:203–210.

Karlin, A. (1980). Molecular Properties of Nicotinic Acetylcholine Receptors. In G. Poste, G. L. Nicolson, and C. W. Cotman, eds. *Cell Surface Reviews,* 191–260. Amsterdam: Elsevier.

Karlin, A. (1983). Anatomy of a receptor. *Neurosci. Comment* 1:111-123.

Karlin, A., Holtzman, E., Yodh, N., Lobel, P., Wall, J., and Hainfeld, J. (1983). The arrangement of the subunits of the acetylcholine receptor of *Torpedo californica. J. Biol. Chem.* 258:6678–6681.

Karlin, A. (1993). Structure of nicotinic acetylcholine receptors. *Curr. Opin. Neurobiol.* 3:299–309.

Karlsson, E., Heilbronn, E., and Widlund, L. (1972). Isolation of the nicotinic acetylcholine receptor by biospecific chromatography on insolubilized *Naja naja* neurotoxin. *FEBS Lett.* 28:107–111.

Kasai, M. and Changeux, J.-P. (1970). Demonstration de l'excitation par des agonistes cholinergiques à partir de fractions de membranes purifiées *in vitro. C. R. Acad. Sci. Paris* 270:1400–1403.

Kasai, M. and Changeux, J.-P. (1971). *In vitro* excitation of purified membrane fragments by cholinergic agonists. I. Pharmacological properties of the excitable membrane fragments. II. The permeability change caused by cholinergic agonists. III. Comparison of the dose response curves to decamethonium with the corresponding binding curves of decamethonium to the cholinergic receptor. *J. Membr. Biol.* 6:1–80.

Kato, G. and Changeux, J.-P. (1976). Studies on the effect of histrionicotoxin on the monocellular electroplax from *Electrophorus electricus* and on the binding of (3H)acetylcholine to membrane fragments from *Torpedo marmorata. Mol. Pharamcol.* 12:92–100.

Katz, B. and Thesleff, S. (1957). A study of "desensitization" produced by acetylcholine at the motor end-plate. *J. Physiol.* 138:63–80.

Katz, B. (1966). *Nerve, Muscle, and Synapse.* New York: McGraw-Hill.

Katz, B. and Miledi, R. (1970). Membrane noise produced by acetylcholine. *Nature* 226:962–963.

Katz, B. and Miledi, R. (1972). The statistical nature of the acetycholine potential and its molecular components. *J. Physiol.* 224:665–699.

Katz, B. and Miledi, R. (1973a). The characteristics of "end-plate noise" produced by different depolarizing drugs. *J. Physiol.* 230:707–717.

Katz, B. and Miledi, R. (1973b). The effect of atropine on acetylcholine action at the neuromuscular junction. *Proc. R. Soc. Lond. B Biol. Sci.* 184:221–226.

Katz, B. and Miledi, R. (1973c). The binding of acetylcholine to receptors and its removal from the synaptic cleft. *J. Physiol.* 231:549–574.

Katz, B. and Miledi, R. (1977). Transmitter leakage from motor nerve endings. *Proc. R. Soc. Lond. B* 196:59–72.

Keiger, C. J., Case, L. D., Kendal-Reed, M., Jones, K. R., Drake, A. F., and Walker, J. C. (2003a). Nicotinic cholinergic receptor expression in the human nasal mucosa. *Ann. Otol. Rhinol. Laryngol.* 112:77–84.

Keiger, C. J., Prevette, D., Conroy, W. G., and Oppenheim, R. W. (2003b). Developmental expression of nicotinic receptors in the chick and human spinal cord. *J. Comp. Neurol.* 455:86–99.

Kelley, S. P., Dunlop, J. I., Kirkness, E. F., Lambert, J. J., and Peters, J. A. (2003). A cytoplasmic region determines single-channel conductance in 5-HT3 receptors. *Nature* 424:321–324.

Kennedy, M. B. (1997). The postsynaptic density at glutamatergic synapses. *Trends Neurosci.* 20:264–268.

Keramidas, A., Moorhouse, A. J., French, C. R., Schofield, P. R., and Barry, P. H. (2000). M2 pore mutations convert the glycine receptor channel from being anion- to cation-selective. *Biophys. J.* 79:247–259.

Kerbiriou, D. and Hervé, G. (1977). An aspartate transcarbamylase lacking catalytic subunit interactions: study of conformational changes by ultraviolet absorbance and circular dichroism spectroscopy. *J. Biol. Chem.* 252:2881–2890.

Kerszberg, M. and Changeux, J.-P. (1993). A model for motor endplate morphogenesis: diffusible morphogens, transmembrane signalling and compartmentalized gene expression. *Neural Comput.* 5:341–358.

Kerszberg, M. and Changeux, J.-P. (1994). A model for reading morphogenetic gradients: autocatalysis and competition at the gene level. *Proc. Natl. Acad. Sci. USA* 91:5823–5827.

Kerszberg, M. and Changeux, J.-P. (1998). A simple molecular model of neurulation. *Bioessays* 20:758–770.

Khromov-Borisov, N. V. and Michelson, M. J. (1966). The mutual disposition of cholino-receptors of locomotor muscles, and the changes in their disposition in the course of evolution. *Pharmacol. Rev.* 18:1051–1090.

Khurana, T. S., Rosmarin, A. G., Shang, J., Krag, T. O., Das, S., and Gammeltoft, S. (1999). Activation of utrophin promoter by heregulin via the ets-related transcription factor complex GA-binding protein alpha/beta. *Mol. Biol. Cell* 10:2075–2086.

Kienker, P., Tomaselli, G., Jurman, M., and Yellen, G. (1994). Conductance mutations of the nicotinic acetylcholine receptor do not act by a simple electrostatic mechanism. *Biophys. J.* 66:325–334.

Kim, J. H., Liao, D., Lau, L. F., and Huganir, R. L. (1998). SynGAP: a synaptic RasGAP that associates with the PSD-95/SAP90 protein family. *Neuron* 20:683–691.

Kirsch, J. and Betz, H. (1998). Glycine-receptor activation is required for receptor clustering in spinal neurons. *Nature* 392:717–720.

Kirschner, K., Eigen, M., Bittman, R., and Voigt, B. (1966). The binding of nicotinamide-adenine dinucleotide to yeast D-phosphoglyceraldehyde-3-phosphate dehydrogenase: temperature-jump relaxation studies on the mechanism of an allosteric enzyme. *Proc. Natl. Acad. Sci. USA* 56:1661–1166.

Kistler, J. and Stroud, R. M. (1981). Crystalline arrays of membrane-bound acetylcholine receptor. *Proc. Natl. Acad. Sci. USA* 78:3678–3682.

Klarsfeld, A. and Changeux, J.-P. (1985). Activity regulates the levels of acetylcholine receptor alpha-subunit mRNA in cultured chicken myotubes. *Proc. Natl. Acad. Sci. USA* 82:4558–4562.

Klarsfeld, A., Daubas, P., Bourachot, B., and Changeux, J.-P. (1987). A 5′-flanking region of the chicken acetylcholine receptor alpha-subunit gene confers tissue specificity and developmental control of expression in transfected cells. *Mol. Cell Biol.* 7:951–955.

Klarsfeld, A., Laufer, R., Fontaine, B., Devillers-Thiéry, A., Dubreuil, C., and Changeux, J.-P. (1989). Regulation of muscle AChR alpha subunit gene expression by electrical activity: involvement of protein kinase C and Ca2$^+$. *Neuron* 2:1229–1236.

Klarsfeld, A., Bessereau, J.-L., Salmon, A. M., Triller, A., Babinet, C., and Changeux, J.-P. (1991). An acetylcholine receptor alpha-subunit promoter conferring preferential synaptic expression in muscle of transgenic mice. *EMBO J.* 10:625–632.

Klett, R. P., Fulpius, B. W., Cooper, D., Smith, M., Reich, E., and Posani, L. D. (1973). The acetylcholine receptor. I. Purification and characterization of a macromolecule isolated from *Electrophorus electricus*. *J. Biol. Chem.* 248:6841–6853.

Klink, R., De Kerchove D'Exaerde, A., Zoli, M., and Changeux, J.-P. (2001). Molecular and physiological diversity of nicotinic acetylcholine receptors in the midbrain dopaminergic nuclei. *J. Neurosci.* 21:1452–1463.

Klymkowsky, M. W. and Stroud, R. M. (1979). Immunospecific identification and three-dimensional structure of a membrane-bound acetylcholine receptor from *Torpedo marmorata*. *J. Mol. Biol.* 128:319–334.

Klymkowsky, M. W., Heuser, J. E., and Stroud, R. M. (1980). Protease effects on the structure of acetylcholine receptor membranes from *Torpedo californica*. *J. Cell Biol.* 85:823–838.

Koblin, D. D. and Lester, H. A. (1979). Voltage-dependent and voltage-independent blockade of acetylcholine receptors by local anesthetics in *Electrophorus* electroplaques. *Mol. Pharmacol.* 15:559–580.

Koike, S., Schaeffer, L., and Changeux, J.-P. (1995). Identification of a DNA element determining synaptic expression of the mouse acetylcholine receptor delta-subunit gene. *Proc. Natl. Acad. Sci. USA* 92:10624–10628.

Konno, T., Busch, C., Von Kitzing, E., Imoto, K., Wang, F., Nakai, J., Mishina, M., Numa, S., and Sakmann, B. (1991). Rings of anionic amino acids as structural determinants of ion selectivity in the acetylcholine receptor channel. *Proc. R. Soc. Lond. B Biol. Sci.* 244:69–79.

Koob, G. F. (1996). Drug addiction: the yin and yang of hedonic homeostasis. *Neuron* 16:893–896.

Kordeli, E., Cartaud, J., Nghiem, H. O., Pradel, L. A., Dubreuil, C., Paulin, D., and Changeux, J.-P. (1986). Evidence for a polarity in the distribution of proteins from the cytoskeleton in *Torpedo marmorata* electrocytes. *J. Cell Biol.* 102:748–761.

Kordeli, E., Cartaud, J., Nghiem, H. O., Devillers-Thiéry, A., and Changeux, J.-P. (1989). Asynchronous assembly of the acetylcholine receptor and of the 43-kD nu1 protein in the postsynaptic membrane of developing *Torpedo marmorata* electrocyte. *J. Cell Biol.* 108:127–139.

Kornau, H. C., Schenker, L. T., Kennedy, M. B., and Seeburg, P. H. (1995). Domain interaction between NMDA receptor subunits and the postsynaptic density protein PSD-95. *Science* 269:1737–1740.

Koshland, D. E. (1963). The role of flexibility in enzyme action. *Cold Spring Harb. Symp. Quant. Biol.* 28:473–480.

Koshland, D. E., Némethy, G., and Filmer, D. (1966). Comparison of experimental binding data and theoretical models in proteins containing subunits. *Biochemistry* 5:365–385.

Krause, R. M., Buisson, B., Bertrand, S., Corringer, P.-J., Galzi, J.-L., Changeux, J.-P., and Bertrand, D. (1998). Ivermectin: a positive allosteric effector of the alpha7 neuronal nicotinic acetylcholine receptor. *Mol. Pharmacol.* 53:283–294.

Kreienkamp, H. J., Sine, S. M., Maeda, R. K., and Taylor, P. (1994). Glycosylation sites selectively interfere with alpha-toxin binding to the nicotinic acetylcholine receptor. *J. Biol. Chem.* 269:8108–8114.

Kreis, T. E. (1990). Role of microtubules in the organisation of the Golgi apparatus. *Cell Motil. Cytoskel.* 15:67–70.

Krodel, E. K., Beckman, R. A., and Cohen, J. B. (1979). Identification of a local anesthetic binding site in nicotinic post-synaptic membranes isolated from *Torpedo marmorata* electric tissue. *Mol. Pharmacol.* 15:294–312.

Kubalek, E., Ralston, S., Lindstrom, J., and Unwin, N. (1987). Location of subunits within the acetylcholine receptor by electron image analysis of tubular crystals from *Torpedo marmorata*. *J. Cell Biol.* 105:9–18.

Kuffler, S. W. and Yoshikami, D. (1975). The distribution of acetylcholine sensitivity at the post-synaptic membrane of vertebrate skeletal twitch muscles: iontophoretic mapping in the micron range. *J. Physiol.* 244:703–730.

Kühne, W. (1862). *Ueber die peripherischen Endorgane der motorischen Nerven.* Leipzig: Engelmann.

Kuryatov, A., Gerzanich, V., Nelson, M., Olale, F., and Lindstrom, J. (1997). Mutation causing autosomal dominant nocturnal frontal lobe epilepsy alters Ca^{2+} permeability, conductance, and gating of human alpha4beta2 nicotinic acetylcholine receptors. *J. Neurosci.* 17:9035–9047.

Kuryatov, A., Olale, F., Cooper, J., Choi, C., and Lindstrom, J. (2000). Human alpha6 AChR subtypes: subunit composition, assembly, and pharmacological responses. *Neuropharmacology* 39:2570–2590.

Labarca, C., Schwarz, J., Deshpande, P., Schwarz, S., Nowak, M. W., Fonck, C., Nashmi, R., Kofuji, P., Dang, H., Shi, W., et al. (2001). Point mutant mice with hypersensitive alpha4 nicotinic receptors show dopaminergic deficits and increased anxiety. *Proc. Natl. Acad. Sci. USA* 98:2786–2791.

Lacazette, E., Le Calvez, S., Gajendran, N., and Brenner, H. R. (2003). A novel pathway for MuSK to induce key genes in neuromuscular synapse formation. *J. Cell Biol.* 161:727–736.

Lander, E. S., Linton, L. M., Birren, B., Nusbaum, C., Zody, M. C., Baldwin, J., Devon, K., Dewar, K., Doyle, M., FitzHugh, W., et al. (2001). Initial sequencing and analysis of the human genome. *Nature* 409:860–921.

Lang, B. and Vincent, A. (1996). Autoimmunity to ion channels and other proteins in paraneoplastic disorders. *Curr. Opin. Immunol.* 8:865–871.

Langenbuch-Cachat, J., Bon, C., Goeldner, M., Hirth, C., and Changeux, J.-P. (1988). Photoaffinity labeling by aryldiazonium derivatives of *Torpedo marmorata* acetylcholine receptor. *Biochemistry* 27:2337–2345.

Langley, J. N. (1905). On the reaction of cells and of nerve-endings to certain poisons, chiefly as regards the reactions of striated muscle to nicotine and to curare. *J. Physiol.* 33:374–413.

Langley, J. N. (1906). Croonian Lecture, 1906. On nerve endings and on special excitable substances in cells. *Proc. Roy. Soc. Ser. B.* 78:170–194.

Langley, J. N. (1907). On the contraction of muscle, chiefly in relation to the presence of "receptive" substances. *J. Physiol.* 36:347–384.

Langmuir, I. (1918). The adsorption of gasses on plastine surfaces: glass, mica, and platinum. *J. Ann. Chem. Soc.* 40:1361–1403.

Langosh, D., Laube, B., Rundström, N., Schmieden, V., Bormann, J., and Betz, H. (1994). Decreased agonist affinity and chloride conductance of mutant glycine receptors associated with human hereditary hyperekplexia. *EMBO J.* 13:4223–4228.

Lapa, A. J., Albuquerque, E. X., Sarvey, J. M., Daly, J., and Witkop, B. (1975). Effects of histri-

Works Cited

onicotoxin on the chemosensitive and electrical properties of skeletal muscle. *Exp. Neurol.* 47:558–580.

Larmie, E. T. and Webb, G. D. (1973). Desensitization in the electroplax. *J. Gen. Physiol.* 61:263.

LaRochelle, W. J. and Froehner, S. C. (1986). Determination of the tissue distributions and relative concentrations of the postsynaptic 43-kDa protein and the acetylcholine receptor in *Torpedo*. *J. Biol. Chem.* 261:5270–5274.

Larsson, A. and Engel, J. A. (2004). Neurochemical and behavioral studies on ethanol and nicotine interactions. *Neurosci. Biobehav. Rev.* 27:713–720.

Lassar, A. B., Buskin, J. N., Lockshon, D., Davis, R. L., Apone, S., Hauschka, S. D., and Weintraub, H. (1989). MyoD is a sequence-specific DNA binding protein requiring a region of myc homology to bind to the muscle creatine kinase enhancer. *Cell* 58:823–831.

Lau, F. T. and Fersht, A. R. (1987). Conversion of allosteric inhibition to activation in phosphofructokinase by protein engineering. *Nature* 326:811–812.

Laube, B., Langosch, D., Betz, H., and Schmieden, V. (1995). Hyperekplexia mutations of the glycine receptor unmask the inhibitory subsite for beta-amino-acids. *Neuroreport* 6:897–900.

Laudenbach, V., Medja, F., Zoli, M., Rossi, F. M., Evrard, P., Changeux, J.-P., and Gressens, P. (2002). Selective activation of central subtypes of the nicotinic acetylcholine receptor has opposite effects on neonatal excitotoxic brain injuries. *Faseb J.* 16:423–425.

Laufer, R. and Changeux, J.-P. (1989). Activity-dependent regulation of gene expression in muscle and neuronal cells. *Mol. Neurobiol.* 3:1–53.

Laviolette, S. R. and van der Kooy, D. (2004). The neurobiology of nicotine addiction: bridging the gap from molecules to behaviour. *Nat. Rev. Neurosci.* 5:55–65

Le Novère, N. and Changeux, J.-P. (1995). Molecular evolution of the nicotinic acetylcholine receptor: an example of multigene family in excitable cells. *J. Mol. Evol.* 40:155–172.

Le Novère, N., Zoli, M., and Changeux, J.-P. (1996). Neuronal nicotinic receptor α6 subunit RNA is selectively concentrated in catecholaminergic nuclei of the rat brain. *Eur. J. Neurosci.* 8:2428–2439.

Le Novère, N. (1998). Contribution à l'étude de la relation structure-fonction dans la famille des sous-unités des récepteurs nicotiniques de l'acétylcholine. Ph.D. thesis, University of Paris-6.

Le Novère, N., Corringer, P.-J., and Changeux, J.-P. (1999). Improved secondary structure predictions for a nicotinic receptor subunit: incorporation of solvent accessibility and experimental data into a two-dimensional representation. *Biophys. J.* 76:2329–2345.

Le Novère, N., Corringer, P.-J., and Changeux, J.-P. (2002a). The diversity of subunit composition in nAChRs: evolutionary origins, physiologic and pharmacologic consequences. *J. Neurobiol.* 53:447–456.

Le Novère, N., Grutter, T., and Changeux, J.-P. (2002b). Models of the extracellular domain of the nicotinic receptors and of agonist- and Ca2+-binding sites. *Proc. Natl. Acad. Sci. USA* 99:3210–3215.

Lee, C. Y. and Chang, C. C. (1966). Modes of actions of purified toxins from elapid venoms on neuromuscular transmission. *Mem. Inst. Butantan São Paulo* 33:555–572.

Lee, C. Y. and Tseng, L. F. (1966). Distribution of *Bungarus multicinctus* venom following envenomation. *Toxicon* 3:281–290.

Lee, S. H. and Sheng, M. (2000). Development of neuron-neuron synapses. *Curr. Opin. Neurobiol.* 10:125–131.

Lee, T., Witzemann, V., Schimerlik, M., and Raftery, M. A. (1977). Cholinergic ligand-induced affinity changes in *Torpedo californica* acetylcholine receptor. *Arch. Biochem. Biophys.* 183:57–63.

Lefkowitz, R. J., Cotecchia, S., Samama, P., and Costa, T. (1993). Constitutive activity of receptors coupled to guanine nucleotide regulatory proteins. *Trends Pharmacol. Sci.* 14:303–307.

Léna, C., Changeux, J.-P., and Mulle, C. (1993). Evidence for "preterminal" nicotinic receptors on GABAergic axons in the rat interpeduncular nucleus. *J. Neurosci.* 13:2680–2688.

Léna, C. and Changeux, J.-P. (1997). Role of Ca^{2+} ions in nicotinic facilitation of GABA release in mouse thalamus. *J. Neurosci.* 17:576–585.

Léna, C., Popa, D., Grailhe, R., Escourrou, P., Changeux, J.-P., and Adrien, J. (2004). Beta2-containing nicotinic receptors contribute to the organization of sleep and regulate putative micro-arousals in mice. *J. Neurosci.* 24:5711–5718.

Leonard, R. J., Labarca, C. G., Charnet, P., Davidson, N., and Lester, H. A. (1988). Evidence that the M2 membrane-spanning region lines the ion channel pore of the nicotinic receptor. *Science* 242:1578–1581.

Lepore, M. and Franklin, K. B. (1996). N-methyl-D-aspartate lesions of the pedunculopontine nucleus block acquisition and impair maintenance of responding reinforced with brain stimulation. *Neuroscience* 71:147–155.

Lester, H. (1970). Postsynaptic action of cobra toxin at the myoneural junction. *Nature* 227:727–728.

Lester, H., Changeux, J.-P., and Sheridan, R. E. (1975). Conductance increases produced by bath application of cholinergic agonists to *Electrophorus* electroplaques. *J. Gen. Physiol.* 65:797–816.

Levin, E. D. (1992). Nicotinic systems and cognitive function. *Psychopharmacology (Berl.)* 108:417–431.

Levin, E. D., Briggs, S. J., Christopher, N. C., and Rose, J. E. (1993). Prenatal nicotine exposure and cognitive performance in rats. *Neurotoxicol. Teratol.* 15:251–260.

Levin, E. D., Mead, T., Rezvani, A. H., Rose, J. E., Gallivan, C., and Gross, R. (2000). The nicotinic antagonist mecamylamine preferentially inhibits cocaine vs. food self-administration in rats. *Physiol. Behav.* 71:565–570.

Levin, E. D. and Rezvani, A. H. (2002). Nicotinic treatment for cognitive dysfunction. *Curr. Drug Targets CNS Neurol. Disord.* 1:423–431.

Levitan, I. B. (1994). Modulation of ion channels by protein phosphorylation and dephosphorylation. *Ann. Rev. Physiol.* 56:193–212.

Lichtman, J. W. and Sanes, J. R. (2003). Watching the neuromuscular junction. *J. Neurocytol.* 32:767–775.

Lillie, R. S. (1909). On the connection between changes of permeability and stimulation and on the significance of changes in permeability to carbon dioxide. *Am. J. Physiol.* 24:14–44.

Lin, W., Burgess, R. W., Dominguez, B., Pfaff, S. L., Sanes, J. R., Lee, K. F., and Terrado, J. (2001). Distinct roles of nerve and muscle in postsynaptic differentiation of the neuromuscular synapse. *Nature* 410:1057–1064.

Lindstrom, J. and Patrick, J. (1974). Purification of the Acetylcholine Receptor by Affinity Chromatography. In M.V.L. Bennett, ed. *Synaptic Transmission and Neuronal Interaction*, 191–216. New York: Raven Press.

Lindstrom, J., Walter, B., and Einarson, B. (1979). Immunochemical similarities between subunits of acetylcholine receptors from *Torpedo, Electrophorus*, and mammalian muscle. *Biochemistry* 18:4470–4480.

Lindstrom, J. and Dau, P. (1980). Biology of myasthenia gravis. *Ann. Rev. Pharmacol. Toxicol.* 20:337–362.

Lindstrom, J. (1986). Probing nicotinic acetylcholine receptors with monoclonal antibodies. *Trends Neurosci.* 9:401–407.

Lindstrom, J. M. (2000). Acetylcholine receptors and myasthenia. *Muscle Nerve* 23:453–477.

Lindstrom, J. (2002). Autoimmune diseases involving nicotinic receptors. *J. Neurobiol.* 53:656–665.

Lipscomb, W. N. (1994). Aspartate transcarbamylase from *Escherichia coli:* activity and regulation. *Adv. Enzymol. Relat. Areas Mol. Biol.* 68:67–151.

Littleton, J. T. and Ganetzky, B. (2000). Ion channels and synaptic organization: analysis of the *Drosophila* genome. *Neuron* 26:35–43.

Liu, Q., Melnikova, I. N., Hu, M., and Gardner, P. D. (1999). Cell type-specific activation of neuronal nicotinic acetylcholine receptor subunit genes by Sox10. *J. Neurosci.* 19:9747–9755.

Liu, Q. S., Kawai, H., and Berg, D. K. (2001). β-amyloid peptide blocks the response of α7-containing nicotinic receptors on hippocampal neurons. *Proc. Natl. Acad. Sci. USA* 98:4734–4739.

Lo, M. M., Garland, P. B., Lamprecht, J., and Barnard, E. A. (1980). Rotational mobility of the membrane-bound acetylcholine receptor of *Torpedo* electric organ measured by phosphorescence depolarisation. *FEBS Lett.* 111:407–412.

Lo, M. M., Barnard, E. A., and Dolly, J. O. (1982). Size of acetylcholine receptors in the membrane. An improved version of the radiation inactivation method. *Biochemistry* 21:2210–2217.

Lomo, T. and Rosenthal, J. (1972). Control of ACh sensitivity by muscle activity in the rat. *J. Physiol.* 221:493–513.

Lomo, T. and Westgaard, R. H. (1976). Control of ACh sensitivity in rat muscle fibers. *Cold Spring Harb. Symp. Quant. Biol.* 40:263–274.

Works Cited

Lomo, T. and Slater, C. R. (1980). Control of junctional acetylcholinesterase by neural and muscular influences in the rat. *J. Physiol.* 303:191–202.

Loring, R. H. and Salpeter, M. M. (1980). Denervation increases turnover rate of junctional acetylcholine receptors. *Science* 210:550–551.

Lu, J. T., Son, Y. J., Lee, J., Jetton, T. L., Shiota, M., Moscoso, L., Niswender, K. D., Loewy, A. D., Magnuson, M. A., Sanes, J. R., and Emeson, R. B. (1999). Mice lacking alpha-calcitonin gene-related peptide exhibit normal cardiovascular regulation and neuromuscular development. *Mol. Cell Neurosci.* 14:99–120.

Luyten, W. H. (1986). A model for the acetylcholine binding site of the nicotinic acetylcholine receptor. *J. Neurosci. Res.* 16:51–73.

Machold, J., Weise, C., Utkin, Y., Tsetlin, V., and Hucho, F. (1995). The handedness of the subunit arrangement of the nicotinic acetylcholine receptor from *Torpedo californica*. *Eur. J. Biochem.* 234:427–430.

MacKinnon, R. (1995). Pore loops: an emerging theme in ion channel structure. *Neuron* 14:889–892.

Macol, C. P., Tsuruta, H., Stec, B., and Kantrowitz, E. R. (2001). Direct structural evidence for a concerted allosteric transition in *Escherichia coli* aspartate transcarbamoylase. *Nat. Struct. Biol.* 8:423–426.

Maelicke, A. and Reich, E. (1976). On the interaction between cobra alpha-neurotoxin and the acetylcholine receptor. *Cold Spring Harb. Symp. Quant. Biol.* 40:231–235.

Maelicke, A., Fulpius, B. W., Klett, R. P., and Reich, E. (1977). Acetylcholine receptor: responses to drug binding. *J. Biol. Chem.* 252:4811–4830.

Maelicke, A., Coban, T., Storch, A., Schrattenholz, A., Pereira, E. F., and Albuquerque, E. X. (1997). Allosteric modulation of *Torpedo* nicotinic acetylcholine receptor ion channel activity by noncompetitive agonists. *J. Recept. Signal Transduct. Res.* 17:11–28.

Maeno, T. (1966). Analysis of sodium and potassium conductances in the procaine endplate potential. *J. Physiol.* 183:592–606.

Magleby, K. L. and Pallotta, B. S. (1981). A study of desensitization of acetylcholine receptors using nerve-released transmitter in the frog. *J. Physiol.* 316:225–250.

Mannervik, M., Nibu, Y., Zhang, H., and Levine, M. (1999). Transcriptional coregulators in development. *Science* 284:606–609.

Mansvelder, H. D. and McGehee, D. S. (2000). Long-term potentiation of excitatory inputs to brain reward areas by nicotine. *Neuron* 27:349–357.

Mansvelder, H. D., Keath, J. R., and McGehee, D. S. (2002). Synaptic mechanisms underlie nicotine-induced excitability of brain reward areas. *Neuron* 33:905–919.

Marchand, S., Stetzkowski-Marden, F., and Cartaud, J. (2001). Differential targeting of components of the dystrophin complex to the postsynaptic membrane. *Eur. J. Neurosci.* 13:221–229.

Marchand, S. and Cartaud, J. (2002). Targeted trafficking of neurotransmitter receptors to synaptic sites. *Mol. Neurobiol.* 26:117–135.

Marchand, S., Devillers-Thiéry, A., Pons, S., Changeux, J.-P., and Cartaud, J. (2002). Rapsyn escorts the nicotinic acetylcholine receptor along the exocytic pathway via association with lipid rafts. *J. Neurosci.* 22:8891–8901.

Marden, M. C., Kiger, L., Poyart, C., and Edelstein, S. J. (1998). Identifying the conformational state of bi-liganded haemoglobin. *Cell Mol. Life Sci.* 54:1365–1384.

Marks, M. J., Pauly, J. R., Gross, S. D., Deneris, E. S., Hermans-Borgmeyer, I., Heinemann, S. F., and Collins, A. C. (1992). Nicotine binding and nicotinic receptor subunit RNA after chronic nicotine treatment. *J. Neurosci.* 12:2765–2784.

Marshall, L. M., Sanes, J. R., and McMahan, U. J. (1977). Reinnervation of original synaptic sites on muscle fiber basement membrane after disruption of the muscle cells. *Proc. Natl. Acad. Sci. USA* 74:3073–3077.

Martin, M., Czajkowski, C., and Karlin, A. (1996). The contributions of aspartyl residues in the acetylcholine receptor gamma and delta subunits to the binding of agonists and competitive antagonists. *J. Biol. Chem.* 271:13497–13503.

Martinez, K. L., Corringer, P.-J., Edelstein, S. J., Changeux, J.-P., and Merola, F. (2000). Structural differences in the two agonist binding sites of the *Torpedo* nicotinic acetylcholine receptor revealed by time-resolved fluorescence spectroscopy. *Biochemistry* 39:6979–6990.

Martinez-Carrion, M. and Raftery, M. A. (1973). Use of a fluorescent probe for the study of ligand binding by the isolated cholinergic receptor of *Torpedo californica*. *Biochem. Biophys. Res. Commun.* 55:1156–1164.

Martin-Ruiz, C., Lawrence, S., Piggott, M., Kuryatov, A., Lindstrom, J., Gotti, C., Cookson, M. R., Perry, R. H., Jaros, E., Perry, E. K., and Court, J. A. (2002). Nicotinic receptors in the putamen of patients with dementia with Lewy bodies and Parkinson's disease: relation to changes in alpha-synuclein expression. *Neurosci. Lett.* 335:134–138.

Martin-Ruiz, C. M., Lee, M., Perry, R. H., Baumann, M., Court, J. A., and Perry, E. K. (2004). Molecular analysis of nicotinic receptor expression in autism. *Brain Res. Mol. Brain Res.* 123:81–90.

Marubio, L. M., del Mar Arroyo-Jimenez, M., Cordero-Erausquin, M., Léna, C., Le Novère, N., De Kerchove D'Exaerde, A., Huchet, M., Damaj, M. I., and Changeux, J.-P. (1999). Reduced antinociception in mice lacking neuronal nicotinic receptor subunits. *Nature* 398:805–810.

Marubio, L. M., Gardier, A. M., Durier, S., David, D., Klink, R., Arroyo-Jimenez, M. M., McIntosh, J. M., Rossi, F., Champtiaux, N., Zoli, M., and Changeux, J.-P. (2003). Effects of nicotine in the dopaminergic system of mice lacking the alpha4 subunit of neuronal nicotinic acetylcholine receptors. *Eur. J. Neurosci.* 17:1329–1337.

Maskos, U., Molles, B. E., Pons, S., Besson, M., Guiard, B. P., Guilloux, J.-P., Evrard, A., Cazala, P., Cormier, A., Mameli-Engvall, M., Dufour, N., Cloëz-Tayarani, I., Bemelmans, A.-P., Mallet, J., Gardier, A. M., David, V., Faure, P., Granon, S., and Changeux, J.-P. (2005). Recovery of nicotine reinforcement and cognitive functions by targeted expression of nicotinic receptors. *Nature* 436:103–107.

Massoulie, J., Rieger, F., and Tsuji, S. (1970). [Solubilization of electric organ acetylcholinesterase from eels: action of trypsin]. *Eur. J. Biochem.* 14:430–439.

Massoulie, J. and Bon, S. (1982). The molecular forms of cholinesterase and acetylcholinesterase in vertebrates. *Ann. Rev. Neurosci.* 5:57–106.

Massoulie, J., Sussman, J., Bon, S., and Silman, I. (1993). Structure and functions of acetylcholinesterase and butyrylcholinesterase. *Prog. Brain Res.* 98:139–146.

McClatchey, A. I., Van den Bergh, P., Pericak-Vance, M. A., Raskind, W., Verellen, C., McKenna-Yasek, D., Rao, K., Haines, J. L., Bird, T., Brown, R. H., Jr., and et al. (1992). Temperature-sensitive mutations in the III-IV cytoplasmic loop region of the skeletal muscle sodium channel gene in paramyotonia congenita. *Cell* 68:769–774.

McConnell, H. C., Devaux, P., and Scandella, C. J. (1972). Lateral Diffusion and Phase Separations in Biological Membranes. In C. F. Fox, ed. *Membrane Research,* 27-37. New York: Academic Press.

McConville, J. and Vincent, A. (2002). Diseases of the neuromuscular junction. *Curr. Opin. Pharmacol.* 2:296–301.

McCrea, P. D., Popot, J. L., and Engelman, D. M. (1987). Transmembrane topography of the nicotinic acetylcholine receptor delta subunit. *EMBO J.* 6:3619–3626.

McDonough, J. and Deneris, E. (1997). Beta43′: an enhancer displaying neural-restricted activity is located in the 3′-untranslated exon of the rat nicotinic acetylcholine receptor beta4 gene. *J. Neurosci.* 17:2273–2283.

McDonough, J., Francis, N., Miller, T., and Deneris, E. S. (2000). Regulation of transcription in the neuronal nicotinic receptor subunit gene cluster by a neuron-selective enhancer and ETS domain factors. *J. Biol. Chem.* 275:28962–28970.

McGehee, D. S., Heath, M. J., Gelber, S., Devay, P., and Role, L. W. (1995). Nicotine enhancement of fast excitatory synaptic transmission in CNS by presynaptic receptors. *Science* 269:1692–1696.

McHugh, T. J., Blum, K. I., Tsien, J. Z., Tonegawa, S., and Wilson, M. A. (1996). Impaired hippocampal representation of space in CA1-specific NMDAR1 knockout mice. *Cell* 87: 1339–1349.

McMahan, U. J., Sanes, J. R., and Marshall, L. M. (1978). Cholinesterase is associated with the basal lamina at the neuromuscular junction. *Nature* 271:172–174.

McMahon, L. L., Yoon, K. W., and Chiappinelli, V. A. (1994a). Nicotinic receptor activation facilitates GABAergic neurotransmission in the avian lateral spiriform nucleus. *Neuroscience* 59:689–698.

McMahon, L. L., Yoon, K. W., and Chiappinelli, V. A. (1994b). Electrophysiological evidence for presynaptic nicotinic receptors in the avian ventral lateral geniculate nucleus. *J. Neurophysiol.* 71:826–829.

Mechawar, N., Saghatelyan, A., Grailhe, R., Scoriels, L., Gheusi, G., Gabellec, M. M., Lledo, P. M., and Changeux, J.-P. (2004). Nicotinic receptors regulate the survival of newborn neurons in the adult olfactory bulb. *Proc. Natl. Acad. Sci. USA* 101:9822–9826.

Mega, M. S. (2000). The cholinergic deficit in Alzheimer's disease: impact on cognition, behaviour and function. *Int. J. Neuropsychopharmacol.* 3:3–12.

Meiniel, R. and Bourgeois, J. P. (1982). Appearance and distribution "in situ" of nicotinic acetylcholine receptors in cervical myotomes of young chick embryos. Radioautographic studies by light and electron microscopy. *Anat. Embryol.* 164:349–368.

Mendelzon, D., Changeux, J.-P., and Nghiem, H. O. (1994). Phosphorylation of myogenin in chick myotubes: regulation by electrical activity and by protein kinase C. Implications for acetylcholine receptor gene expression. *Biochemistry* 33:2568–2575.

Mendez, B., Valenzuela, P., Martial, J. A., and Baxter, J. D. (1980). Cell-free synthesis of acetylcholine receptor polypeptides. *Science* 209:695–697.

Menez, A., Morgat, J. L., Fromageot, P., Ronseray, A. M., Boquet, P., and Changeux, J.-P. (1971). Tritium labelling of the alpha-neurotoxin of *Naja nigricollis*. *FEBS Lett.* 17:333–338.

Merlie, J.-P., Sobel, A., Changeux, J.-P., and Gros, F. (1975). Synthesis of acetylcholine receptor during differentiation of cultured embryonic muscle cells. *Proc. Natl. Acad. Sci. USA* 72:4028–4032.

Merlie, J.-P., Changeux, J.-P., and Gros, F. (1976). Acetylcholine receptor degradation measured by pulse chase labelling. *Nature* 264:74–76.

Merlie, J.-P., Changeux, J.-P., and Gros, F. (1978). Skeletal muscle acetylcholine receptor: purification, characterization, and turnover in muscle cell cultures. *J. Biol. Chem.* 253:2882–2891.

Merlie, J.-P., Sebbane, R., Gardner, S., Olson, E., and Lindstrom, J. (1983). The regulation of acetylcholine receptor expression in mammalian muscle. *Cold Spring Harb. Symp. Quant. Biol.* 48 (1):135–146.

Merlie, J.-P. and Sanes, J. R. (1985). Concentration of acetylcholine receptor mRNA in synaptic regions of adult muscle fibres. *Nature* 317:66–68.

Merlie, J.-P. and Smith, M. M. (1986). Synthesis and assembly of acetylcholine receptor, a multisubunit membrane glycoprotein. *J. Membr. Biol.* 91:1–10.

Merlie, J.-P. and Kornhauser, J. M. (1989). Neural regulation of gene expression by an acetylcholine receptor promoter in muscle of transgenic mice. *Neuron* 2:1295–1300.

Mesulam, M. (2000). Brain, mind, and the evolution of connectivity. *Brain Cogn.* 42:4–6.

Methot, N. and Baenziger, J. E. (1998). Secondary structure of the exchange-resistant core from the nicotinic acetylcholine receptor probed directly by infrared spectroscopy and hydrogen/deuterium exchange. *Biochemistry* 37:14815–14822.

Methot, N., Ritchie, B. D., Blanton, M. P., and Baenziger, J. E. (2001). Structure of the pore-forming transmembrane domain of a ligand-gated ion channel. *J. Biol. Chem.* 276:23726–23732.

Meunier, J. C., Huchet, M., Boquet, P., and Changeux, J.-P. (1971a). [Separation of the receptor protein of acetylcholine and acetylcholinesterase]. *C. R. Acad. Sci. Hebd. Séances Acad. Sci. D Sci. Nat.* 272:117–120.

Meunier, J. C., Olsen, R., Menez, A., Morgat, J. L., Fromageot, P., Ronseray, A. M., Boquet, P., and Changeux, J.-P. (1971b). [Some physical properties of the acetylcholine protein receptor studied with a radioactive neurotoxin]. *C. R. Acad. Sci. Hebd. Séances Acad. Sci. D Sci. Nat.* 273:595–598.

Meunier, J. C., Olsen, R. W., and Changeux, J.-P. (1972a). Studies on the cholinergic receptor protein from *Electrophorus electricus*: effect of detergents on some hydrodynamic properties of the receptor protein in solution. *FEBS Lett.* 24:63–68.

Meunier, J. C., Olsen, R. W., Menez, A., Fromageot, P., Boquet, P., and Changeux, J.-P. (1972b). Some physical properties of the cholinergic receptor protein from *Electrophorus electricus* revealed by a tritiated alpha-toxin from *Naja nigricollis* venom. *Biochemistry* 11:1200–1210.

Meunier, J. C. and Changeux, J.-P. (1973). Comparison between the affinities for reversible cholinergic ligands of a purified and membrane bound state of the acetylcholine-receptor protein from *Electrophorus electricus*. *FEBS Lett.* 32:143–148.

Meunier, J. C., Sealock, R., Olsen, R., and Changeux, J.-P. (1974). Purification and properties of the cholinergic receptor protein from *Electrophorus electricus* electric tissue. *Eur. J. Biochem.* 45:371–394.

Michaelis, L. and Menten, M. L. (1913). Die Kinetik der inverten Wirkung. *Biochem Z.* 49:333–369.

Middleton, R. E. and Cohen, J. B. (1991). Mapping of the acetylcholine binding site of the nicotinic acetylcholine receptor: [3H] nicotine as an agonist photoaffinity label. *Biochemistry* 30:6987–6997.

Middleton, R. E., Strnad, N. P., and Cohen, J. B. (1999). Photoaffinity labeling the *Torpedo* nicotinic acetylcholine receptor with [(3)H]tetracaine, a nondesensitizing noncompetitive antagonist. *Mol. Pharmacol.* 56:290–299.

Mihovilovic, M. and Richman, D. P. (1984). Modification of alpha-bungarotoxin and cholinergic ligand-binding properties of *Torpedo* acetylcholine receptor by a monoclonal anti-acetylcholine receptor antibody. *J. Biol. Chem.* 259:15051–15059.

Miledi, R. (1960). The acetylcholine sensitivity of frog muscle fibers after complete or partial denervation. *J. Physiol.* 151:1–23.

Miledi, R., Molinoff, P., and Potter, L. T. (1971). Isolation of the cholinergic receptor protein of *Torpedo* electric tissue. *Nature* 229:554–557.

Miledi, R. and Potter, L. T. (1971). Acetylcholine receptors in muscle fibres. *Nature* 233:599–603.

Mileo, A. M., Monaco, L., Palma, E., Grassi, F., Miledi, R., and Eusebi, F. (1995). Two forms of acetylcholine receptor gamma subunit in mouse muscle. *Proc. Natl. Acad. Sci. USA* 92:2686–2690.

Miller, D. L., Moore, H. P. H., Hartig, P. R., and Raftery, M. A. (1978). Fast cation flux from *Torpedo californica* membrane preparations: implications for a functional role for acetylcholine receptor dimers. *Biochem. Biophys. Res. Commun.* 85:632–640.

Miller, J., Witzemann, V., Quast, U., and Raftery, M. A. (1979). Proton magnetic resonance studies of cholinergic ligand binding to the acetylcholine receptor in its membrane environment. *Proc. Natl. Acad. Sci. USA* 76:3580–3584.

Milone, M., Wang, H. L., Ohno, K., Fukudome, T., Pruitt, J. N., Bren, N., Sine, S. M., and Engel, A. G. (1997). Slow-channel myasthenic syndrome caused by enhanced activation, desensitization, and agonist binding affinity attributable to mutation in the M2 domain of the acetylcholine receptor alpha subunit. *J. Neurosci.* 17:5651–5665.

Milone, M., Wang, H. L., Ohno, K., Prince, R., Fukudome, T., Shen, X. M., Brengman, J. M., Griggs, R. C., Sine, S. M., and Engel, A. G. (1998). Mode switching kinetics produced by a naturally occurring mutation in the cytoplasmic loop of the human acetylcholine receptor epsilon subunit. *Neuron* 20:575–588.

Milton, N. G., Bessis, A., Changeux, J.-P., and Latchman, D. S. (1995). The neuronal nicotinic acetylcholine receptor alpha2 subunit gene promoter is activated by the Brn-3b POU family transcription factor and not by Brn-3a or Brn-3c. *J. Biol. Chem.* 270:15143–15147.

Mishina, M., Kurosaki, T., Tobimatsu, T., Morimoto, Y., Noda, M., Yamamoto, T., Terao, M., Lindstrom, J., Takahashi, T., Kuno, M., and et al. (1984). Expression of functional acetylcholine receptor from cloned cDNAs. *Nature* 307:604–608.

Mishina, M., Tobimatsu, T., Imoto, K., Tanaka, K., Fujita, Y., Fukuda, K., Kurasaki, M., Takahashi, H., Morimoto, Y., and Hirose, T. (1985). Location of functional regions of acetylcholine receptor alpha-subunit by site-directed mutagenesis. *Nature* 313:364–369.

Mishina, M., Takai, T., Imoto, K., Noda, M., Takahashi, T., Numa, S., Methfessel, C., and Sakmann, B. (1986). Molecular distinction between fetal and adult forms of muscle acetylcholine receptor. *Nature* 312:406–411.

Mitra, A. K., McCarthy, M. P., and Stroud, R. M. (1989). Three-dimensional structure of the nicotinic acetylcholine receptor and location of the major associated 43-kD cytoskeletal protein, determined at 22 A by low dose electron microscopy and x-ray diffraction to 12.5 A. *J. Cell Biol.* 109:755–774. [Erratum published in 109:1185.]

Miyazawa, A., Fujiyoshi, Y., Stowell, M., and Unwin, N. (1999). Nicotinic acetylcholine receptor at 4.6 Å resolution: transverse tunnels in the channel wall. *J. Mol. Biol.* 288:765–786.

Miyazawa, A., Fujiyoshi, Y., and Unwin, N. (2003). Structure and gating mechanism of the acetylcholine receptor pore. *Nature* 424:949–955.

Mohler, H., Crestani, F., and Rudolph, U. (2001). GABA(A)-receptor subtypes: a new pharmacology. *Curr. Opin. Pharmacol.* 1:22–25.

Mongan, N. P., Baylis, H. A., Adcock, C., Smith, G. R., Sansom, M. S., and Sattelle, D. B. (1998). An extensive and diverse gene family of nicotinic acetylcholine receptor alpha subunits in *Caenorhabditis elegans*. *Recept. Chann.* 6:213–228.

Monod, J. and Jacob, F. (1961). General conclusions: telenomic mechanisms in cellular metabolism, growth, and differentiation. *Cold Spring Harb. Symp. Quant. Biol.* 26:389–401.

Monod, J., Changeux, J.-P., and Jacob, F. (1963). Allosteric proteins and cellular control systems. *J. Mol. Biol.* 6:306–329.

Monod, J., Wyman, J., and Changeux, J.-P. (1965). On the nature of allosteric transitions: a plausible model. *J. Mol. Biol.* 12:88–118.

Montal, M. (1986). Functional Reconstitution of Membrane Proteins in Planar Lipid Bilayer Membranes. In E. I. Ragan and R. J. Cherry, eds. *Techniques for the Analysis of Membrane Proteins*, 98–128. London: Chapman and Hall.

Works Cited

Montal, M., Anholt, R., and Labarca, P. (1986). The Reconstituted Receptor. In C. Miller, ed. *Ion Channel Reconstitution*, 157–204. New York: Plenum Press.

Montal, M. and Opella, S. J. (2002). The structure of the M2 channel-lining segment from the nicotinic acetylcholine receptor. *Biochem. Biophys. Acta* 1565:287–293.

Montal, M. O., Iwamoto, T., Tomich, J. M., and Montal, M. (1993). Design, synthesis and functional characterization of a pentameric channel protein that mimics the presumed pore structure of the nicotinic cholinergic receptor. *FEBS Lett.* 320:261–266.

Moody, T. W., Schmidt, J., and Raftery, M. A. (1973). Binding of acetylcholine and related compounds to purified acetylcholine receptor from *Torpedo californica* electroplax. *Biochem. Biophys. Res. Commun.* 53:761–772.

Moore, H. and Raftery, M. A. (1980). Direct spectroscopic studies of cation translocation by *Torpedo* acetylcholine receptor on a time scale of physiological relevance. *Proc. Natl. Acad. Sci. USA* 77:4509–4513.

Moreau, M. and Changeux, J.-P. (1976). Studies on the electrogenic action of acetylcholine with *Torpedo marmorata* electric organ. I. Pharmacological properties of the electroplaque. *J. Mol. Biol.* 106:457–467.

Moriyoshi, K., Masu, M., Ishii, T., Shigemoto, R., Mizuno, N., and Nakanishi, S. (1991). Molecular cloning and characterization of the rat NMDA receptor. *Nature* 354:31–37.

Mosckovitz, R. and Gershoni, J. M. (1988). Three possible disulfides in the acetylcholine receptor alpha-subunit. *J. Biol. Chem.* 263:1017–1022.

Muhn, P., Fahr, A., and Hucho, F. (1984). Photoaffinity labeling of acetylcholine receptor in millisecond time scale. *FEBS Lett.* 166:146–150.

Muir-Robinson, G., Hwang, B. J., and Feller, M. B. (2003). Retinogeniculate axons undergo eye-specific segregation in the absence of eye-specific layers. *J. Neurosci.* 22:5259–5264.

Mulac-Jericevic, B. and Atassi, M. Z. (1986). Segment alpha 182–198 of *Torpedo californica* acetylcholine receptor contains second toxin-binding region and binds anti-receptor antibodies. *FEBS Lett.* 199:68–74.

Mulle, C. and Changeux, J.-P. (1990). A novel type of nicotinic receptor in the rat central nervous system characterized by patch-clamp techniques. *J. Neurosci.* 10:169–175.

Mulle, C., Vidal, C., Benoit, P., and Changeux, J.-P. (1991). Existence of different subtypes of nicotinic acetylcholine receptors in the rat habenulo-interpeduncular system. *J. Neurosci.* 11:2588–2597.

Mulle, C., Choquet, D., Korn, H., and Changeux, J.-P. (1992a). Calcium influx through nicotinic receptor in rat central neurons: its relevance to cellular regulation. *Neuron* 8:135–143.

Mulle, C., Léna, C., and Changeux, J.-P. (1992b). Potentiation of nicotinic receptor response by external calcium in rat central neurons. *Neuron* 8:937–945.

Musil, L. S., Carr, C., Cohen, J. B., and Merlie, J.-P. (1988). Acetylcholine receptor-associated 43K protein contains covalently bound myristate. *J. Cell Biol.* 107:1113–1121.

Nachmansohn, D. (1959). *The Chemical and Molecular Basis of Nerve Activity*. New York: Academic Press.

Nakazawa, K., Mikawa, S., Hashikawa, T., and Ito, M. (1995). Transient and persistent phosphorylation of AMPA-type glutamate receptor subunits in cerebellar Purkinje cells. *Neuron* 15:697–709.

Narahashi, T., Aistrup, G. L., Marszalec, W., and Nagata, K. (1999). Neuronal nicotinic acetylcholine receptors: a new target site of ethanol. *Neurochem. Int.* 35:131–141.

Nastuk, W. L. (1953). The electrical activity of the muscle cell membrane at the neuromuscular junction. *J. Cell Comp. Physiol.* 42:249–272.

Nef, P., Mauron, A., Stalder, R., Alliod, C., and Ballivet, M. (1984). Structure linkage, and sequence of the two genes encoding the delta and gamma subunits of the nicotinic acetylcholine receptor. *Proc. Natl. Acad. Sci. USA* 81:7975–7979.

Neher, E. and Sakmann, B. (1976). Single channel currents recorded from membrane of denervated frog muscle fibers. *Nature* 260:799–802.

Neher, E. and Steinbach, J. H. (1978). Local anaesthetics transiently block currents through single acetylcholine-receptor channels. *J. Physiol.* 277:153–176.

Neher, E. (1983). The charge carried by single-channel currents of rat cultured muscle cells in the presence of local anaesthetics. *J. Physiol.* 339:663–678.

Nelson, M. E. and Lindstrom, J. (1999). Single channel properties of human alpha3 AChRs: impact of beta2, beta4 and alpha5 subunits. *J. Physiol.* 516:657–678.

Nelson, N., Anholt, R., Lindstrom, J., and Montal, M. (1980). Reconstitution of purified acetylcholine receptors with functional ion channels in planar lipid bilayers. *Proc. Natl. Acad. Sci. USA* 77:3057–3061.

Neubig, R. R. and Cohen, J. B. (1979). Equilibrium binding of [3H]tubocurarine and [3H]acetylcholine by *Torpedo* post-synaptic membranes: stoichiometry and ligand interactions. *Biochemistry* 18:5464–5475.

Neubig, R. R., Krodel, E. K., Boyd, N. D., and Cohen, J. B. (1979). Acetylcholine and local anesthetic binding to *Torpedo* nicotinic postsynaptic membranes after removal of nonreceptor peptides. *Proc. Natl. Acad. Sci. USA* 76:690–694.

Neubig, R. R. and Cohen, J. B. (1980). Permeability control by cholinergic receptors in *Torpedo* post-synaptic membranes: agonist dose-response relations measured at second and millisecond times. *Biochemistry* 19:2770–2779.

Neubig, R. R., Boyd, N. D., and Cohen, J. B. (1982). Conformations of *Torpedo* acetylcholine receptor associated with ion transport and desensitization. *Biochemistry* 21:3460–3467.

Neumann, D., Gershoni, J. M., Fridkin, M., and Fuchs, S. (1985). Antibodies to synthetic peptides as probes for the binding site on the alpha subunit of the acetylcholine receptor. *Proc. Natl. Acad. Sci. USA* 82:3490–3493.

Neumann, D., Barchan, D., Safran, A., Gershoni, J. M., and Fuchs, S. (1986). Mapping of the alpha-bungarotoxin binding site within the alpha subunit of the acetylcholine receptor. *Proc. Natl. Acad. Sci. USA* 83:3008–3011.

New, H. V. and Mudge, A. W. (1986). Calcitonin gene-related peptide regulates muscle acetylcholine receptor synthesis. *Nature* 323:809–811.

Nghiem, H. O., Cartaud, J., Dubreuil, C., Kordeli, C., Buttin, G., and Changeux, J.-P. (1983). Production and characterization of a monoclonal antibody directed against the 43,000-dalton v1 polypeptide from *Torpedo marmorata* electric organ. *Proc. Natl. Acad. Sci. USA* 80:6403–6407.

Nghiem, H. O., Hill, J., and Changeux, J.-P. (1991). Developmental changes in the subcellular distribution of the 43K (v1) polypeptides in *Torpedo marmorata* electrocyte: support for a role in acetylcholine receptor stabilization. *Development* 113:1059–1067.

Nghiem, H. O., Bettendorff, L., and Changeux, J.-P. (2000). Specific phosphorylation of *Torpedo* 43K rapsyn by endogenous kinase(s) with thiamine triphosphate as the phosphate donor. *Faseb J.* 14:543–554.

Nichols, P., Croxen, R., Vincent, A., Rutter, R., Hutchinson, M., Newsom-Davis, J., and Beeson, D. (1999). Mutation of the acetylcholine receptor epsilon-subunit promoter in congenital myasthenic syndrome. *Ann. Neurol.* 45:439–443.

Nickel, E. and Potter, L. T. (1973). Ultrastructure of isolated membranes of *Torpedo* electric tissue. *Brain Res.* 57:508–517.

Niethammer, M., Valtschanoff, J. G., Kapoor, T. M., Allison, D. W., Weinberg, T. M., Craig, A. M., and Sheng, M. (1998). CRIPT, a novel postsynaptic protein that binds to the third PDZ domain of PSD-95/SAP90. *Neuron* 20:693–707.

Nisell, M., Nomikos, G. G., and Svensson, T. H. (1994). Systemic nicotine-induced dopamine release in the rat nucleus accumbens is regulated by nicotinic receptors in the ventral tegmental area. *Synapse* 16:36–44.

Nitkin, R. M., Smith, M. A., Magill, C., Fallon, J. R., Yao, Y. M., Wallace, B. G., and McMahan, U. J. (1987). Identification of agrin, a synaptic organizing protein from *Torpedo* electric organ. *J. Cell Biol.* 105:2471–2478.

Noakes, P. G., Phillips, W. D., Hanley, T. A., Sanes, J. R., and Merlie, J.-P. (1993). 43K protein and acetylcholine receptors colocalize during the initial stages of neuromuscular synapse formation *in vivo*. *Dev. Biol.* 155:275–280.

Noda, M., Takahashi, H., Tanabe, T., Toyosato, M., Furutani, Y., Hirose, T., Asai, M., Inayama, S., Miyata, T., and Numa, S. (1982). Primary structure of alpha-subunit precursor of *Torpedo californica* acetylcholine receptor deduced from cDNA sequence. *Nature* 299:793–797.

Noda, M., Takahashi, H., Tanabe, T., Toyosato, M., Kikyotani, S., Furutani, Y., Hirose, T., Takashima, H., Inayama, S., Miyata, T., and Numa, S. (1983a). Structural homology of *Torpedo californica* acetylcholine receptor subunits. *Nature* 302:528–532.

Noda, M., Takahashi, H., Tanabe, T., Toyosato, M., Kikyotani, S., Hirose, T., Asai, M., Takashima, H., Inayama, S., Miyata, T., and Numa, S. (1983b). Primary structures of beta- and delta-subunit precursors of *Torpedo californica* acetylcholine receptor deduced from cDNA sequences. *Nature* 301:251–255.

Nomikos, G. G., Schilstrom, B., Hildebrand, B. E., Panagis, G., Grenhoff, J., and Svensson, T. H. (2000). Role of alpha7 nicotinic receptors in nicotine dependence and implications for psychiatric illness. *Behav. Brain Res.* 113:97–103.

Nomoto, H., Takahashi, N., Nagaki, Y., Endo, S., Arata, Y., and Hayashi, K. (1986). Carbohydrate structures of acetylcholine receptor from *Torpedo californica* and distribution of oligosaccharides among the subunits. *Eur. J. Biochem.* 157:233–242.

Nordberg, A. (2001). Nicotinic receptor abnormalities of Alzheimer's disease: therapeutic implications. *Biol. Psychiatry* 49:200–210.

Nowak, M. W., Keatney, P. C., Sampson, J. R., Saks, M. E., Labarca, C. G., Silverman, S. K., Zhong, W., Thorson, J., Abelson, J. N., Davidson, N., et al. (1995). Nicotinic receptor binding site probed with unnatural amino acid incorporation in intact cells. *Science* 268:439–442.

Numa, S., Noda, M., Takahashi, H., Tanabe, T., Toyosato, M., Furutani, Y., and Kikyotani, S. (1983). Molecular structure of the nicotinic acetylcholine receptor. *Cold Spring Harb. Symp. Quant. Biol.* 48:57–69.

Numa, S. (1989). A molecular view of neurotransmitter receptors and ionic channels. *Harvey Lectures* 83:121–165.

Oberthur, W., Muhn, P., Baumann, H., Lottspeich, F., Wittmann-Liebold, B., and Hucho, F. (1986). The reaction site of a non-competitive antagonist in the delta-subunit of the nicotinic acetylcholine receptor. *EMBO J.* 5:1815–1819.

Oblas, B., Singer, R. H., and Boyd, N. D. (1986). Location of a polypeptide sequence within the alpha-subunit of the acetylcholine receptor containing the cholinergic binding site. *Mol. Pharmacol.* 29:649–656.

Ochoa, E. L., Chattopadhyay, A., and McNamee, M. G. (1989). Desensitization of the nicotinic acetylcholine receptor: molecular mechanisms and effect of modulators. *Cell Mol. Neurobiol.* 9:141–178.

Ohno, K., Hutchinson, D. O., Milone, M., Brengman, J. M., Bouzat, C., Sine, S. M., and Engel, A. G. (1995). Congenital myasthenic syndrome caused by prolonged acetylcholine receptor channel openings due to a mutation in the M2 domain of the epsilon subunit. *Proc. Natl. Acad. Sci. USA* 92:758–762.

Ohno, K., Wang, H. L., Milone, M., Bren, N., Brengman, J. M., Nakano, S., Quiram, P., Pruitt, J. N., Sine, S. M., and Engel, A. G. (1996). Congenital myasthenic syndrome caused by decreased agonist binding affinity due to a mutation in the acetylcholine receptor epsilon subunit. *Neuron* 17:157–170.

Ohno, K., Quiram, P. A., Milone, M., Wang, H. L., Harper, M. C., Pruitt, J. N., 2nd, Brengman, J. M., Pao, L., Fischbeck, K. H., Crawford, T. O., et al. (1997). Congenital myasthenic syndromes due to heteroallelic nonsense/missense mutations in the acetylcholine receptor epsilon subunit gene: identification and functional characterization of six new mutations. *Hum. Mol. Genet.* 6:753–766.

Ohno, K., Brengman, J., Tsujino, A., and Engel, A. G. (1998). Human endplate acetylcholinesterase deficiency caused by mutations in the collagen-like tail subunit (ColQ) of the asymmetric enzyme. *Proc. Natl. Acad. Sci. USA* 95:9654–9659.

Ohno, K., Anlar, B., and Engel, A. G. (1999). Congenital myasthenic syndrome caused by a mutation in the Ets-binding site of the promoter region of the acetylcholine receptor epsilon subunit gene. *Neuromuscul. Disord.* 9:131–135.

Ohno, K., Tsujino, A., Brengman, J. M., Harper, C. M., Bajzer, Z., Udd, B., Beyring, R., Robb, S., Kirkham, F. J., and Engel, A. G. (2001). Choline acetyltransferase mutations cause myasthenic syndrome associated with episodic apnea in humans. *Proc. Natl. Acad. Sci. USA* 98:2017–2022.

Ohno, K., Engel, A. G., Shen, X. M., Selcen, D., Brengman, J., Harper, C. M., Tsujino, A., and Milone, M. (2002). Rapsyn mutations in humans cause endplate acetylcholine-receptor deficiency and myasthenic syndrome. *Am. J. Hum. Genet.* 70:875–885.

Ohno, K., Sadeh, M., Blatt, I., Brengman, J. M., and Engel, A. G. (2003). E-box mutations in the RAPSN promoter region in eight cases with congenital myasthenic syndrome. *Hum. Mol. Genet.* 12:739–748.

Olausson, P., Akesson, P., Engel, J. A., and Soderpalm, B. (2001). Effects of 5-HT1A and 5-HT2 receptor agonists on the behavioral and neurochemical consequences of repeated nicotine treatment. *Eur. J. Pharmacol.* 420:45–54.

O'Leary, M. E. and White, M. M. (1992). Mutational analysis of ligand-induced activation of the *Torpedo* acetylcholine receptor. *J. Biol. Chem.* 267:8360–8365.

Olmstead, M. C. and Franklin, K. B. (1994). Lesions of the pedunculopontine tegmental nucleus block drug-induced reinforcement but not amphetamine-induced locomotion. *Brain Res.* 638:29–35.

Olmstead, M. C., Inglis, W. L., Bordeaux, C. P., Clarke, E. J., Wallum, N. P., Everitt, B. J., and Robbins, T. W. (1999). Lesions of the pedunculopontine tegmental nucleus increase

sucrose consumption but do not affect discrimination or contrast effects. *Behav. Neurosci.* 113:732–743.
Olsen, R. W., Meunier, J.-C., and Changeux, J.-P. (1972). Progress in the purification of the cholinergic receptor protein from *Electrophorus electricus* by affinity chromatography. *FEBS Lett.* 28:96–100.
Olson, E. N. (1990). MyoD family: a paradigm for development? *Genes Dev.* 4:1454–1461.
Ono, F., Shcherbatko, A., Higashijima, S., Mandel, G., and Brehm, P. (2002). The Zebrafish motility mutant twitch once reveals new roles for rapsyn in synaptic function. *J. Neurosci.* 22:6491–6498.
Opella, S. J., Marassi, F. M., Gesell, J. J., Valente, A. P., Kim, Y., Oblatt-Montal, M., and Montal, M. (1999). Structures of the M2 channel-lining segments from nicotinic acetylcholine and NMDA receptors by NMR spectroscopy. *Nat. Struct. Biol.* 6:374–379.
Orr-Urtreger, A., Goldner, F. M., Saeki, M., Lorenzo, I., Goldberg, L., De Biasi, M., Dani, J. A., Patrick, J. W., and Beaudet, A. L. (1997). Mice deficient in the alpha7 neuronal nicotinic acetylcholine receptor lack alpha-bungarotoxin binding sites and hippocampal fast nicotinic currents. *J. Neurosci.* 17:9165–9171.
Orr-Urtreger, A., Broide, R. S., Kasten, M. R., Dang, H., Dani, J. A., Beaudet, A. L., and Patrick, J. W. (2000). Mice homozygous for the L250T mutation in the alpha7 nicotinic acetylcholine receptor show increased neuronal apoptosis and die within 1 day of birth. *J. Neurochem.* 74:2154–2166.
Ortells, M. O. and Lunt, G. G. (1995). Evolutionary history of the ligand-gated ion-channel superfamily of receptors. *Trends Neurosci.* 18:121–127.
Ortells, M. O., Barrantes, G. E., Wood, C., Lunt, G. G., and Barrantes, F. J. (1997). Molecular modelling of the nicotinic acetylcholine receptor transmembrane region in the open state. *Prot. Eng.* 10:511–517.
Osaka, H., Malany, S., Kanter, J. R., Sine, S. M., and Taylor, P. (1999). Subunit interface selectivity of the alpha-neurotoxins for the nicotinic acetylcholine receptor. *J. Biol. Chem.* 274:9581–9586.
Osterlund, M., Fontaine, B., Devillers-Thiéry, A., Geoffroy, B., and Changeux, J.-P. (1989). Acetylcholine receptor expression in primary cultures of embryonic chick myotubes. I. Discoordinate regulation of alpha-, gamma- and delta- subunit gene expression by calcitonin gene-related peptide and by muscle electrical activity. *Neuroscience* 32:279–287.
Oswald, R., Sobel, A., Waksman, G., Roques, B., and Changeux, J.-P. (1980). Selective labelling by [3H]trimethisoquin azide of polypeptide chains present in acetylcholine receptor-rich membranes from *Torpedo marmorata*. *FEBS Lett.* 111:29–34.
Oswald, R. and Changeux, J.-P. (1981a). Ultraviolet light-induced labeling by noncompetitive blockers of the acetylcholine receptor from *Torpedo marmorata*. *Proc. Natl. Acad. Sci. USA* 78:3925–3929.
Oswald, R. E. and Changeux, J.-P. (1981b). Selective labeling of the delta subunit of the acetylcholine receptor by a covalent local anesthetic. *Biochemistry* 20:7166–7174.
Oswald, R. E. and Changeux, J.-P. (1982). Crosslinking of α-bungarotoxin to the acetylcholine receptor from *Torpedo marmorata* by ultraviolet light irradiation. *FEBS Lett.* 139:225–229.
Paas, Y. (1998). The macro- and microarchitectures of the ligand-binding domain of glutamate receptors. *Trends Neurosci.* 21:117–125.
Padlan, E. A., Davies, D. R., Rudikoff, S., and Potter, M. (1976). Structural basis for the specificity of phosphorylcholine-binding immunoglobulins. *Immunochemistry* 13:945–949.
Palade, G. E. and Palay, S. L. (1954). Electron microscope observations of interneuronal and neuromuscular synapses. *Anat. Rec.* 118:335–336.
Palanche, T., Ilien, B., Zoffmann, S., Reck, M. P., Bucher, B., Edelstein, S. J., and Galzi, J. L. (2001). The neurokinin a receptor activates calcium and cAMP responses through distinct conformational states. *J. Biol. Chem.* 276:34853–34861.
Palma, E., Bertrand, S., Binzoni, T., and Bertrand, D. (1996). Neuronal nicotinic alpha 7 receptor expressed in *Xenopus* oocytes presents five putative binding sites for methyllycaconitine. *J. Physiol.* 491:151–161.
Paterson, D. and Nordberg, A. (2000). Neuronal nicotinic receptors in the human brain. *Prog. Neurobiol.* 61:75–111.
Paton, W. and Zaimis, E. (1949). The pharmacological actions of polymethylenebistrimethylammonium salts. *Br. J. Pharmacol. Chemother.* 4:381–400.
Paton, W. (1961). A theory of drug action based on rate of drug-receptor combination. *Proc. R. Soc. (London) Ser. B* 154:21–69.

Works Cited

Paton, W. (1970). Receptors as Defined by their Pharmacological Properties: Molecular Properties of Drug Receptors. In R. Porter and M. O'Connor, eds. *Molecular Properties of Drug Receptors*, 3–32. London: Churchill.

Patrick, J. and Lindstrom, J. (1973). Autoimmune response to acetylcholine receptor. *Science* 180:871–872.

Patrick, J., Lindstrom, J., Culp, B., and McMillan, J. (1973). Studies on purified eel acetylcholine receptor and anti-acetylcholine receptor antibody. *Proc. Natl. Acad. Sci. USA* 70:3334–3338.

Pauling, L. (1935). The oxygen equilibrium of hemoglobin and its structural interpretation. *Proc. Natl. Acad. Sci. USA* 21:186–191.

Paylor, R., Nguyen, M., Crawley, J. N., Patrick, J., Beaudet, A., and Orr-Urtreger, A. (1998). Alpha7 nicotinic receptor subunits are not necessary for hippocampal-dependent learning or sensorimotor gating: a behavioral characterization of Acra7-deficient mice. *Learn Mem.* 5:302–316.

Pedersen, S. E., Dreyer, E. B., and Cohen, J. B. (1986). Location of ligand-binding sites on the nicotinic acetylcholine receptor alpha-subunit. *J. Biol. Chem.* 261:13735–13743.

Pedersen, S. E. and Cohen, J. B. (1988). Photoaffinity labelling of the high and low affinity d-tubocurare binding sites of the nicotinic acetylcholine receptor (AChR) by [3H]d-tubocurare (d-Tc). *Biophys. J.* 53:35la.

Pedersen, S. E. and Cohen, J. B. (1990). d-Tubocurarine binding sites are located at α-γ and α-δ subunit interfaces of the nicotinic acetylcholine receptor. *Proc. Natl. Acad. Sci. USA* 87:2785–2789.

Pedersen, S. E., Sharp, S. D., Liu, W. S., and Cohen, J. B. (1992). Structure of the noncompetitive antagonist-binding site of the *Torpedo* nicotinic acetylcholine receptor. [3H]meproadifen mustard reacts selectively with alpha-subunit Glu-262. *J. Biol. Chem.* 267:10489–10499.

Peng, H. B., Cheng, P. C., and Luther, P. W. (1981). Formation of ACh receptor clusters induced by positively charged latex beads. *Nature* 292:831–834.

Peng, H. B. and Froehner, S. C. (1985). Association of the postsynaptic 43K protein with newly formed acetylcholine receptor clusters in cultured muscle cells. *J. Cell Biol.* 100:1698–1705.

Penn, A. A., Riquelme, P. A., Feller, M. B., and Shatz, C. J. (1998). Competition in retinogeniculate patterning driven by spontaneous activity. *Science* 279:2108–2112.

Pennisi, E. (2003). DNA's cast of thousands. *Science* 300:282–285.

Perrier, A. L., Massoulie, J., and Krejci, E. (2002). PRiMA: the membrane anchor of acetylcholinesterase in the brain. *Neuron* 33:275–285.

Perry, E., Walker, M., Grace, J., and Perry, R. (1999). Acetylcholine in mind: a neurotransmitter correlate of consciousness? *Trends Neurosci.* 22:273–280.

Perry, E. K., Lee, M. L., Martin-Ruiz, C. M., Court, J. A., Volsen, S. G., Merrit, J., Folly, E., Iversen, P. E., Bauman, M. L., Perry, R. H., and Wenk, G. L. (2001). Cholinergic activity in autism: abnormalities in the cerebral cortex and basal forebrain. *Am. J. Psychiatry* 158: 1058–1066.

Perry, E. K. and Perry, R. H. (2004). Neurochemistry of consciousness: cholinergic pathologies in the human brain. *Prog. Brain Res.* 145:287–299.

Perutz, M. F. (1989). Mechanisms of cooperativity and allosteric regulation in proteins. *Quart. Rev. Biophys.* 22:139–236.

Pettit, D. L., Shao, Z., and Yakel, J. L. (2001). Beta-amyloid(1–42) peptide directly modulates nicotinic receptors in the rat hippocampal slice. *J. Neurosci.* 21:RC120.

Phillips, H. A., Favre, I., Kirkpatrick, M., Zuberi, S. M., Goudie, D., Heron, S. E., Scheffer, I. E., Sutherland, G. R., Berkovic, S. F., Bertrand, D., and Mulley, J. C. (2001). CHRNB2 is the second acetylcholine receptor subunit associated with autosomal dominant nocturnal frontal lobe epilepsy. *Am. J. Hum. Genet.* 68:225–231.

Phillips, W. D., Maimone, M. M., and Merlie, J.-P. (1991). Mutagenesis of the 43-kD postsynaptic protein defines domains involved in plasma membrane targeting and AChR clustering. *J. Cell Biol.* 115:1713–1723.

Picciotto, M. R., Zoli, M., Léna, C., Bessis, A., Lallemand, Y., Le Novère, N., Vincent, P., Pich, E. M., Brulet, P., and Changeux, J.-P. (1995). Abnormal avoidance learning in mice lacking functional high-affinity nicotine receptor in the brain. *Nature* 374:65–67.

Picciotto, M. R., Zoli, M., Rimondini, R., Léna, C., Marubio, L. M., Pich, E. M., Fuxe, K., and Changeux, J.-P. (1998). Acetylcholine receptors containing the beta2 subunit are involved in the reinforcing properties of nicotine. *Nature* 391:173–177.

Picciotto, M. R. and Zoli, M. (2002). Nicotinic receptors in aging and dementia. *J. Neurobiol.* 53:641–655.

Picones, A. and Korenbrot, J. I. (1995). Spontaneous, ligand-independent activity of the cGMP-gated ion channel in cone photoreceptors of fish. *J. Physiol.* 485:699–714.

Pidoplichko, V. I., DeBiasi, M., Williams, J. T., and Dani, J. A. (1997). Nicotine activates and desensitizes midbrain dopamine neurons. *Nature* 390:401–404.

Piette, J., Bessereau, J. L., Huchet, M., and Changeux, J.-P. (1990). Two adjacent MyoD1-binding sites regulate expression of the acetylcholine receptor alpha-subunit gene. *Nature* 345:353–355.

Piette, J., Huchet, M., Houzelstein, D., and Changeux, J.-P. (1993). Compartmentalized expression of the alpha- and gamma-subunits of the acetylcholine receptor in recently fused myofibers. *Dev. Biol.* 157:205–213.

Pittenger, C. and Kandel, E. R. (2003). In search of general mechanisms for long-lasting plasticity: aplysia and the hippocampus. *Philos. Trans. R. Soc. Lond. B Biol. Sci.* 358:757–763.

Podleski, T. R. and Bartels, E. (1963). Difference between tetracaine and d-tubocurarine in the competition with carbamylcholine. *Biochem. Biophys. Acta* 75:387–396.

Podleski, T. R. and Nachmansohn, D. (1966). Similarities between active sites of acetylcholine receptor and acetylcholinesterase tested with quinolinium ions. *Proc. Natl. Acad. Sci. USA* 56:1034–1039.

Podleski, T. R. (1967). Distinction between the active sites of acetylcholine-receptor and acetylcholinesterase. *Proc. Natl. Acad. Sci. USA* 58:268–273.

Pontieri, F. E., Tanda, G., Orzi, F., and Di Chiara, G. (1996). Effects of nicotine on the nucleus accumbens and similarity to those of addictive drugs. *Nature* 382:255–257.

Popot, J. L., Sugiyama, H., and Changeux, J.-P. (1974). [*In vitro* demonstration of drug desensitization of the acetylcholine receptor with fragments of excitable membrane from the torpedo fish]. *C. R. Hebd. Séances Acad. Sci. D Sci. Nat.* 279:1721–1724.

Popot, J. L., Demel, R. A., Sobel, A., Van Deenen, L. L., and Changeux, J.-P. (1978). Interaction of the acetylcholine (nicotinic) receptor protein from *Torpedo marmorata* electric organ with monolayers of pure lipids. *Eur. J. Biochem.* 85:27–42.

Popot, J. L., Cartaud, J., and Changeux, J.-P. (1981). Reconstitution of a functional acetylcholine receptor: incorporation into artificial lipid vesicles and pharmacology of the agonist-controlled permeability changes. *Eur. J. Biochem.* 118:203–214.

Popot, J. L. and Changeux, J.-P. (1984). Nicotinic receptor of acetylcholine: structure of an oligomeric integral membrane protein. *Physiol. Rev.* 64:1162–1239.

Porter, C. W. and Barnard, E. (1976). Ultrastructural studies on the acetylcholine receptor at motor end plates of normal and pathologic muscles. *Ann. N. Y. Acad. Sci.* 274:85–107.

Potter, L. T. (1973). Acetylcholine Receptors in Vertebrate Skeletal Muscle and Electric Tissue. In H. P. Rang, ed. *Drug Receptors*, 295–312. London: Macmillan.

Prince, R. J. and Sine, S. M. (1996). Molecular dissection of subunit interfaces in the acetylcholine receptor: identification of residues that determine agonist selectivity. *J. Biol. Chem.* 271:25770–25777.

Pun, S., Sigrist, M., Santos, A. F., Ruegg, M. A., Sanes, J. R., Jessell, T. M., Arber, S., and Caroni, P. (2002). An intrinsic distinction in neuromuscular junction assembly and maintenance in different skeletal muscles. *Neuron* 34:357–370.

Quast, U., Schimerlik, M., Lee, T., Witzemann, T. L., Blanchard, S., and Raftery, M. A. (1978a). Ligand-induced conformation changes in *Torpedo californica* membrane-bound acetylcholine receptor. *Biochemistry* 17:2405–2414.

Quast, U., Schimerlik, M., and Raftery, M. A. (1978b). Stopped flow kinetics of carbamylcholine binding to membrane bound acetylcholine receptor. *Biochem. Biophys. Res. Commun.* 81:955–964.

Quiram, P. A., Ohno, K., Milone, M., Patterson, M. C., Pruitt, N. J., Brengman, J. M., Sine, S. M., and Engel, A. G. (1999). Mutation causing congenital myasthenia reveals acetylcholine receptor beta/delta subunit interaction essential for assembly. *J. Clin. Invest.* 104:1403–1410.

Radding, W., Corfield, P. W., Levinson, L. S., Hashim, G. A., and Low, B. W. (1988). Alpha-toxin binding to acetylcholine receptor alpha 179–191 peptides: intrinsic fluorescence studies. *FEBS Lett.* 231:212–216.

Raftery, M. A., Schmidt, J., Clark, D. G., and Wolcott, R. G. (1971). Demonstration of a specific α-bungarotoxin binding component in *Electrophorus electricus* electroplax membranes. *Biochem. Biophys. Res. Commun.* 45:1622–1629.

Raftery, M. A., Schmidt, J., and Clark, D. G. (1972). Specificity of α-bungarotoxin binding to *Torpedo californica* electroplax. *Arch. Biochem. Biophys.* 152:882–886.

Raftery, M. A., Vandlen, R., Michaelson, D., Bode, J., Moody, T., Chao, Y., Reed, K., Deutsch,

Works Cited

J., and Duguid, J. (1974). The biochemistry of an acetylcholine receptor. *J. Supramol. Struct.* 2:582–592.

Raftery, M. A., Bode, J., Vandlen, R., Micahelson, D., Deutsch, J., Moody, T., Ross, M. J., and Stroud, R. M. (1975). Structural and Functional Studies of an Acetylcholine Receptor. In H. Sund., ed. *Protein-Ligand Interactions,* 328–352. Berlin: Walter de Gruyter.

Raftery, M. A., Hunkapiller, M. W., Strader, C. D., and Hood, L. E. (1980). Acetylcholine receptor: complex of homologous subunits. *Science* 208:1454–1456.

Raggenbass, M. and Bertrand, D. (2002). Nicotinic receptors in circuit excitability and epilepsy. *J. Neurobiol.* 53:580–589.

Raimondi, E., Rubboli, F., Moralli, D., Chini, B., Fornasari, D., Tarroni, P., De Carli, L., and Clementi, F. (1992). Chromosomal localization and physical linkage of the genes encoding the human alpha3, alpha5, and beta4 neuronal nicotinic receptor subunits. *Genomics* 12:849–850.

Rajendra, S., Lynch, J., Pierce, K. D., French, C. R., Barry, P. H., and Schofield, P. R. (1994). Startle disease mutations reduce the agonist sensitivity of the human inhibitory glycine receptor. *J. Biol. Chem.* 269:18739–18742.

Ralston, E., Lu, Z., and Ploug, T. (1999). The organization of the Golgi complex and microtubules in skeletal muscle is fiber type-dependent. *J. Neurosci.* 19:10694–10705.

Ralston, S., Sarin, V., Thanh, H. L., Rivier, J., Fox, J. L., and Lindstrom, J. (1987). Synthetic peptides used to locate the alpha-bungarotoxin binding site and immunogenic regions on alpha subunits of the nicotinic acetylcholine receptor. *Biochemistry* 26:3261–3266.

Ramarao, M. K. and Cohen, J. B. (1998). Mechanism of nicotinic acetylcholine receptor cluster formation by rapsyn. *Proc. Natl. Acad. Sci. USA* 95:4007–4012.

Ramarao, M. K., Bianchetta, M. J., Lanken, J., and Cohen, J. B. (2001). Role of rapsyn tetratricopeptide repeat and coiled-coil domains in self-association and nicotinic acetylcholine receptor clustering. *J. Biol. Chem.* 276:7475–7483.

Ramirez-Latorre, J., Yu, C. R., Qu, X., Perin, F., Karlin, A., and Role, L. (1996). Functional contributions of alpha5 subunit to neuronal acetylcholine receptor channels. *Nature* 380:347–351.

Ramón y Cajal, S. (1909–1911). *Histologie du système nerveux de l'homme et des vertébrés.* 2 vols., trans. L. Azoulay. Paris: Maloine.

Rang, H. P. and Ritter, J. M. (1970). On the mechanism of desensitization of cholinergic receptors. *Mol. Pharmacol.* 6:357–382.

Rang, H. P., Dale, M. M., Ritter, J. M., Moore, P. K., and Lamb, P. (2003). *Pharmacology.* 5th edition. Edinburgh: Churchill Livingstone.

Ranvier, L. (1875). *Traité technique d'histologie.* Paris: Savy.

Ratnam, M., Gullick, W., Spiess, J., Wan, K., Criado, M., and Lindstrom, J. (1986). Structural heterogeneity of the alpha subunits of the nicotinic acetylcholine receptor in relation to agonist affinity alkylation and antagonist binding. *Biochemistry* 25:4268–4275.

Rauer, B., Neumann, E., Widengren, J., and Rigler, R. (1996). Fluorescence correlation spectrometry of the interaction kinetics of tetramethylrhodamin α-bungarotoxin with *Torpedo californica* acetylcholine receptor. *Biophys. Chem.* 58:3–12.

Raymond, L. A., Blackstone, C. D., and Huganir, R. L. (1993). Phosphorylation of amino acid neurotransmitter receptors in synaptic plasticity. *Trends Neurosci.* 16:147–153.

Reed, K., Vandlen, R., Bode, J., Duguid, J., and Raftery, M. A. (1975). Characterization of acetylcholine receptor-rich and acetylcholinesterase-rich membrane particles from *Torpedo california* electroplax. *Arch. Biochem. Biophys.* 167:138–144.

Reid, M. S., Mickalian, J. D., Delucchi, K. L., and Berger, S. P. (1999). A nicotine antagonist, mecamylamine, reduces cue-induced cocaine craving in cocaine-dependent subjects. *Neuropsychopharmacology* 20:297–307.

Reiter, M. J., Cowburn, D. A., Prives, J. M., and Karlin, A. (1972). Affinity labeling of the acetylcholine receptor in the electroplax: electrophoretic separation in sodium dodecyl sulfate. *Proc. Natl. Acad. Sci. USA* 69:1168–1172.

Renshaw, G., Rigby, P., Self, G., Lamb, A., and Goldie, R. (1993). Exogenously administered alpha-bungarotoxin binds to embryonic chick spinal cord: implications for the toxin-induced arrest of naturally occurring motoneuron death. *Neuroscience* 53:1163–1172.

Révah, F., Galzi, J. L., Giraudat, J., Haumont, P. Y., Lederer, F., and Changeux, J.-P. (1990). The noncompetitive blocker [3H]chlorpromazine labels three amino acids of the acetylcholine receptor gamma subunit: implications for the alpha-helical organization of regions MII and for the structure of the ion channel. *Proc. Natl. Acad. Sci. USA* 87:4675–4679.

Révah, F., Bertrand, D., Galzi, J. L., Devillers-Thiéry, A., Mulle, C., Hussy, N., Bertrand, S.,

Ballivet, M., and Changeux, J.-P. (1991). Mutations in the channel domain alter desensitization of a neuronal nicotinic receptor. *Nature* 353:846–849.

Reynolds, J. A. and Karlin, A. (1978). Molecular weight in detergent solution of acetylcholine receptor from *Torpedo californica. Biochemistry* 17:2035–2038.

Richmond, J. E. and Jorgensen, E. M. (1999). One GABA and two acetylcholine receptors function at the *C. elegans* neuromuscular junction. *Nat. Neurosci.* 2:791–797.

Robbins, T. W. (2000). Chemical neuromodulation of frontal-executive functions in humans and other animals. *Exp. Brain Res.* 133:130–138.

Roberts, A. C., Robbins, T. W., and Weiskrantz, L. (1998). *The Prefrontal Cortex: Executive and Cognitive Functions.* New York: Oxford University Press.

Robertson, J. D. (1956). The ultrastructure of a reptilian myoneural junction. *J. Biophys. Biochem. Cytol.* 2:381–394.

Robertson, J. D. (1960). Electron microscopy of the motor end-plate and the neuromuscular spindle. *Am. J. Phys. Med.* 39:1–43.

Rodbell, M. (1995). Nobel Lecture. Signal transduction: evolution of an idea. *Biosci. Rep.* 15:117–133.

Rogers, S. W., Andrews, P. I., Gahring, L. C., Whisenand, T., Cauley, K., Crain, B., Hughes, T. E., Heinemann, S. F., and McNamara, J. O. (1994). Autoantibodies to glutamate receptor GluR3 in Rasmussen's encephalitis. *Science* 265:648–651.

Role, L. W. and Berg, D. K. (1996). Nicotinic receptors in the development and modulation of CNS synapses. *Neuron* 16:1077–1085.

Role, L. W. and McGehee, D. S. (1996). Presynaptic ionotropic receptors. *Nature* 383:670–671.

Rosenberg, M. M., Blitzblau, R. C., Olsen, D. P., and Jacob, M. H. (2002). Regulatory mechanisms that govern nicotinic synapse formation in neurons. *J. Neurobiol.* 53:542–555.

Rosenbluth, J. (1975). Synaptic membrane structure in *Torpedo* electric organ. *J. Neurocytol.* 4:697–712.

Rosenfeld, M. R., Wong, E., Dalmau, J., Manley, G., Posner, J. B., Sher, E., and Furneaux, H. M. (1993). Cloning and characterization of a Lambert-Eaton myasthenic syndrome antigen. *Ann. Neurol.* 33:113–120.

Rosenmund, C., Stern-Bach, Y., and Stevens, C. F. (1998). The tetrameric structure of a glutamate receptor channel. *Science* 280:1596–1599.

Ross, M. J., Klymkowsky, M. W., Agard, D. A., and Stroud, R. M. (1977). Structural studies of a membrane-bound acetylcholine receptor from *Torpedo californica. J. Mol. Biol.* 116:635–659.

Ross, S. A., Wong, J. Y., Clifford, J. J., Kinsella, A., Massalas, J. S., Horne, M. K., Scheffer, I. E., Kola, I., Waddington, J. L., Berkovic, S. F., and Drago, J. (2000). Phenotypic characterization of an alpha4 neuronal nicotinic acetylcholine receptor subunit knock-out mouse. *J. Neurosci.* 20:6431–6441.

Rossi, F. M., Pizzorusso, T., Porciatti, V., Marubio, L. M., Maffei, L., and Changeux, J.-P. (2001). Requirement of the nicotinic acetylcholine receptor beta2 subunit for the anatomical and functional development of the visual system. *Proc. Natl. Acad. Sci. USA* 98:6453–6458.

Rotenberg, A., Mayford, M., Hawkins, R. D., Kandel, E. R., and Muller, R. U. (1996). Mice expressing activated CaMKII lack low frequency LTP and do not form stable place cells in the CA1 region of the hippocampus. *Cell* 87:1351–1361.

Rothlin, C. V., Katz, E., Verbitsky, M., and Elgoyhen, A. B. (1999). The alpha9 nicotinic acetylcholine receptor shares pharmacological properties with type A gamma-aminobutyric acid, glycine, and type 3 serotonin receptors. *Mol. Pharmacol* 55:248–254.

Rotzler, S., Schramek, H., and Brenner, H. R. (1991). Metabolic stabilization of endplate acetylcholine receptors regulated by Ca^{2+} influx associated with muscle activity. *Nature* 349:337–339.

Rouget, C. (1862). Note sur la terminaison des nerfs moteurs dans les muscles chez les reptiles, les oiseaux et les mammifères. *C. R. Acad. Sci.* 55:548–551.

Rousselet, A., Cartaud, J., Devaux, P. F., and Changeux, J.-P. (1982). The rotational diffusion of the acetylcholine receptor in *Torpeda marmorata* membrane fragments studied with a spin-labelled alpha-toxin: importance of the 43 000 protein(s). *EMBO J.* 1:439–445.

Rowell, P. P. and Li, M. (1997). Dose-response relationship for nicotine-induced up-regulation of rat brain nicotinic receptors. *J. Neurochem.* 68:1982–1989.

Roztocil, T., Matter-Sadzinski, L., Gomez, M., Ballivet, M., and Matter, J. M. (1998). Functional properties of the neuronal nicotinic acetylcholine receptor beta3 promoter in the developing central nervous system. *J. Biol. Chem.* 273:15131–15137.

Works Cited

Rubin, G. M., Yandell, M. D., Wortman, J. R., Gabor Miklos, G. L., Nelson, C. R., Hariharan, I. K., Fortini, M. E., Li, P. W., Apweiler, R., Fleischmann, W., et al. (2000). Comparative genomics of the eukaryotes. *Science* 287:2204–2215.

Rubin, M. M. and Changeux, J.-P. (1966). On the nature of allosteric transitions: implications of non-exclusive ligand binding. *J. Mol. Biol.* 21:265–274.

Rudolph, U. and Mohler, H. (2004). Analysis of GABAA receptor function and dissection of the pharmacology of benzodiazepines and general anesthetics through mouse genetics. *Ann. Rev. Pharmacol. Toxicol.* 44:475–498.

Ruegg, M. A. and Briguet, A. (2000). The Ets transcription factor GABP is required for post-synaptic differentiation *in vivo*. *J. Neurosci.* 20:5989–5996.

Ruiz, M. and Karpen, J. W. (1999). Opening mechanism of a cyclic nucleotide-gated channel based on analysis of single channels locked in each liganded state. *J. Gen. Physiol.* 113:873–895.

Rust, G., Burgunder, J. M., Lauterburg, T. E., and Cachelin, A. B. (1994). Expression of neuronal nicotinic acetylcholine receptor subunit genes in the rat autonomic nervous system. *Eur. J. Neurosci.* 6:478–485.

Saitoh, T., Wennogle, L. P., and Changeux, J.-P. (1979). Factors regulating the susceptibility of the acetylcholine receptor protein to heat inactivation. *FEBS Lett.* 108:489–494.

Saitoh, T. and Changeux, J.-P. (1980). Phosphorylation *in vitro* of membrane fragments from *Torpedo marmorata* electric organ: effect on membrane solubilization by detergents. *Eur. J. Biochem.* 105:51–62.

Saitoh, T., Oswald, R., Wennogle, L. P., and Changeux, J.-P. (1980). Conditions for the selective labelling of the 66 000 dalton chain of the acetylcholine receptor by the covalent non-competitive blocker 5-azido-[3H]trimethisoquin. *FEBS Lett.* 116:30–36.

Saitoh, T. and Changeux, J.-P. (1981). Change in state of phosphorylation of acetylcholine receptor during maturation of the electromotor synapse in *Torpedo marmorata* electric organ. *Proc. Natl. Acad. Sci. USA* 78:4430–4434.

Sakmann, B., Patlak, J., and Neher, E. (1980). Single acetylcholine-activated channels show burst-kinetics in presence of desensitizing concentrations of agonist. *Nature* 286:71–73.

Sallette, J., Bohler, S., Benoit, P., Soudant, M., Pons, S., Le Novere, N., Changeux, J.-P., and Corringer, P.-J. (2004). An extracellular protein microdomain controls up-regulation of neuronal nicotinic acetylcholine receptors by nicotine. *J. Biol. Chem.* 279:18767–18775.

Sallette, J., Pons, S., Devillers-Thiéry, A., Soudant, M., Changeux, J.-P., and Corringer, P.-J. (2005). Nicotine enhances intracellular nicotinic receptor maturation: a novel mechanism of neural plasticity? *Neuron* 46:595–607.

Salmon, A. M. and Changeux, J.-P. (1992). Regulation of an acetylcholine receptor LacZ transgene by muscle innervation. *Neuroreport* 3:973–976.

Salmon, A. M., Damaj, I., Sekine, S., Picciotto, M. R., Marubio, L., and Changeux, J.-P. (1999). Modulation of morphine analgesia in alphaCGRP mutant mice. *Neuroreport* 10:849–854.

Salpeter, M. M. and Loring, R. H. (1985). Nicotinic acetylcholine receptors in vertebrate muscle: properties, distribution and neural control. *Prog. Neurobiol.* 25:297–325.

Sanchez, C. and Changeux, J.-P. (1966). On the properties of biosynthetic L-threonine deaminase of a mutant of *E. coli* K 12. *Bull. Soc. Chim. Biol.* 48:705–713.

Sander, A., Hesser, B. A., and Witzemann, V. (2001). MuSK induces *in vivo* acetylcholine receptor clusters in a ligand-independent manner. *J. Cell. Biol.* 155:1287–1296.

Sandrock, A. W., Jr., Dryer, S. E., Rosen, K. M., Gozani, S. N., Kramer, R., Theill, L. E., and Fischbach, G. D. (1997). Maintenance of acetylcholine receptor number by neuregulins at the neuromuscular junction *in vivo*. *Science* 276:599–603.

Sanes, J. R. and Lichtman, J. W. (1999). Development of the vertebrate neuromuscular junction. *Ann. Rev. Neurosci.* 22:389–442.

Sanes, J. R. and Lichtman, J. W. (2001). Induction, assembly, maturation and maintenance of a postsynaptic apparatus. *Nat. Rev. Neurosci.* 2:791–805.

Sansom, M. S., Adcock, C., and Smith, G. R. (1998). Modelling and simulation of ion channels: applications to the nicotinic acetylcholine receptor. *J. Struct. Biol.* 121:246–262.

Sapru, M. K., Florance, S. K., Kirk, C., and Goldman, D. (1998). Identification of a neuregulin and protein-tyrosine phosphatase response element in the nicotinic acetylcholine receptor epsilon subunit gene: regulatory role of an Rts transcription factor. *Proc. Natl. Acad. Sci. USA* 95:1289–1294.

Sargent, P. B. (1993). The diversity of neuronal nicotinic acetylcholine receptors. *Ann. Rev. Neurosci.* 16:403–443.

Sasco, A. J., Secretan, M. B., and Straif, K. (2004). Tobacco smoking and cancer: a brief review of recent epidemiological evidence. *Lung Cancer* 45 (Suppl. 2):S3–9.

Sassoon, D. A. (1993). Myogenic regulatory factors: dissecting their role and regulation during vertebrate embryogenesis. *Dev. Biol.* 156:11–23.

Sato, S., Abe, T., and Tamiya, N. (1970). Binding of iodinated erabutoxin b, a sea snake toxin, to the endplates of the mouse diaphragm. *Toxicon* 8:313–314.

Savi, P. (1844). *Études anatomiques sur le système nerveux et sur l'organe éléctrique de la Torpille*. Paris: Fortin Masson.

Schaeffer, L., Duclert, N., Huchet-Dymanus, M., and Changeux, J.-P. (1998). Implication of a multisubunit Ets-related transcription factor in synaptic expression of the nicotinic acetylcholine receptor. *EMBO J.* 17:3078–3090.

Schaeffer, L., De Kerchove d'Exaerde, A., and Changeux, J.-P. (2001). Targeting transcription to the neuromuscular synapse. *Neuron* 31:15–22.

Schiebler, W. (1977). Acetylcholine receptor enriched membranes: acetylcholine binding and excitability after reduction *in vitro*. *Mol. Cell. Biochem.* 18:151–172.

Schilstrom, B., Svensson, H. M., Svensson, T. H., and Nomikos, G. G. (1998). Nicotine and food induced dopamine release in the nucleus accumbens of the rat: putative role of alpha7 nicotinic receptors in the ventral tegmental area. *Neuroscience* 85:1005–1009.

Schilstrom, B., Fagerquist, M. V., Zhang, X., Hertel, P., Panagis, G., Nomikos, G. G., and Svensson, T. H. (2000). Putative role of presynaptic alpha7* nicotinic receptors in nicotine stimulated increases of extracellular levels of glutamate and aspartate in the ventral tegmental area. *Synapse* 38:375–383.

Schindler, H. and Quast, U. (1980). Functional acetylcholine receptor from *Torpedo marmorata* in planar membranes. *Proc. Natl. Acad. Sci. USA* 77:3052–3056.

Schindler, H., Spillecke, F., and Neumann, E. (1984). Different channel properties of *Torpedo* acetylcholine receptor monomers and dimers reconstituted in planar membranes. *Proc. Natl. Acad. Sci. USA* 81:6222–6226.

Schirmer, T. and Evans, P. R. (1990). Structural basis of the allosteric behavior of phosphofructokinase. *Nature* 343:140–145.

Schmidt, E. K., Liebermann, T., Kreiter, M., Jonczyk, A., Naumann, R., Offenhausser, A., Neumann, E., Kukol, A., Maelicke, A., and Knoll, W. (1998). Incorporation of the acetylcholine receptor dimer from *Torpedo californica* in a peptide supported lipid membrane investigated by surface plasmon and fluorescence spectroscopy. *Biosens. Bioelectron.* 13:585–591.

Schmidt, T. J. and Raftery, M. A. (1972). Use of affinity chromatography for acetylcholine receptor purification. *Biochem. Biophys. Res. Commun.* 49:572–758.

Schmidt, T. J. and Raftery, M. A. (1973). Purification of acetylcholine receptors from *Torpedo californica* electroplax by affinity chromatography. *Biochemistry* 12:852–856.

Schneider, H. J., Güttes, D., and Schneider, U. (1986). A macrobicyclic polyphenoxide as receptor analogue for choline and related ammonium compounds. *Angew. Chem. Int. Ed. Engl.* 25:647–649.

Schneider, H. J., Güttes, D., and Schneider, U. (1988). Host-guest complexes with water-soluble macrocyclic polyphenolates including induced fit and simple elements of a proton pump. *J. Am. Chem. Soc.* 110:6449–6454.

Schoffeniels, E. and Nachmansohn, D. (1957). An isolated single electroplax preparation. 1. New data on the effect of acetylcholine and related compounds. *Biochem. Biophys. Acta* 26:1–15.

Schofield, P. R. (2001). Genetics, an alternative way to discover, characterize and understand ion channels. *Clin. Exp. Pharmacol. Physiol.* 28:84–88.

Schuetze, S. M. and Role, L. W. (1987). Developmental regulation of nicotinic acetylcholine receptors. *Ann. Rev. Neurosci.* 10:403–457.

Schultz, W., Dayan, P., and Montague, P. R. (1997). A neural substrate of prediction and reward. *Science* 275:1593–1599.

Schulz, R., Sawruk, E., Mulhardt, C., Bertrand, S., Baumann, A., Phannavong, B., Betz, H., Bertrand, D., Gundelfinger, E. D., and Schmitt, B. (1998). D alpha3, a new functional alpha-subunit of nicotinic acetylcholine receptors from *Drosophila*. *J. Neurochem.* 71:853–862.

Schwarz, T. L., Umbach, J. A., Gundersen, C. B., Kidokoro, Y., and Saitoe, M. (2002). Absence of junctional glutamate receptor clusters in *Drosophila* mutants lacking spontaneous transmitter release. *Trends Neurosci.* 25:385–386.

Schwille, P., Meyer-Almes, F.-J., and Rigler, R. (1997). Dual-color fluorescence cross-correlation spectroscopy for multicomponent diffusional analysis in solution. *Biophys. J.* 72:1878–1886.

Sealock, R. (1982). Cytoplasmic surface structure in postsynaptic membranes from electric tissue visualized by tannic-acid-mediated negative contrasting. *J. Cell Biol.* 92:514–522.

Sealock, R., Wray, B. E., and Froehner, S. C. (1984). Ultrastructural localization of the Mr 43,000 protein and the acetylcholine receptor in *Torpedo* postsynaptic membranes using monoclonal antibodies. *J. Cell Biol.* 98:2239–2244.

Seguela, P., Wadiche, J., Dineley-Miller, K., Dani, J. A., and Patrick, J. W. (1993). Molecular cloning, functional properties, and distribution of rat brain alpha7: a nicotinic cation channel highly permeable to calcium. *J. Neurosci.* 13:596–604.

Shainberg, A. and Burstein, M. (1976). Decrease of acetylcholine receptor synthesis in muscle cultures by electrical stimulation. *Nature* 264:368–369.

Shallice, T., Fletcher, P., Frith, C. D., Grasby, P., Frackowiak, R. S., and Dolan, R. J. (1982). Brain regions associated with acquisition and retrieval of verbal episodic memory. *Phil. Trans. R. Soc. Lond. B Biol. Sci.* 298:199–209.

Sharples, C. G., Kaiser, S., Soliakov, L., Marks, M. J., Collins, A. C., Washburn, M., Wright, E., Spencer, J. A., Gallagher, T., Whiteaker, P., and Wonnacott, S. (2000). UB-165: a novel nicotinic agonist with subtype selectivity implicates the alpha4beta2* subtype in the modulation of dopamine release from rat striatal synaptosomes. *J. Neurosci.* 20:2783–2791.

Sheng, M. (1997). Excitatory synapses. Glutamate receptors put in their place. *Nature* 386:221–223.

Sheng, M. and Wysznyski, M. (1997). Ion channel targeting in neurons. *BioEssays* 19:847–853.

Shiang, R., Ryan, S. G., Zhu, Y. Z., Hahn, A. F., O'Connell, P., and Wasmuth, J. J. (1993). Mutations in the alpha 1 subunit of the inhibitory glycine receptor cause the dominant neurologic disorder, hyperekplexia. *Nat. Genet.* 5:351–358.

Shieh, B. H., Ballivet, M., and Schmidt, J. (1987). Quantitation of an alpha subunit splicing intermediate: evidence for transcriptional activation in the control of acetylcholine receptor expression in denervated chick skeletal muscle. *J. Cell Biol.* 104:1337–1341.

Shillito, P., Molenaar, P. C., Vincent, A., Leys, K., Zheng, W., van den Berg, R. J., Plomp, J. J., van Kempen, G. T., Chauplannaz, G., Wintzen, A. R., and et al. (1995). Acquired neuromyotonia: evidence for autoantibodies directed against K^+ channels of peripheral nerves. *Ann. Neurol.* 38:714–722.

Shulman, R. G. (2001). Spectroscopic contributions to the understanding of hemoglobin function: implications for structural biology. *IUBMB Life* 51:351–357.

Si, J., Wang, Q., and Mei, L. (1999). Essential roles of c-JUN and c-JUN N-terminal kinase (JNK) in neuregulin-increased expression of the acetylcholine receptor epsilon-subunit. *J. Neurosci.* 19:8498–8508.

Sieb, J. P., Milone, M., and Engel, A. G. (1996). Effects of the quinoline derivatives quinine, quinidine, and chloroquine on neuromuscular transmission. *Brain Res.* 712:179–189.

Siegwart, R., Krahenbuhl, K., Lambert, S., and Rudolph, U. (2003). Mutational analysis of molecular requirements for the actions of general anaesthetics at the gamma-aminobutyric acidA receptor subtype, alpha1beta2gamma2. *BMC Pharmacol.* 3:13.

Sigman, D. S. and Young, A. P. (1981). Volatile anesthetic facilitation of *in vitro* desensitization of membrane-bound acetylcholine receptor from *Torpedo californica*. *Mol. Pharmacol.* 20:498–505.

Sigman, D. S. and Young, A. P. (1983). Allosteric effects of volatile anesthetics on the membrane-bound acetylcholine receptor protein. *Biochemistry* 22:2155–2162.

Simon, A. M., Hoppe, P., and Burden, S. J. (1992). Spatial restriction of AChR gene expression to subsynaptic nuclei. *Development* 114:545–553.

Sine, S. and Taylor, P. (1979). Functional consequences of agonist-mediated state transitions in the cholinergic receptor: studies in cultured muscle cells. *J. Biol. Chem.* 254:3315–3325.

Sine, S. M. (1993). Molecular dissection of subunit interfaces in the acetylcholine receptor: identification of residues that determine curare selectivity. *Proc. Natl. Acad. Sci. USA* 90:9436–9440.

Sine, S. M., Quiram, P., Papanikolaou, F., Kreienkamp, H.-J., and Taylor, P. (1994). Conserved tyrosines in the alpha-subunit of the nicotinic acetylcholine receptor stabilize quaternary ammonium groups of agonists and curariform antagonists. *J. Biol. Chem.* 269:8808–8816.

Sine, S. M., Kreienkamp, H. J., Bren, N., Maeda, R., and Taylor, P. (1995a). Molecular dissection of subunit interfaces in the acetylcholine receptor: identification of determinants of alpha-conotoxin M1 selectivity. *Neuron* 15:205–211.

Sine, S. M., Kreienkamp, H.-J., Bren, N., Maeda, R., and Taylor, P. (1995b). Molecular dissection of subunit interfaces in the acetylcholine receptor: identification of determinants of α-conotoxin M1 selectivity. *Neuron* 15:205–211.

Sine, S. M., Ohno, K., Bouzat, C., Auerbach, A., Milone, M., Pruitt, J. N., and Engel, A. G. (1995c). Mutation of the acetylcholine receptor alpha subunit causes a slow-channel myasthenic syndrome by enhancing agonist binding affinity. *Neuron* 15:229–239.

Singer, S. J. and Doolittle, R. F. (1966). Antibody active sites and immunoglobulin molecules: recent studies give more details of the structure and function of antibodies and pathological immunoglobulins. *Science* 153:13–25.

Skou, J. C. (1965). Enzymatic basis for active transport of Na^+ and K across cell membranes. *Physiol. Rev.* 45:596–617.

Smit, A. B., Syed, N. I., Schaap, D., van Minnen, J., Klumperman, J., Kits, K. S., Lodder, H., van der Schors, R. C., van Elk, R., Sorgedrager, B., et al. (2001). A glia-derived acetylcholine-binding protein that modulates synaptic transmission. *Nature* 411:261–268.

So, K. F., Campbell, G., and Lieberman, A. R. (1990). Development of the mammalian retinogeniculate pathway: target finding, transient synapses and binocular segregation. *J. Exp. Biol.* 153:85–104.

Sobel, A., Heidmann, T., and Changeux, J.-P. (1977a). [Purification of a protein binding quinacrine and histrionicotoxin from membrane fragments rich in cholinergic receptors in *Torpedo marmorata*]. *C. R. Acad. Sci. Hebd. Séances Acad. Sci. D Sci. Nat.* 285:1255–1258.

Sobel, A., Weber, M., and Changeux, J.-P. (1977b). Large-scale purification of the acetylcholine-receptor protein in its membrane-bound and detergent-extracted forms from *Torpedo marmorata* electric organ. *Eur. J. Biochem.* 80:215–224.

Sobel, A., Heidmann, T., Hofler, J., and Changeux, J.-P. (1978). Distinct protein components from *Torpedo marmorata* membranes carry the acetylcholine receptor site and the binding site for local anesthetics and histrionicotoxin. *Proc. Natl. Acad. Sci. USA* 75:510–514.

Sobel, A., Hofler, J., Heidmann, T., and Changeux, J.-P. (1979). Structural and functional properties of the acetylcholine regulator. *Adv. Cytopharmacol.* 3:191–196.

Sobel, A., Heidmann, T., Cartaud, J., and Changeux, J.-P. (1980). Reconstitution of a functional acetylcholine receptor: polypeptide chains, ultrastructure, and binding sites for acetylcholine and local anesthetics. *Eur. J. Biochem.* 110:13–33.

St. John, P. A., Froehner, S. C., Goodenough, D. A., and Cohen, J. B. (1982). Nicotinic postsynaptic membranes from *Torpedo*: sidedness, permeability to macromolecules, and topography of major polypeptides. *J. Cell Biol.* 92:333–342.

Starace, D. M., Stefani, E., and Bezanilla, F. (1997). Voltage-dependent proton transport by the voltage sensor of the Shaker K^+ channel. *Neuron* 19:1319–1327.

Stefurak, T. L. and van der Kooy, D. (1994). Tegmental pedunculopontine lesions in rats decrease saccharin's rewarding effects but not its memory-improving effect. *Behav. Neurosci.* 108:972–980.

Steinbach, A. B. (1968). Alteration by xylocaine (lidocaine) and its derivatives of the time course of the end plate potential. *J. Gen. Physiol.* 52:144–161.

Steinlein, O. K., Mulley, J. C., Propping, P., Wallace, R. H., Phillips, H. A., Sutherland, G. R., Scheffer, I. E., and Berkovic, S. F. (1995). A missense mutation in the neuronal nicotinic receptor α4 subunit is associated with autosomal dominant nocturnal frontal lobe epilepsy. *Nat. Genet.* 11:201–203.

Steinlein, O. K. (2000). Neuronal nicotinic receptors in human epilepsy. *Eur. J. Pharmacol.* 393:243–247.

Steinlein, O.K. (2004). Genetic mechanisms that underlie epilepsy. *Nat. Rev. Neurosci.* 5:400–408.

Stoltzfus, A., Spencer, D. F., Zuker, M., Logsdon, J. M., Jr., and Doolittle, W. F. (1994). Testing the exon theory of genes: the evidence from protein structure. *Science* 265:202–207.

Strader, C. B., Revel, J. P., and Raftery, M. A. (1979). Demonstration of the transmembrane nature of the acetylcholine receptor by labeling with anti-receptor antibodies. *J. Cell Biol.* 83:499–510.

Strader, C. B. and Raftery, M. A. (1980). Topographic studies of *Torpedo* acetylcholine receptor subunits as a transmembrane complex. *Proc. Natl. Acad. Sci. USA* 77:5807–5811.

Straub, V. and Campbell, K. P. (1997). Muscular dystrophies and the dystrophin-glycoprotein complex. *Curr. Opin. Neurobiol.* 10:168–175.

Stroud, R. M. (1983). Acetylcholine receptor structure. *Neurosci. Comm.* 1:124–138.

Suarez-Isla, B. A. and Hucho, F. (1977). Acetylcholine receptor: SH group reactivity as indicator of conformational changes and functional states. *FEBS Lett.* 75:65–69.

Sugiyama, H., Benda, P., Meunier, J. C., and Changeux, J.-P. (1973). Immunological characterisation of the cholinergic receptor protein from *Electrophorus electricus*. *FEBS Lett.* 35:124–128.

Sugiyama, H. and Changeux, J.-P. (1975). Interconversion between different states of affinity for acetylcholine of the cholinergic receptor protein from *Torpedo marmorata*. *Eur. J. Biochem.* 55:505–515.

Sugiyama, N., Boyd, A. E., and Taylor, P. (1996). Anionic residue in the α-subunit of the nicotinic acetylcholine receptor contributing to subunit assembly and ligand binding. *J. Biol. Chem.* 271:26575–26581.

Sumikawa, K., Houghton, M., Emtage, J. S., Richards, B. M., and Barnard, E. A. (1981). Active multi-subunit ACh receptor assembled by translation of heterologous mRNA in *Xenopus* oocytes. *Nature* 292:862–864.

Sumikawa, K., Houghton, M., Smith, J. C., Bell, L., Richards, B. M., and Barnard, E. A. (1982). The molecular cloning and characterisation of cDNA coding for the alpha subunit of the acetylcholine receptor. *Nucl. Acids Res.* 10:5809–5822.

Sunesen, M., Huchet-Dymanus, M., Christensen, M. O., and Changeux, J.-P. (2003). Phosphorylation-elicited quaternary changes of GABP in transcriptional activation. *Mol. Cell. Biol.* 23:8008–8018.

Suppes, P. (2002). *Representation and Invariance in Scientific Structures*. Stanford, Cal.: CSLI Publications.

Sussman, J. L., Harel, M., Frolow, F., Oefner, C., Goldman, A., Toker, L., and Silman, I. (1991). Atomic structure of acetylcholinesterase from *Torpedo californica:* a prototypic acetylcholine-binding protein. *Science* 253:872–879.

Swanson, G. T., Kamboj, S. K., and Cull-Candy, S. G. (1997). Single-channel properties of recombinant AMPA receptors depend on RNA editing, splice variation, and subunit composition. *J. Neurosci.* 17:58–69.

Swope, S. L., Qu, Z., and Huganir, R. L. (1995). Phosphorylation of the nicotinic acetylcholine receptor by protein tyrosine kinases. *Ann. N. Y. Acad. Sci.* 757:197–214.

Takai, T., Noda, M., Mishina, M., Shimizu, S., Furutani, Y., Kayano, T., Ikeda, T., Kubo, T., Takahashi, H., Takahashi, T., and et al. (1985). Cloning, sequencing and expression of cDNA for a novel subunit of acetylcholine receptor from calf muscle. *Nature* 315:761–764.

Takeuchi, A. and Takeuchi, N. (1960). On the permeability of endplate membrane during the action of transmitter. *J. Physiol.* 154:52–67.

Talib, S., Okarma, T. B., and Lebkowski, J. S. (1993). Differential expression of human nicotinic acetylcholine receptor alpha subunit variants in muscle and non-muscle tissues. *Nucl. Acids Res.* 21:233–237.

Taly, A., Delarue, M., Grutter, T., Nilges, M., Le Novère, N., Corringer, P.-J., and Changeux, J.-P. (2005). Normal mode analysis suggests a quaternary twist model for the nicotinic receptor gating mechanism. *Biophys. J.* 88:3954–3965.

Tang, J., Jo, S. A., and Burden, S. J. (1994). Separate pathways for synapse-specific and electrical activity-dependent gene expression in skeletal muscle. *Development* 120:1799–1804.

Tanji, J. and Hoshi, E. (2001). Behavioral planning in the prefrontal cortex. *Curr. Opin. Neurobiol.* 11:164–170.

Tansey, M. G., Chu, G. C., and Merlie, J.-P. (1996). ARIA/HRG regulates AChR epsilon subunit gene expression at the neuromuscular synapse via activation of phosphatidylinositol 3-kinase and Ras/MAPK pathway. *J. Cell Biol.* 134:465–476.

Tarrab-Hazdai, R., Bercovici, T., Goldfarb, V., and Gitler, C. (1980). Identification of the acetylcholine receptor subunit in the lipid bilayer of *Torpedo* electric organ excitable membranes. *J. Biol. Chem.* 255:1204–1209.

Tassonyi, E., Charpantier, E., Muller, D., Dumont, L., and Bertrand, D. (2002). The role of nicotinic acetylcholine receptors in the mechanisms of anesthesia. *Brain Res. Bull.* 57:133–150.

Teichberg, V. I. and Changeux, J.-P. (1977). Evidence for protein phosphorylation and dephosphorylation in membrane fragments isolated from the electric organ of *Electrophorus electricus*. *FEBS Lett.* 74:71–76.

Teichberg, V. I., Sobel, A., and Changeux, J.-P. (1977). *In vitro* phosphorylation of the acetylcholine receptor. *Nature* 267:540–542.

Thayer, M. J., Tapscott, S. J., Davis, R. L., Wright, W. E., Lassar, A. B., and Weintraub, H. (1989). Positive autoregulation of the myogenic determination gene MyoD1. *Cell* 58:241–248.

Thesleff, S. (1955). The mode of neuromuscular block caused by actylcholine, nicotine, decamethonium and succinylcholine. *Acta Physiol. Scand.* 34:218–231.

Thom, R. (1979). Modélisation et scientificité. In P. Delattre and M. Thellier, eds. *Élaboration et justification des modèles,* 21–30. Paris: Maloine.

Thorndike, E. L. (1911). *Animal Intelligence.* New York: Macmillan.

Tibbs, G. R., Goulding, E. H., and Siegelbaum, S. A. (1997). Allosteric activation and tuning of ligand efficacy in cyclic-nucleotide-gated channels. *Nature* 386:612–615.

Tikhonov, D. B. and Zhorov, B. S. (1998). Kinked-helices model of the nicotinic acetylcholine receptor ion channel and its complexes with blockers: simulation by Monte Carlo-with-energy-minimization method. *Biophys. J.* 74:242–256.

Tilson, H. A., McLamb, R. L., Shaw, S., Rogers, B. C., Pediaditakis, P., and Cook, L. (1988). Radial-arm maze deficits produced by colchicine administered into the area of the nucleus basalis are ameliorated by cholinergic agents. *Brain Res.* 438:83–94.

Tomaselli, G. F., McLaughlin, J. T., Jurman, M., Hawrot, E., and Yellen, G. (1991). Mutations affecting agonist sensitivity of the nicotinic acetylcholine receptor. *Biophys. J.* 60:721–727.

Toyoshima, C. and Unwin, N. (1988). Ion channel of acetylcholine receptor reconstructed from images of postsynaptic membranes. *Nature* 336:247–250.

Treinin, M., Gillo, B., Liebman, L., and Chalfie, M. (1998). Two functionally dependent acetylcholine subunits are encoded in a single *Caenorhabditis elegans* operon. *Proc. Natl. Acad. Sci. USA* 95:15492–15495.

Triggle, D. J. (1980). Desensitization. *Trends Pharmacol. Sci.* 1:395–398.

Trinidad, J. C., Fischbach, G. D., and Cohen, J. B. (2000). The Agrin/MuSK signaling pathway is spatially segregated from the neuregulin/ErbB receptor signaling pathway at the neuromuscular junction. *J. Neurosci.* 20:8762–8770.

Trussell, L. O. and Fischbach, G. D. (1989). Glutamate receptor desensitization and its role in synaptic transmission. *Neuron* 3:209–218.

Tsay, H. J. and Schmidt, J. (1989). Skeletal muscle denervation activates acetylcholine receptor genes. *J. Cell Biol.* 108:1523–1526.

Tsay, H. J., Neville, C. M., and Schmidt, J. (1990). Protein synthesis is required for the denervation-triggered activation of acetylcholine receptor genes. *FEBS Lett.* 274:69–72.

Tseng, J., Kwitek-Black, A. E., Erbe, C. B., Popper, P., Jacob, H. J., and Wackym, P. A. (2001). Radiation hybrid mapping of 11 alpha and beta nicotinic acetylcholine receptor genes in *Rattus norvegicus. Brain Res. Mol. Brain. Res.* 91:169–173.

Tsigelny, I., Sugiyama, N., Sine, S. M., and Taylor, P. (1997). A model of the nicotinic receptor extracellular domain based on sequence identity and residue location. *Biophys. J.* 73:52–66.

Tsui, H. C., Cohen, J. B., and Fischbach, G. D. (1990). Variation in the ratio of acetylcholine receptors and the Mr 43,000 receptor-associated protein in embryonic chick myotubes and myoblasts. *Dev. Biol.* 140:437–446.

Turecek, R., Vlachova, V., and Vyklicky, L., Jr. (1997). Spontaneous openings of NMDA receptor channels in cultured rat hippocampal neurons. *Eur. J. Neurosci.* 9:1999–2008.

Twyman, R. E. and Macdonald, R. L. (1991). Kinetic properties of the glycine receptor main- and sub-conductance states of mouse spinal cord neurons in culture. *J. Physiol.* 435:303–331.

Tzartos, S. J. and Lindstrom, J. M. (1980). Monoclonal antibodies used to probe acetylcholine receptor structure: localization of the main immunogenic region and detection of similarities between subunits. *Proc. Natl. Acad. Sci. USA* 77:755–759.

Tzartos, S. J. and Changeux, J.-P. (1983). High affinity binding of alpha-bungarotoxin to the purified alpha-subunit and to its 27,000-dalton proteolytic peptide from *Torpedo marmorata* acetylcholine receptor: requirement for sodium dodecyl sulfate. *EMBO J.* 2:381–387.

Tzartos, S. J. and Changeux, J.-P. (1984). Lipid-dependent recovery of alpha-bungarotoxin and monoclonal antibody binding to the purified alpha-subunit from *Torpedo marmorata* acetylcholine receptor: enhancement by noncompetitive channel blockers. *J. Biol. Chem.* 259:11512–11519.

Umbarger, H. E. (1956). Evidence for a negative feedback mechanism in the biosynthesis of isoleucine. *Science* 123:848.

Unwin, N., Toyoshima, C., and Kubalek, E. (1988). Arrangement of the acetylcholine receptor subunits in the resting and desensitized states, determined by cryoelectron microscopy of crystallized *Torpedo* postsynaptic membranes. *J. Cell Biol.* 107:1123–1138.

Unwin, N. (1993). The nicotinic acetylcholine receptor at 9 Å resolution. *J. Mol. Biol.* 229:1101–1124.

Unwin, N. (1995). Acetylcholine receptor channel imaged in the open state. *Nature* 373: 37–43.

Unwin, N. (2000). The Croonian Lecture 2000. Nicotinic acetylcholine receptor and the structural basis of fast synaptic transmission. *Philos. Trans. R. Soc. Lond. B Biol. Sci.* 355:1813–1829.

Unwin, N., Miyazawa, A., Li, J., and Fujiyoshi, Y. (2002). Activation of the nicotinic acetylcholine receptor involves a switch in conformation of the alpha subunits. *J. Mol. Biol.* 319:1165–1176.

Valera, S., Ballivet, M., and Bertrand, D. (1992). Progesterone modulates a neuronal nicotinic acetylcholine receptor. *Proc. Natl. Acad. Sci. USA* 89:9949–9953.

Valor, L. M., Campos-Caro, A., Carrasco-Serrano, C., Ortiz, J. A., Ballesta, J. J., and Criado, M. (2002). Transcription factors NF-Y and Sp1 are important determinants of the promoter activity of the bovine and human neuronal nicotinic receptor beta4 subunit genes. *J. Biol. Chem.* 277:8866–8876.

van Hooft, J. A., Spier, A. D., Yakel, J. L., Lummis, S. C., and Vijverberg, H. P. (1998). Promiscuous coassembly of serotonin 5-HT3 and nicotinic alpha4 receptor subunits into Ca^{2+}-permeable ion channels. *Proc. Natl. Acad. Sci. USA* 95:11456–11461.

Vandlen, R. L., Wu, W. C., Eisenach, J. C., and Raftery, M. A. (1979). Studies of the composition of purified *Torpedo californica* acetylcholine receptor and of its subunits. *Biochemistry* 18:1845–1854.

Vannier, C. and Triller, A. (1997). Biology of the postsynaptic glycine receptor. *Int. Rev. Cytol.* 176:201–244.

Venter, J. C., Adams, M. D., Myers, E. W., Li, P. W., Mural, R. J., Sutton, G. G., Smith, H. O., Yandell, M., Evans, C. A., Holt, R. A., et al. (2001). The sequence of the human genome. *Science* 291:1304–1351.

Vetter, D. E., Liberman, M. C., Mann, J., Barhanin, J., Boulter, J., Brown, M. C., Saffiote-Kolman, J., Heinemann, S. F., and Elgoyhen, A. B. (1999). Role of alpha9 nicotinic ACh receptor subunits in the development and function of cochlear efferent innervation. *Neuron* 23:93–103.

Vidal, C. and Changeux, J.-P. (1989). Pharmacological profile of nicotinic acetylcholine receptors in the rat prefrontal cortex: an electrophysiological study in a slice preparation. *Neuroscience* 29:261–270.

Vidal, C. and Changeux, J.-P. (1993). Nicotinic and muscarinic modulations of excitatory synaptic transmission in the rat prefrontal cortex *in vitro*. *Neuroscience* 56:23–32.

Villarroel, A., Herlitze, S., Koenen, M., and Sakmann, B. (1991). Location of a threonine residue in the alpha-subunit M2 transmembrane segment that determines the ion flow through the acetylcholine receptor channel. *Proc. R. Soc. Lond. B Biol. Sci.* 243:69–74.

Villarroel, A. and Sakmann, B. (1992). Threonine in the selectivity filter of the acetylcholine receptor channel. *Biophys. J.* 62:196–205.

Vincent, A., Lang, B., and Newsom-Davis, J. (1989). Autoimmunity to the voltage-gated calcium channel underlies the Lambert-Eaton myasthenic syndrome, a paraneoplastic disorder. *Trends Neurosci.* 12:496–502.

von Euler, U. S. (1981). Historical Perspective: Growth and Impact of the Concept of Chemical Neurotransmission. In L. Stjärne, P. Hedqvist, H. Lagercrantz, and A. Wennmalm, eds. *Chemical Neurotransmission: 75 Years*, 3–12. London: Academic Press.

Wada, E., Wada, K., Boulter, J., Deneris, E., Heinemann, S., Patrick, J., and Swanson, L. W. (1989). Distribution of alpha2, alpha3, alpha4, and beta2 neuronal nicotinic receptor subunit mRNAs in the central nervous system: a hybridization histochemical study in the rat. *J. Comp. Neurol.* 284:314–335.

Waksman, G., Changeux, J.-P., and Roques, B. P. (1980a). Structural requirements for agonist and noncompetitive blocking action of acylcholine derivatives on *Electrophorus electricus* electroplaque. *Mol. Pharmacol.* 18:20–27.

Waksman, G., Oswald, R., Changeux, J.-P., and Roques, B. P. (1980b). Synthesis and pharmacological activity on *Electrophorus electricus* electroplaque of photoaffinity labelling derivatives of the non-competitive blockers di- and tri-methisoquin. *FEBS Lett.* 111:23–28.

Walker, J. W., Takeyasu, K., and McNamee, M. G. (1982). Activation and inactivation kinetics of *Torpedo californica* acetylcholine receptor in reconstituted membranes. *Biochemistry* 21:5384–5389.

Walker, J. W., Richardson, C. A., and McNamee, M. G. (1984). Effects of thio-group modifications of *Torpedo californica* acetylcholine receptor on ion flux activation and inactivation kinetics. *Biochemistry* 23:2329–2338.

Wallace, R. H., Marini, C., Petrou, S., Harkin, L. A., Bowser, D. N., Panchal, R. G., Williams, D. A., Sutherland, G. R., Mulley, J. C., Scheffer, I. E., and Berkovic, S. F. (2001). Mutant GABA(A) receptor gamma2-subunit in childhood absence epilepsy and febrile seizures. *Nat. Genet.* 28:49–52.

Wang, C. T., Zhang, H. G., Rocheleau, T. A., ffrench-Constant, R. H., and Jackson, M. B. (1999). Cation permeability and cation-anion interactions in a mutant GABA-gated chloride channel from *Drosophila. Biophys. J.* 77:691–700.

Wang, D., Chiara, D. C., Xie, Y., and Cohen, J. B. (2000). Probing the structure of the nicotinic acetylcholine receptor with 4-benzoylbenzoylcholine, a novel photoaffinity competitive antagonist. *J. Biol. Chem.* 275:28666–28674.

Wang, F. and Imoto, K. (1992). Pore size and negative charge as structural determinants of permeability in the *Torpedo* nicotinic acetylcholine receptor channel. *Proc. R Soc. Lond. B Biol. Sci.* 250:11–17.

Wang, F., Gerzanich, V., Wells, G. B., Anand, R., Peng, X., Keyser, K., and Lindstrom, J. (1996). Assembly of human neuronal nicotinic receptor alpha5 subunits with alpha3, beta2, and beta4 subunits. *J. Biol. Chem.* 271:17656–17665.

Wang, F., Nelson, M. E., Kuryatov, A., Olale, F., Cooper, J., Keyser, K., and Lindstrom, J. (1998). Chronic nicotine treatment up-regulates human alpha3 beta2 but not alpha3 beta4 acetylcholine receptors stably transfected in human embryonic kidney cells. *J. Biol. Chem.* 273:28721–28732.

Wang, H. L., Auerbach, A., Bren, N., Ohno, K., Engel, A. G., and Sine, S. M. (1997). Mutation in the M1 domain of the acetylcholine receptor alpha subunit decreases the rate of agonist dissociation. *J. Gen. Physiol.* 109:757–766.

Wang, H. L., Milone, M., Ohno, K., Shen, X. M., Tsujino, A., Batocchi, A. P., Tonali, P., Brengman, J., Engel, A. G., and Sine, S. M. (1999). Acetylcholine receptor M3 domain: stereochemical and volume contributions to channel gating. *Nat. Neurosci.* 2:226–233.

Wang, N., Orr-Urtreger, A., Chapman, J., Rabinowitz, R., Nachman, R., and Korczyn, A. D. (2002). Autonomic function in mice lacking alpha5 neuronal nicotinic acetylcholine receptor subunit. *J. Physiol.* 542:347-354.

Watanabe, H., Zoli, M., and Changeux, J.-P. (1998). Promoter analysis of the neuronal nicotinic acetylcholine receptor alpha4 gene: methylation and expression of the transgene. *Eur. J. Neurosci.* 10:2244–2253.

Watters, D. and Maelicke, A. (1983). Organization of ligand binding sites at the acetylcholine receptor: a study with monoclonal antibodies. *Biochemistry* 22:1811–1819.

Weber, I. T., Johnson, L. N., Wilson, K. S., Yeates, D. G., Wild, D. L., and Jenkins, J. A. (1978). Crystallographic studies on the activity of glycogen phosphorylase b. *Nature* 274:433–437.

Weber, M., Menez, A., Fromageot, P., Boquet, P., and Changeux, J.-P. (1972). [Effect of cholinergic agents and local anesthetics on the kinetics of binding of tritiated toxin of *Naja nigricollis* to the cholinergic receptor]. *C. R. Acad. Sci. Hebd. Séances Acad. Sci. D Sci. Nat.* 274:1575–1578.

Weber, M. and Changeux, J.-P. (1974a). Binding of *Naja nigricollis* (3H)alpha-toxin to membrane fragments from *Electrophorus* and *Torpedo* electric organs. I. Binding of the tritiated alpha-neurotoxin in the absence of effector. *Mol. Pharmacol.* 10:1–14.

Weber, M. and Changeux, J.-P. (1974b). Binding of *Naja nigricollis* (3H)alpha-toxin to membrane fragments from *Electrophorus* and *Torpedo* electric organs. III. Effects of local anaesthetics on the binding of the tritiated alpha-neurotoxin. *Mol. Pharmacol.* 10:35–40.

Weber, M. and Changeux, J.-P. (1974c). Binding of *Naja nigricollis* (3H)alpha-toxin to membrane fragments from *Electrophorus* and *Torpedo* electric organs. II. Effect of cholinergic agonists and antagonists on the binding of the tritiated alpha-neurotoxin. *Mol. Pharmacol.* 10:15–34.

Weber, M., David-Pfeuty, T., and Changeux, J.-P. (1975). Regulation of binding properties of the nicotinic receptor protein by cholinergic ligands in membrane fragments from *Torpedo marmorata. Proc. Natl. Acad. Sci. USA* 72:3443–3447.

Weiland, G., Georgia, B., Wee, V. T., Chignell, C. F., and Taylor, P. (1976). Ligand interactions with cholinergic receptor-enriched membranes from *Torpedo:* influence of agonist exposure on receptor properties. *Mol. Pharmacol.* 12:1091–1105.

Weiland, G., Georgia, B., Lappi, S., Chignell, C. F., and Taylor, P. (1977). Kinetics of agonist-mediated transitions in state of the cholinergic receptor. *J. Biol. Chem.* 252:7648–7656.

Weiland, G., Frisman, D., and Taylor, P. (1979). Affinity labeling of the subunits of the membrane associated cholinergic receptor. *Mol. Pharmacol.* 15:213–226.

Weiland, S., Witzemann, V., Villarroel, A., Propping, P., and Steinlein, O. K. (1996). An amino acid exchange in the second transmembrane segment of a neuronal nicotinic receptor causes partial epilepsy by altering its desensitization kinetics. *FEBS Lett.* 398:91–96.

Weill, C. L., McNamee, M. G., and Karlin, A. (1974). Affinity-labeling of purified acetylcholine receptor from *Torpedo californica*. *Biochem. Biophys. Res. Commun.* 61:997–1003.

Wennogle, L. P. and Changeux, J.-P. (1980). Transmembrane orientation of proteins present in acetylcholine receptor-rich membranes from *Torpedo marmorata* studied by selective proteolysis. *Eur. J. Biochem.* 106:381–393.

Wennogle, L. P., Oswald, R., Saitoh, T., and Changeux, J.-P. (1981). Dissection of the 66,000-dalton subunit of the acetylcholine receptor. *Biochemistry* 20:2492–2497.

White, B. H. and Cohen, J. B. (1992). Agonist-induced changes in the structure of the acetylcholine receptor M2 regions revealed by photoincorporation of an uncharged nicotinic noncompetitive antagonist. *J. Biol. Chem.* 267:15770–15783.

Whiting, P., Vincent, A., and Newsom-Davis, J. (1985). Monoclonal antibodies to *Torpedo* acetylcholine receptor: characterisation of antigenic determinants within the cholinergic binding site. *Eur. J. Biochem.* 150:533–539.

Wilson, G. G. and Karlin, A. (1998). The location of the gate in the acetylcholine receptor channel. *Neuron* 20:1269–1281.

Wilson, P. T., Gershoni, J. M., Hawrot, E., and Lentz, T. L. (1984). Binding of alpha-bungarotoxin to proteolytic fragments of the alpha subunit of *Torpedo* acetylcholine receptor analyzed by protein transfer on positively charged membrane filters. *Proc. Natl. Acad. Sci. USA* 81:2553–2557.

Wilson, P. T., Lentz, T. L., and Hawrot, E. (1985). Determination of the primary amino acid sequence specifying the alpha-bungarotoxin binding site on the alpha subunit of the acetylcholine receptor from *Torpedo californica*. *Proc. Natl. Acad. Sci. USA* 82:8790–8794.

Wise, D. S., Karlin, A., and Schoenborn, B. P. (1979). An analysis by low-angle neutron scattering of the structure of the acetylcholine receptor from *Torpedo californica* in detergent solution. *Biophys. J.* 28:473–496.

Wise, D. S., Schoenborn, B. P., and Karlin, A. (1981). Structure of acetylcholine receptor dimer determined by neutron scattering and electron microscopy. *J. Biol. Chem.* 256:4124–4126.

Witzemann, V. and Raftery, M. (1978). Specific molecular aggregates of *Torpedo californica* acetylcholine receptor. *Biochem. Biophys. Res. Commun.* 81:1025–1031.

Witzemann, V., Barg, B., Nishikawa, Y., Sakmann, B., and Numa, S. (1987). Differential regulation of muscle acetylcholine receptor gamma- and epsilon-subunit mRNAs. *FEBS Lett.* 223:104–112.

Witzemann, V., Barg, B., Criado, M., Stein, E., and Sakmann, B. (1989). Developmental regulation of five subunit specific mRNAs encoding acetylcholine receptor subtypes in rat muscle. *FEBS Lett.* 242:419–424.

Witzemann, V. and Sakmann, B. (1991). Differential regulation of MyoD and myogenin mRNA levels by nerve induced muscle activity. *FEBS Lett.* 282:259–264.

Wonnacott, S., Kaiser, S., Mogg, A., Soliakov, L., and Jones, I. W. (2000). Presynaptic nicotinic receptors modulating dopamine release in the rat striatum. *Eur. J. Pharmacol.* 393:51–58.

Woolf, N. J. (1991). Cholinergic systems in mammalian brain and spinal cord. *Prog. Neurobiol.* 37:475–524.

Wu, W. C. S. and Raftery, M. (1979). Carbamycholine induced rapid cation efflux from reconstituted membrane vesicles containing purified acetylcholine receptor. *Biochem. Biophys. Res. Commun.* 89:26–35.

Wu, W. C. S. and Raftery, M. A. (1981a). Functional properties of acetylcholine receptor monomeric and dimeric forms in reconstituted membranes. *Biochem. Biophys. Res. Commun.* 99:436–444.

Wu, W. C. S. and Raftery, M. A. (1981b). Reconstitution of acetylcholine receptor function, using purified receptor protein. *Biochemistry* 20:694–701.

Wyllie, D. J., Traynelis, S. F., and Cull-Candy, S. G. (1993). Evidence for more than one type of non-NMDA receptor in outside-out patches from cerebellar granule cells of the rat. *J. Physiol.* 463:193–226.

Wyman, J. (1948). Heme proteins. *Adv. Prot. Chem.* 4:407–531.

Xu, W., Gelber, S., Orr-Urtreger, A., Armstrong, D., Lewis, R. A., Ou, C. N., Patrick, J., Role, L., De Biasi, M., and Beaudet, A. L. (1999a). Megacystis, mydriasis, and ion channel defect in mice lacking the alpha3 neuronal nicotinic acetylcholine receptor. *Proc. Natl. Acad. Sci. USA* 96:5746–5751.

Xu, W., Orr-Urtreger, A., Nigro, F., Gelber, S., Sutcliffe, C. B., Armstrong, D., Patrick, J. W., Role, L. W., Beaudet, A. L., and De Biasi, M. (1999b). Multiorgan autonomic dysfunction in mice lacking the beta2 and the beta4 subunits of neuronal nicotinic acetylcholine receptors. *J. Neurosci.* 19:9298–9305.

Yakel, J. L., Lagrutta, A., Adelman, J. P., and North, R. A. (1993). Single amino acid substitution affects desensitization of the 5-hydroxytryptamine type 3 receptor expressed in *Xenopus* oocytes. *Proc. Natl. Acad. Sci. USA* 90:5030–5033.

Yang, X., McDonough, J., Fyodorov, D., Morris, M., Wang, F., and Deneris, E. S. (1994). Characterization of an acetylcholine receptor alpha 3 gene promoter and its activation by the POU domain factor SCIP/Tst-1. *J. Biol. Chem.* 269:10252–10264.

Yang, X., Arber, S., William, C., Li, L., Tanabe, Y., Jessell, T. M., Birchmeier, C., and Burden, S. J. (2001). Patterning of muscle acetylcholine receptor gene expression in the absence of motor innervation. *Neuron* 30:399–410.

Yang, X. Y., Schmidt, T., Chinenov, Y., Wang, R., Martin, M. E., and Yang, X. (1997). Elements between the protein-coding regions of the adjacent beta4 and alpha3 acetylcholine receptor genes direct neuron-specific expression in the central nervous system. *J. Biol. Chem.* 272:29060–29067.

Yang, Y. R. and Schachman, H. K. (1987). Hybridization as a technique for studying interchain interactions in the catalytic trimers of aspartate transcarbamoylase. *Anal. Biochem.* 163:188–195.

Yates, R. A. and Pardee, A. B. (1956). Control of pyrimidine biosynthesis in *Escherichia coli* by a feedback mechanism. *J. Biol. Chem.* 221:757–770.

Yeramian, E., Trautmann, A., and Claverie, P. (1986). Acetylcholine receptors are not functionally independent. *Biophys. J.* 50:253–263.

Yu, Y., Shu, L., and Karlin, A. (2003). Structural effects of quinacrine binding in the open channel of the acetylcholine receptor. *Proc. Natl. Acad. Sci. USA* 100:3907–3912.

Zachariou, V., Caldarone, B. J., Weathers-Lowin, A., George, T. P., Elsworth, J. D., Roth, R. H., Changeux, J.-P., and Picciotto, M. R. (2001). Nicotine receptor inactivation decreases sensitivity to cocaine. *Neuropsychopharmacology* 24:576–589.

Zagotta, W. N. and Siegelbaum, S. A. (1996). Structure and function of cyclic nucleotide-gated channels. *Ann. Rev. Neurosci.* 19:235–263.

Zhong, W., Gallivan, J. P., Zhang, Y., Li, L., Lester, H. A., and Dougherty, D. A. (1998). From *ab initio* quantum mechanics to molecular neurobiology: a cation-π binding site in the nicotinic receptor. *Proc. Natl. Acad. Sci. USA* 95:12088–12093.

Zhou, M., Engel, A. G., and Auerbach, A. (1999). Serum choline activates mutant acetylcholine receptors that cause slow channel congenital myasthenic syndromes. *Proc. Natl. Acad. Sci. USA* 96:10466–10471.

Zhou, Y., Morais-Cabral, J. H., Kaufman, A., and MacKinnon, R. (2001). Chemistry of ion coordination and hydration revealed by a K^+ channel-Fab complex at 2.0 Å resolution. *Nature* 414:43–48.

Ziebell, M. R., Nirthanan, S., Husain, S. S., Miller, K. W., and Cohen, J. B. (2004). Identification of binding sites in the nicotinic acetylcholine receptor for [3H]azietomidate, a photoactivatable general anesthetic. *J. Biol. Chem.* 279:17640–17649.

Zingsheim, H. P., Neugebauer, D. C., Barrantes, F. J., and Frank, J. (1980). Structural details of membrane-bound acetylcholine receptor from *Torpedo marmorata*. *Proc. Natl. Acad. Sci. USA* 77:952–956.

Zingsheim, H. P., Neugebauer, D. C., Frank, J., Hanicke, W., and Barrantes, F. J. (1982). Dimeric arrangement and structure of the membrane-bound acetylcholine receptor studied by electron microscopy. *EMBO J.* 1:541–547.

Zoli, M., Le, N. N., Hill, J. A. J., and Changeux, J.-P. (1995). Developmental regulation of nicotinic ACh receptor subunit mRNAs in the rat central and peripheral nervous systems. *J. Neurosci.* 15:1912–1939.

Zoli, M., Léna, C., Picciotto, M. R., and Changeux, J.-P. (1998). Identification of four classes of brain nicotinic receptors using beta2 mutant mice. *J. Neurosci.* 18:4461–4472.

Zoli, M., Picciotto, M. R., Ferrari, R., Cocchi, D., and Changeux, J.-P. (1999). Increased neurodegeneration during ageing in mice lacking high-affinity nicotine receptors. *EMBO J.* 18:1235–1244.

Zupancic, A. O. (1952). The modification of acetylcholine. *Acta Physiol. Scand.* 29:63–71.

INDEX

acetylcholine, 52, 58–59, 84, 154; action of, 8, 142, 177–78, 191; [^{3}H], 86; response of microsacs to, 23–25
acetylcholine-binding pocket. See binding pocket
acetylcholine-binding protein. See AChBP (acetylcholine-binding protein)
acetylcholine-binding site, 74. See binding site
acetylcholine mustard, 55
acetylcholine receptor: biosynthesis during endplate formation, 144–46; conformational states, 85–88; in postsynaptic membrane, 142–44. See also nicotinic receptor
acetylcholine receptor-inducing activity. See ARIA
acetylcholinesterase, 8, 29, 117, 127, 143, 164
acetylcholine synthesis, 127
AChBP (acetylcholine-binding protein), 54, 108; crystal structure of, 115–19, 130; "desensitized" state of, 122; molluscan, 108, 115–19, 121, 124, 126
activation, 201
activation, of ion channel, 83–85, 90; modeling, 89–95
activity, effect on receptor stability, 174
addiction, role of nicotinic receptor in, 196–200
ADNFLE. See under epilepsy
affinity, preferential order of, for receptors, 61
affinity chromatography, 30
aging, role of nicotinic receptor in, 193–96
agonist, use of term, 3
agrin, 155, 157, 164, 169, 173, 175
allosteric, use of term, 13
"allosteric diseases," 128–31, 139
allosteric effectors, and modulation of conformational transitions, 104–6
allosteric enzyme model, 83
allosteric enzymes, 122. See also names
allosteric interactions, 12–19, 82, 138

allosteric mechanism, 83
allosteric membrane proteins, 211
allosteric models, 103; extended, 101–2; generalized, 91; for slow-channel myasthenic mutant, 131–34
allosteric protein theory, 64–65
allosteric schema, 100–101; experimental evidence for, 95–100
allosteric site, 64; secondary, 66
allosteric transition, 124, 201; structural basis of, 121–23
α-bungarotoxin, 28, 173–74, 181, 193
α-cobratoxin, 123
α-conotoxin, 182
α-dystrobrevin, 165
α-helical structures, in transmembrane domain, 115
α-toxin, from snake venom, 22, 28–29, 31, 52, 86, 112, 142
Alzheimer's disease, 63, 192, 195–96, 208–9, 211
amino acids, 65, 72–74, 77, 81, 122; aromatic, 57–59; of binding site, 52–53, 55–57, 61–62; involved in ion channel conductance, 74–75. See also names
analgesia, nicotinic, 200–201
anionic rings, 75–76, 81, 129
anions, carboxylate, 57
ankyrin, 165
antagonist, use of term, 3
antibody induction mechanism, 126–27
Aplysia, 163
APV (2-amino-5-phosphonopentanoic acid), 187
ARIA (acetylcholine receptor-inducing activity), 155–57
arousal response, 203
asolectin, 34
aspartates, 56, 59, 191
aspartate transcarbamylase, 13–15, 18, 122
aspartic acid residues, 56, 57, 62
attention deficit hyperactivity disorders, 192

277

Index

Augustinsson, K. B., 8
autism, 208–9, 211
autocatalytic switching, 148
autosomal dominant nocturnal frontal lobe epilepsy (ADNFLE). *See under* epilepsy
axonal compartment, "preterminal," 189–90
axons, 189–90
axon terminals, nicotinic receptors in, 187–89

Bacon, Francis, 20
Baldwin, J., 18
Barnard, E. A., 74
basal lamina, 173
BCNI (bis(choline)-N-(4-nitrobenzo-2-oxa-1,3-diazol-7-yl)-iminodipropionate), 80
benzodiazepines, 106, 136
Bernard, Claude, 2–3
Bernstein, Julius, 9
β-dystroglycan, 170, 172
binding, 13–19, 118; channel-blocker, 70–72; chemical interactions responsible for, 57–59; and flux dynamics, 85–88; nonexclusive, 95–96
binding pocket, 54–57, 117–19
binding site, 83; amino acids belonging to, 52–53; at boundary between subunits, 53–54; distance to ion channel, 81–82; location of, 61, 119–21, 124, 180–82; loops of, 54–57; nonequivalence of, 102–3
biosynthesis, of nicotinic acetylcholine receptor, 144–46, 155, 174
"blooming," 122
Blount, P., 53
botulinum toxin, 173
brain: distribution of cholinergic neurons in, 177–78; distribution of nicotinic receptor oligomers in, 179–82
brain development, role of nicotinic receptor in, 193–96
brain function, and nicotinic receptor, 134–35, 192–93, 208–11
bridging concepts, 21
bungarotoxin, 204
Bungarus (snakes), 22; *B. multicinctus*, 27

Ca^{2+} ion, 104, 161, 175
Caenorhabditis, 48
calcitonin gene-related peptide (CGRP), 140, 155
calf receptor protein, 74
CANNTG sites, 159
carbamylcholine, 69–70, 72
CAT (chloramphenicol acetyltransferase), 150–51
cations, of acetylcholine, 57
caveolin, 172

CGRP (calcitonin gene-related peptide), 140, 155
Changeux, J.-P., 13, 30, 53, 67, 90, 115, 162
channel blocker binding, and functional state of receptor, 70–72
channel blockers, 65–69, 81
channel gate, 78–79, 81
channelopathies, 137–38
chemical interactions responsible for binding, 57–59
chemical transmission, 2–6
chickens, research use of, 148–49, 160–61, 169, 183–84
chimera experiments, 77, 202
chlorpromazine, 66–67, 74, 96, 129; [^{3}H], 70–72, 81
choline acetyltransferase, 127
cholinergic ligands, 52
Chothia, C., 18
chromosome localization of subunit genes, 47–48
clustering: of acetylcholine receptors, 154–55, 176; of nicotinic receptor genes, 185–86; of nicotinic receptors, 168–69, 173; of receptor proteins, 168
CNQX, 187
"coalitions," 21
cognition, nicotinic receptor and, 204–8
cognitive nodes, 21
Cold Spring Harbor Symposium 1961, 13
collagen, 164
combination of subunits, 50. *See also* subunits, of nicotinic receptor
communication: intercellular, 177, 211; intracellular, 140–42
compartmentalization: of gene expression, 146; of nicotinic receptor oligomers in brain, 186–91
compartmentalized transcription, 146–50, 162
concerted model. *See* MWC model (concerted model)
conductance, ion channel, 74–76, 84, 111
conductance states, 107; multiple, 100–102
conformational states, of acetylcholine receptor, 89, 91, 93, 99; high-affinity, 88; hybrid, 100–102; intermediate-affinity, 87–88; low-affinity, 87; multiple, 100–102, 107
conformational transitions, 124; modulation by multiple allosteric effectors, 104–6
congenital myasthenia, 137, 153, 160, 175
congenital myasthenic syndromes, 98, 100, 125–28, 137, 153, 160, 169, 175
consciousness, and nicotinic receptor, 202–4
Conus (marine snail), 182
cooperativity, 96
coupling, of binding entity and ion channel, 64–65

Couteaux, R., 141
Cox, D. H., 101
"crown," extracellular, 122
cryoelectron microscopy, 40
crystallographic analysis, 57, 78, 119–21
crystal structure of AChBP, 115–19, 130; compared to electron microscopy, 119–21
cupredoxin superfamily, 113
curare, 2–3, 65
cyanogen bromide, 72
cyclothiazides, 106
cysteine, 52–53, 55
cysteine scanning, 121
cytoplasmic boundary, 109
cytoplasmic domain, 106, 111–12, 175
cytoskeletal proteins, 165

Dale, Sir Henry, 4, 6, 22
dansyl-C_6 choline, 60, 67–69, 80, 86–88
dantrolene, 161
DDF (p-N, N-dimethylamino benzene diazonium fluoroborate), 27, 53–55, 122; [^{3}H], 57
decamethonium, 29, 52
Del Castillo, J., 7–8, 111
delipidation, 34–35
dementia, 196
denervation, 143, 148–49, 171, 173
Descartes, René, 5, 20
desensitization, 11–13, 19, 83–85, 88–89, 99–100, 201; modeling of, 89–95; recovery from, 99–100
detergents, effect on acetylcholine binding, 85
detergents, nondenaturing, 28, 35
DHβE (dihydro-beta-erythroidine), 97–98, 204
diseases, directly associated with nicotinic receptor, 125–26. See also congenital myasthenic syndromes; epilepsy
distance, between binding site and ion channel, 79–83
distributed chimera, 77
diversity: of receptor sites, 60–63; of subunit genes, 50
dopaminergic system, 197–200
Doyère, M., 140
Drosophila (fruit fly), 44, 47, 77, 154
drug action, early investigations of, 2–3
drugs of abuse, 197
d-tubocurarine, 27, 52, 55–56, 59, 193; [^{3}H], 53, 60
Du Bois-Reymond, Emil, 5
dynamics, of binding and flux, 85–88
dystrobrevin, 165, 172
dystroglycans, 165, 173
dystrophin, 165, 172
dystrophin-glycoprotein complex, 173, 175–76

E box, 159–62, 169, 175
Edelstein, S. J., 94
Ehrlich, Paul, 3, 7
Eigen, M., 101
electrical activity, as regulatory mechanism, 159–62
electric organ, of fish, 30; as source of receptors, 22–25. See also Torpedo (fish)
electric potentials, 8–12
electron microscope, 5
electron microscopy, 39, 76, 79, 108, 112, 124, 140, 142–43, 167; compared to X-ray crystallography, 119–21; and structure of receptor protein, 109–11
Electrophorus (fish), 23, 54–55, 85, 89; E. electricus, 23, 25, 28, 30, 65, 67, 70, 96, 109, 126, 142–43
electrophysiology, 74, 84, 187, 198, 203
electroplaque, of fish electric organ, 23, 27–28
Elliott, Thomas R., 5
endplate formation, and acetylcholine receptor biosynthesis, 144–46
enzyme activity, 6–8. See also regulatory enzymes
enzyme catalysis, analogy of, 84
enzymology, 6–8
epibatidine, 182
epilepsy, 134–37; autosomal dominant nocturnal frontal lobe (ADNFLE), 125, 134–35, 192, 204; frontal lobe, 100, 211
ErB, 165
Escherichia coli, 160
ethanol, as allosteric effector, 105
ethidium bromide, 80
exons, in subunit genes, 46–47
experiment, role of, 20–21
experimental evidence, for allosteric schema, 95–100, 107
expression patterns, of nicotinic receptor subunits in brain, 182–83
extraction, of nicotinic receptor, 28–30

fast channel block, 66
feedback inhibition, 13
Feldberg, W., 9
Fessard, A., 9
Filmer, D., 90
Fischbach, G. D., 155
Fischer, Emil, 6–7
flaxedil, 8
flotation gradient centrifugation, 172
flotilin, 172
fluorescence spectroscopy, 52, 60, 79, 103, 122
fluoxetine, 137
43K-rapsyn, 64, 106, 109, 127–28, 165–73, 175–76
four-state model, 91
frog, South American, 66, 182

Index

GABA, 191
GABA currents, 188
GABP factors, 153–54, 158, 175
Galvani, Luigi and Lucia, 8
Galzi, J.-L., 78
gating, 34–35
gating mechanism, 121
gene inactivation studies, 209
general anesthetics, as allosteric effectors, 105
gephyrin, 106, 176
Giraudat, J., 75–76, 113
glutamate, 56, 191
glutamate receptor family, 136
glyceraldehyde-3-phosphate dehydrogenase, 18
glycine receptor, 100–101, 111, 136–37
glycogen phosphorylase, 14
Golgi apparatus, 171–72, 175
Göpfert, H., 9

Heidmann, T., 70, 73, 75, 80
HEK cells, 131
hemoglobin, 12–19, 82, 89–91, 102, 107, 122, 138
hemoglobinopathies, 125
Henry, E. R., 19
HEPES (4-(2-hydroxyethyl)-1-piperazineethanesulfonic acid), 117
heregulin, 175
Herz, J. M., 80
high-affinity binding, and receptor desensitization, 88
high-affinity sites, for channel blockers, 69–74, 80–81
high-performance liquid chromatography (HPLC), 72
hippocampus, vertebrate, 163
Hippocratic school, 2
histidines, 168
histrionicotoxin, 66, 69–70
HLH (helix-loop-helix) family, 159–60
Hodgkin, A. L., 9
HPLC (high-performance liquid chromatography), 72
5-HT$_3$ receptor, 44
Hucho, F., 73, 75–76
Hull, Clark, 196
human genome sequence, 44
Huxley, A. F., 9
hybridization, *in situ*, 148–50, 182
hydrophobic domain, of nicotinic receptor, 80
hydrophobicity, of receptor molecule, 40–41
hyperekplexia (startle disease), 98, 136
hyperkalemic periodic paralysis, 137

ibotenate, 194
identification, of nicotinic receptor, 31–33
immunoglobulin superfamily, 124

immunological experiments, 32–33
induced conformational change model, of allosteric interaction, 16–19
in situ hybridization, 148–50, 182
instructive schema, 90
integrins, 165
intercellular communication, 140–42, 177, 211
intermediate states, 101–2
intracellular signaling pathways, 157–59
introns, in subunit genes, 46–47
ion channel, 8–12, 64–65, 71, 73, 76, 83, 113; distance to acetylcholine-binding site, 81–82; identification of, 66, 81; ion selectivity of, 77–79; M2 segment as component of, 72–74; stereochemical model of, 76–79; structure of, 76–79; voltage-gated, 64, 91, 127
ion gating, 34–35
ionophore, 64
isolation, of nicotinic receptor, 22–28
ivermectin, 105
Iwata, S., 18

Jacob, F., 13

Kao, P. N., 52–53
Karlin, A., 79
Katz, B., 7–8, 11–12, 90–91, 93, 111
Kerszberg, M., 162
Klarsfeld, A., 150–51
KNF model. *See* sequential model (KNF model)
Kornhauser, J. M., 150–51
Koshland, D. E., 90
Kühne, Wilhelm, 3

labeling, of nicotinic receptor, 26–28, 121–22
laminin, 164
Langley, John Newport, 3–4, 11, 20, 84, 148
Lavoisier, Antoine-Laurent, 2
Léna, C., 190
Le Novère, N., 115
leucines, 73
light microscopy autoradiography, 142
Lindstrom, Jon, 126
lipid belt, of nicotinic receptor, 105
lipid rafts, 172–73
liposomes, 35
L-lactate dehydrogenase, 14
local anesthetics, as allosteric effectors, 105
Loewi, Otto, 6
loop component, of nicotinic receptor ion channel, 76–79
loops, multiple, of acetylcholine-binding pocket, 54–57, 117–18
lophotoxin, 55
low-affinity binding, and ion channel opening, 88

Index

low-affinity sites, for channel blockers, 69
Lurcher mutation, in mice, 98
Lymnaea stagnalis, 40

M2 segment, 72–74, 76, 79, 115, 129, 131–34
M4 segment, 115
main immunogenic region (MIR), 126
mapping, 54–55, 105, 152, 184
mathematical models, 20–21; for enzyme activity, 7–8, 11
McMahan, U. J., 155
mecamylamine, 208
medial habenula, 190
[^{14}C]-meproadifen, 69
Merlie, J.-P., 53, 148, 150–51
methanethiosulfonate ethylammonium, 79
methonium, 8
mice, research use of, 47–48, 168–69, 182, 188, 192–93, 195, 200–201, 203–6; "knockin" mice, 193–94; "knockout" mice, 155, 169, 192–93, 198, 209; mutant mice, 157; transgenic mice, 151–53, 159, 185
microsacs, 33, 85; and response to acetylcholine, 23–25
microtubules, 172, 175
MIR (main immunogenic region), 126
mobility, of nicotinic receptor, 173–75
models and modeling, 106–7, 113–15; allosteric enzyme model, 83; allosteric models, 91, 101–3, 131–34; desensitization, 89–95; enzyme activity, 7–8; four-state model, 91; induced conformational change model for allosteric interaction, 16–19; ion channel activation, 89–95; MWC model (concerted model), 90–91, 93, 95–103, 106–7, 133, 139; nicotinic receptor, 91–95; reward processes, 197; sequential model (KNF model), 89–91, 95–97, 99, 107, 133, 139; stereochemical model, of ion channel, 76–79; transmembrane organization, 112–15; two-state model, for allosteric interaction, 13–19
modulation, of conformational transitions by allosteric effectors, 104–6
molecular cloning, 41–42
molecular models, of activation and desensitization, 89–95
molecular phylogeny of nicotinic receptor, 43–47
molecular recognition studies, 57
Monod, J., 13, 90
motor-axon terminal, 140
MPTA (4-(N-maleimido)-phenyl-trimethylammonium iodide), 27
muscarinic agonists, 187
muscle contraction, 126
muscle-specific kinase. *See* MuSK (muscle-specific kinase)
muscular dystrophies, 176

MuSK (muscle-specific kinase), 127–28, 155, 157–58, 165, 169
Mus spretus, 47–48
mutagenesis, site-directed, 80–81, 95–99, 105, 121; and characterization of ion channel, 74–76
mutation: in human population, 125; in 43K-rapsyn, 169, 175–76; of nicotinic receptor genes, 127–31, 137–38; pleiotropic, 96–99
MWC model (concerted model), 90–91, 93, 95–100, 106–7, 133, 139; difficulties and unresolved issues, 100–103
myasthenia gravis, 115, 126–28, 137; acquired, 127. *See also* congenital myasthenic syndromes
myasthenic mutant, slow-channel, 131–34
myasthenic syndrome, slow-channel, 128–31
Mycobacterium tuberculosis, 114
myogenic factor, 160–61
myogenic proteins, 159–60, 162, 175

Na$^+$ channels, 165
Nachmansohn, David, 8–9, 12–13, 28, 65, 189
Naja mossambica mossambica, 62
N box, 150–54, 158, 175
NCAM (neural cell adhesion molecule) factors, 165
nematode *(C. elegans),* 44
Némethy, G., 90
neocortex, 187
nerve-muscle degeneration and regeneration, 173
nervous system, development of, 193
network of transitions, 19
neuregulins, 155–59, 164
neuroblasts, 193
neuromodulator, acetylcholine as, 191
neuromuscular junction, 126–27, 142–44, 164–65, 174, 176–77; pathologies of, 174–76
neuron, use of term, 4
neuronal networks, 176
neuronal workspace, 207–8
neuron-specific expression, of β2-subunit gene, 185
neuroscience, use of term, 1
neurotransmitter, vii, 191; role of, 5–6
neurotransmitter, use of term, 4
NH$_2$-terminal hydrophilic domain, 79
Nicot de Villemain, Jean, 3
nicotine, 3–4, 55–56, 187–89; and cognition, 208; and dopaminergic system, 197–200; neuroprotective effects of, 196
nicotine addiction, 201–2, 211
nicotinic receptor: biosynthesis of, 144–46, 155, 174; and brain function, 134–35, 192–93, 208–11; chemical nature of, 35; and cognition, 204–8; and consciousness,

Index

nicotinic receptor (*continued*) 202–4; and development of visual system, 195; and early development of nervous system, 193; extraction of, without denaturation, 28–30; identification of, 31–33; isolation of, 22–28; "light" and "heavy" forms of, 37; mobility of, 173–75; models for, 91–95; neuronal, 134–35; neuroprotective effect of, 208; and pain, 200–201; pre- and postsynaptic, 187–91; purification of, 29–33, 35; reconstitution of, 33–35; role in addiction, 196–200; role in aging, 193–96; role in brain development, 193–96; role in cognition, 204–8; role in sleep and wakefulness, 202–4; structural predictions for, 113–15; subunit composition of, 37–39; in supramolecular "scaffolding," 170–71; and synaptic connections, 194–95; three-dimensional structure of, 108–9
nicotinic receptor oligomers, distribution in brain, 179–82
nicotinic receptor sites, pharmacological and physiological diversity of, 60–62
noncompetitive blockers. *See* channel blockers
norepinephrine, 197
N-terminal domain, 104, 112, 130, 202
nucleus accumbens, 197

Oberthur, W., 73
okadaic acid, 161–62
opening, of ion channel, 84, 88; prolonged, 128, 130–31; single channel, 95, 103; spontaneous, 95, 97, 99, 128, 131; trimodal distribution of, 131–32
Oswald, R., 53

pain, nicotinic receptor and, 200–201
Paracelsus, 2
paracrine release. *See* volume transmission of acetylcholine
Parkinson's disease, 63, 192, 196, 209, 211
partial agonists, 95–96
passage of receptor oligomers, from nucleus to postsynaptic membrane, 171–74
Pasteur, Louis, 6
patch-clamp techniques, 66, 88–89, 133, 190, 198, 201
pathologies: linked to nicotinic receptors (*see* congenital myasthenic syndromes; epilepsy); of nervous system, 136–37; of neuromuscular junction, 174–76
Patrick, James, 126
patterns of expression of subunit genes, 47–49
PDZ-domain proteins, 106
pentameric molecular structure of nicotinic receptor, 39

peptide fractionation and sequencing techniques, 72
Perutz, M. F., 18
pharmacological agents, 2
pharmacological receptor, 3
pharmacology, novel, 124, 139, 211; for allosteric diseases, 137–38
pharmacology, use of term, 2
phencyclidine, 72, 73
phenotypes, 96–99; "gain of function," 126, 128, 130, 137–39; γ phenotype, 97–98; K phenotype, 97–98; "loss of function," 128, 136; L phenotype, 97–98, 134; produced by mutations, 100, 125, 130
Phillips, David, 8
phosphofructokinase, 14
phosphorylase b, 18
phosphorylation, 112, 162, 168–69
phylogenetic tree, of acetylcholine receptor, 42, 45–47
PKC (protein kinase C), 161
postsynaptic membrane, 140–41, 186–87; acetylcholine receptors in, 142–44; passage of receptor oligomers to, 171–74; structure of, 165
postsynaptic potential, use of term, 9
postsynaptic scaffold, 165–71, 173–74
posttranscriptional processing, of nicotinic receptor subunits in brain, 187
pregnancy, and tobacco smoking, 204
prilocaine, 67
primary structure of receptor molecule, 40–42
promoter approach, 147, 150, 183
promoters, 186
protein kinases, 106
proteolipid, 26
proteolysis, 166
protomers, 90–91
proton magnetic resonance, 52
PSD_{95}, 176
pseudosymmetry of nicotinic receptor molecule, 40, 49
purification, of nicotinic receptor, 29–33, 35

quinacrine, 69, 80–81
quinacrine azide, 72
quinidine, 137
QX-222, 66

Ralston, E., 171
Ramón y Cajal, Santiago, 4–5
Ranvier, L.-A., 141
rapid-mixing techniques, 71, 86
rapsyn. *See* 43K-rapsyn
Rasmussen's encephalitis, 136
rat, research use of, 149, 182–83, 187, 190, 204
"receptor-associated protein at the synapse." *See* 43K-rapsyn

Index

receptor molecule: primary structure of, 40–42; pseudosymmetry of, 40, 49
receptor pharmacology, 2–6
receptor protein, structure of, 109–11
receptors, viii, 20–21; early conceptions of, 3–4
receptor-specific ligands, 26–28
receptor theory, Langley's, 3–4
recombinant DNA technology, 43–44, 65, 74
reconstitution, of nicotinic receptor, 33–35
recovery, from desensitization, 99–100
recovery pathway, 93, 100
regulation: autocatalytic, 161; of expression of nicotinic receptor genes, 145; of extrajunctional receptors, 159–63; of synaptic strength, 167
regulatory enzymes, 12–19
regulatory mechanisms, at gene level, 146–47
regulatory nodes, 186
reinforcement, role of nicotinic receptor in, 197–200
repression, of receptor biosynthesis, 145–46
reproducibility, of purified protein, 34
reward processes, role of nicotinic receptors in, 196–200
rhesus monkey, 183
rotation, of N-terminal domains, 122
Rouget, C., 140

Sanes, J. R., 148
sarcoglycans, 165
scaffolding proteins, 176
SCAM (substituted cysteine accessibility method), 79
Schaefer, H., 9
schizophrenia, 192, 211
scientific cultures, viii, 1–2, 21; allosteric interactions, 12–19; electric potentials and ion channels, 8–12; enzyme activity and stereochemical specificity, 6–8; receptor pharmacology and chemical transmission, 2–6
scopolamine, 204
SDS-polyacrylamide gel electrophoresis, 166
secretory pathway, 172, 174–75, 186–87
selective schema, 90
selectivity, of ion channel, 77–79, 81, 111
selectivity filter, 78–79
sequencing, of subunits, 43–47
sequential binding model, 84
sequential model (KNF model), 89–91, 95–97, 99–100, 107, 133, 139
serines, 73, 75–76
serotonin, 197
Shaker channels, 102
Sherrington, Sir Charles, 5
Shulman, R. G., 90

signaling, neuromodulatory, 177
signaling mechanisms, 147–48, 154–57, 175
signaling pathways, 157–59, 161
signal transduction process, acetylcholine, 83
silence, and desensitization, 99
silencer element, 184
single binding events, 103
single channel recordings, 84, 88–89, 102–3, 131
sites of action, of channel blockers, 67–69
sleep, and nicotinic receptor, 202–4
"slits," between subunits, 111
slow channel block, 66
snake venom, 22, 27–29, 31, 112, 142
Sobel, A., 166
Society for Neuroscience, 1
sodium deoxycholate (detergent), 28
somatodendritic compartment, 190–91
specialization of subunits, 49
spectroscopy, 67–69
stereochemical specificity, 6–8
Streptomyces lividans, 114
structural predictions, for nicotinic receptor, 113–15
structure, of 43K-rapsyn, 168
structure, three-dimensional, of nicotinic receptor, 108–11
subconductance states, 107; multiple, 101–2
subneural domain, 165, 172
substituted cysteine accessibility method (SCAM), 79
subunit assembly rules, 179–82
subunit combination in receptor molecule, 50
subunit composition of nicotinic receptor, 37–39
subunit organization within receptor oligomers, 39–40
subunit patterns of expression, 47–49
subunits, of nicotinic receptor, 41, 49, 61, 79, 81, 109, 125, 128, 171, 178; α subunits, 38–41, 43–44, 113, 119, 121, 126, 128–31, 182–84, 198; β subunits, 38, 44, 167, 182–85, 198, 205–7; boundaries of, 53–54; δ subunit, 38, 40, 72–74; ε subunit, 46, 48, 128, 129, 131–134, 152–153, 154; folding and assembly of, 79–81; γ subunit, 38, 40; interfaces between, 63, 121, 124; labeling of, 55–57; primary structure of, 40–42; transmembrane organization of, 111–13
subunit sequences, 43–47
subunit specialization, 49
subunit transmembrane organization, model of, 112–13
sudden infant death syndrome, 204
Sumikawa, K., 74
supramolecular assembly, of postsynaptic membrane, 164–65
switching mechanism, 150
symmetry, molecular, 49

Index

symmetry axis, of receptor oligomer, 69–70, 76
symmetry principle, 91, 101–2
synapse, vii; chemical *vs.* electrical, 5–6; discovery of, 4–6
synapse, use of term, 4
synaptic cleft, 140
synaptic connections, nicotinic receptor and, 194–95
synaptic plasticity, 174–76, 201–2
synaptic proteins, 164–65
synaptosomes, 187
syntrophin, 165

Taylor, P., 79
TDF (p-(trimethylammonium) benzene diazonium fluoroborate), 27, 54–55
tetracaine, 65, 67
therapeutic chemistry, 51
Thesleff, S., 11–12, 90–91, 93
thiamine triphosphate, 168
Thorndike, Edward L., 196
threonine deaminase, 13–14
time-lapse imaging, 172–74
tobacco addiction, 200–202
tobacco smoking, 127, 192–93, 204
Torpedo (fish), 37, 60, 75, 81, 97, 99, 105, 117, 119, 124, 134, 146, 165–66, 168; electric organ of, 23, 142, 155; electrocytes of, 169, 172; *T. californica*, 30, 41–42, 74, 86, 113; *T. marmorata*, 9, 23, 25, 28, 33, 41–42, 52, 55, 67–72, 74, 85, 109, 113, 126
Tourette's syndrome, 63, 211
TPMP (triphenylmethylphosphonium), 72–74
transcriptional regulation, of brain nicotinic receptor genes, 183–86
transmembrane domain, 122
transmembrane organization, 42, 111–15

transmembrane region, 121
transmembrane segments, of nicotinic receptor, 105; mutations in, 130–31
transsynaptic mechanisms, 165
trimethisoquin, 69–70
Triton X-100 (detergent), 28
tryptophan, 56, 59
tryptophan 57, 56
TTX (tetrodotoxin), 187, 190
"twisting," 122
two-state model, of allosteric interaction, 13–19
tyrosines, 55–56, 58–59

ultraviolet irradiation, 53
unc-29 gene, 44
unc-38 gene, 44
Unwin, N., 40, 121
up-regulation, 200–201
utrophin, 165, 169–70

Vauquelin, Nicholas-Louis, 3
verapamil, 161
volume transmission of acetylcholine, 10–11, 177–78, 190–91
von Euler, Ulf, 6
Vulpian, Alfred, 2–3

wakefulness, and nicotinic receptor, 202–4
Waldeyer, Heinrich, 4
Weber, M., 60, 67
Wilson, G. G., 79
Wyman, J., 90

Xenopus, 44–45, 74–75, 97, 154–55, 168–69
X-ray crystallography, 78, 119–21

zebra fish, 167, 175
zinc-histidine mapping, 121